MULTIUSER DETECTION

Multiuser Detection provides the first comprehensive treatment of the subject of multiuser digital communications. Multiuser detection deals with demodulation of the mutually interfering digital streams of information that occur in areas such as wireless communications, high-speed data transmission, satellite communication, digital television, and magnetic recording. The development of multiuser detection techniques is one of the most important recent advances in communications technology, and this self-contained book gives a comprehensive coverage of the design and analysis of receivers for multiaccess channels, while focusing on fundamental models and algorithms.

The author begins with a review of multiaccess communications, dealing in particular with code-division multiple-access (CDMA) channels. Background material on hypothesis testing and the effect of multiuser interference on single-user receivers are discussed next. This is followed by the design and analysis of optimum and linear multiuser detectors. Also covered in detail are topics such as decision-driven multiuser detection and noncoherent multiuser detection.

The elements of multiuser detection are clearly and systematically presented along with more advanced recent results, some of which are published here for the first time. The extensive set of references and bibliographical notes offer a comprehensive account of the state of the art in the subject.

The only prerequisites assumed are undergraduate-level probability, linear algebra, and introductory digital communications. The book contains over 300 exercises and is a suitable textbook for electrical engineering students. It is also an ideal self-study guide for practicing engineers, as well as a valuable reference volume for researchers in communications and signal processing.

Sergio Verdú is Professor of Electrical Engineering at Princeton University. His contributions to the technology of multiuser detection span his pioneering work in the early 1980s to recent results included in this text. Professor Verdú is also well known for his work on information theory, in which he explores the fundamental limits of data transmission and compression systems. Recipient of a number of awards, he is a Fellow of the IEEE and served as President of the IEEE Information Theory Society in 1997.

MULTIUSER DETECTION

SERGIO VERDÚ

CAMBRIDGE
UNIVERSITY PRESS

PUBLISHED BY THE PRESS SYNDICATE OF THE UNIVERSITY OF CAMBRIDGE
The Pitt Building, Trumpington Street, Cambridge, United Kingdom

CAMBRIDGE UNIVERSITY PRESS
The Edinburgh Building, Cambridge CB2 2RU, UK
40 West 20th Street, New York, NY 10011-4211, USA
10 Stamford Road, Oakleigh, VIC 3166, Australia
Ruiz de Alarcón 13, 28014 Madrid, Spain
Dock House, The Waterfront, Cape Town 8001, South Africa

http://www.cambridge.org

First published 1998
Reprinted 2001

Printed in the United States of America

Typeset in Times Roman 10.5/14 pt. and Futura in LaTeX 2ε [TB]

A catalog record for this book is available from the British Library

Library of Congress Cataloging in Publication Data is available

ISBN 0 521 59373 5 hardback

To Mercedes and Ariana

CONTENTS

LIST OF FIGURES

PREFACE

He that will not apply new remedies
must expect new evils:
for time is the greatest innovator.
Francis Bacon (1561–1626)

Research and development of digital communications systems is undergoing a revolution fueled by rapid advances in technology. With the ever-growing sophistication of signal processing and computation, advances in communication theory have an increasing potential to bridge the gap between practically feasible channel utilization and the fundamental information theoretic limits on channel capacity. If conquering channel capacity is the manifest destiny of communications technology, the need for efficient use of channel bandwidth and transmission power is felt most acutely in wireless communication, where the exponentially growing demand for data rate must be accommodated in a finite segment of the radio spectrum. To add to the challenge, information is transmitted not by a single source but by several uncoordinated, bursty, and geographically separated sources.

Multiuser Detection deals with the demodulation of mutually interfering digital streams of information. Cellular telephony, satellite communication, high-speed data transmission lines, digital radio/television broadcasting, fixed wireless local loops, and multitrack magnetic recording are some of the communication systems subject to multiaccess interference. The superposition of transmitted signals may originate from nonideal characteristics of the transmission medium, or it may be an integral part of the multiplexing method as in the case of Code-Division Multiple-Access (CDMA). Multiuser detection (also known as cochannel interference suppression, multiuser demodulation, interference cancellation, etc.) exploits the considerable structure of the multiuser interference in order to increase the efficiency with which channel resources are employed.

Although isolated generalizations of digital communication models to multi-input multi-output channels had taken place as early as the 1960s, it was not until the mid 1980s that multiuser detection started developing as a cohesive body of analytical results that took into account the specific features of multiuser channels. Since then, the number of researchers working within this discipline has rapidly multiplied, to the point where it is now one of the most active and vibrant branches of digital communications. The extensive set of references collected in this book, although not pretending to be comprehensive in any way, gives evidence of the level of activity in multiuser detection in the past few years. The bibliographical notes at the end of each chapter provide an account of the development of the main results as well as a snapshot of the current state of the art. I can only hope that that part of the book will become quickly obsolete in view of the speed at which the field is currently evolving.

While aiming for a fairly comprehensive coverage of the design and analysis of receivers for multiaccess channels, my goal has been to distill the elements of multiuser detection in the simplest setting that brings out the key concepts. A fertile ground for geometrical intuition, the linearly modulated synchronous multiuser channel proves to be a garden of Euclidean delights. Borrowing from the tradition in multiuser information theory, most of the main ideas are first introduced in the two-user channel, which emerges as a powerful pedagogical tool.

Chapter 1 gives a brief introduction to the main approaches in multiaccess communications. Chapter 2 introduces the basic channel models used throughout the book. The main paradigm is the Code-Division Multiple-Access channel, in which each user modulates its own signature waveform. This channel is general enough to encompass orthogonal and nonorthogonal multiplexing methods, with or without spread-spectrum signaling. Chapter 3 covers background material on hypothesis testing and single-user detection and analyzes the effects of multiaccess interference on the single-user receiver. Chapter 4 is devoted to the design and analysis of optimum multiuser detectors. Linear signal processing for multiuser detection is studied in Chapters 5 and 6, with and without the constraint of complete multiuser interference suppression, respectively. Adaptive linear multiuser detection is covered in Chapter 6. Chapter 7 deals with nonlinear multiuser detectors that use decisions on the interfering digital streams to mitigate their effect.

Whether it is used as a textbook, self-study tool, or research reference, the set of over 300 problems comprises an essential component of this book. They range from simple drill exercises to research results that complement the theory expounded in the text. I hope the reader will draw some sense of accomplishment from solving them.

No prerequisites are assumed beyond undergraduate-level probability, linear algebra, and an introductory course on communications. At Princeton, I have used this text to teach a one-semester course on Multiuser Detection to first- and second-year graduate students with diverse backgrounds. Although previous or concurrent exposure to a conventional detection and estimation course may be beneficial, Chapter 3 gives a self-contained presentation of the required material. A typical "single-user" digital communications course covering the fundamentals of equalization is not required either. In fact, it is my contention that (synchronous) multiaccess channels provide an easier setting for learning many of the fundamentals of equalization in digital communications than the conventional single-user intersymbol interference channel.

The text contains substantial material that can be tailored to serve as the core of various master's and doctoral courses on multiuser communication. In addition, the book can be used as a self-study guide for practicing engineers and as a reference volume for academic and industrial researchers in communications and signal processing.

SPECIAL THANKS

EZIO BIGLIERI • GIUSEPPE CAIRE • BRAD DICKINSON • PHIL MEYLER
JIM FREEBERSYSER • MIKE HONIG • VISI LUCAS • NARAYAN MANDAYAM
ANDY MCKELLIPS • U. MADHOW • TARAGAY OSKIPER • JAY PLETT
LAURIE NELSON • CARL NUZMAN • MERCEDES PARATJE • VINCE POOR
CRAIG RUSHFORTH • SHLOMO SHAMAI • JOHN SMEE • XIAODONG WANG
RAJESH SUNDARESAN • MINERVA YEUNG • MICHELLE YOUNG • BIN YU

> A companion web site for this book can be found at
> http://www.cup.org/Titles/59/0521593735.html

CHAPTER ONE

MULTIACCESS COMMUNICATIONS

1.1 THE MULTIACCESS CHANNEL

The idea of using a communication channel to enable several transmitters to send information simultaneously dates back to Thomas A. Edison's 1873 invention of the *diplex*.[1] This revolutionary system enabled the simultaneous transmission of two telegraphic messages in the same direction through the same wire. One message was encoded by changes of polarity; the other by changing absolute values.

Nowadays, there are numerous examples of *multiaccess communication* in which several transmitters share a common channel: mobile telephones transmitting to a base station, ground stations communicating with a satellite, a bus with multiple taps, multidrop telephone lines, local area networks, packet-radio networks, and interactive cable television networks, to name a few. A common feature of those communication channels is that the receiver obtains (a noisy version of) the superposition of the signals sent by the active transmitters (Figure 1.1). Oftentimes, the superposition of signals sent by different transmitters occurs unintentionally owing to nonideal effects; for example, *crosstalk* in telephony and in multitrack magnetic recording or any time the same radio frequency band is used simultaneously by distant transmitters, as in cellular telephony, radio/television broadcasting, and wireless local loops. Although occasionally the terms *multiplexing* and *multiaccess* are used interchangeably, multiaccess usually refers to situations where the message sources are not collocated and/or operate autonomously. The message sources in a multiaccess channel are referred to as *users*.

The multiaccess communication scenario depicted in Figure 1.1 encompasses not only the case of a common receiver for all users, but the case of several receivers, each of which is interested in the information sent by one

[1] See Conot [60].

1

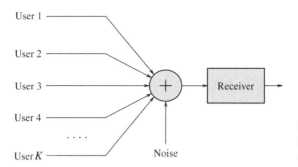

FIGURE 1.1.
Multiaccess
communication.

user (or a subset of users) only. Multiaccess communication is sometimes referred to as *multipoint-to-point* communication. The engineering issues in the dual *point-to-multipoint*[2] channel depend on the commonality of the information transmitted to each destination. At one extreme, the same information is delivered to all recipients, for example, in radio and television broadcasting or in cable television; at the other extreme, the messages transmitted to different recipients are independent, for example, a base station transmitting to mobile units. The latter scenario falls conceptually within the multiaccess channel model; in that case, the receiver (say one of the mobile units) is interested in only one of the information sources transmitted by the base station.

1.2   FDMA AND TDMA

The advent of radio-frequency modulation in the early twentieth century enabled several radio transmissions to coexist in time and space without mutual interference by using different carrier frequencies. The same idea was used in long-distance wire telephony. *Frequency-Division Multiplexing* or *Frequency-Division Multiple Access* (FDMA) assigns a different carrier frequency to each user so that the resulting spectra do not overlap (Figure 1.2). Band-pass filtering (or heterodyning) enables separate demodulation of each channel.

In *Time-Division Multiplexing*, time is partitioned into slots assigned to each incoming digital stream in round-robin fashion (Figure 1.3). Demultiplexing is carried out by simply switching on to the received signal at the appropriate epochs. Time division can be used not only to multiplex collocated message sources but also by geographically separated users who

---

[2] Point-to-multipoint and multipoint-to-point channels are sometimes distinguished either as *downlink* and *uplink* channels, respectively, or as *forward* and *reverse* channels, respectively.

2

FIGURE 1.2.
Frequency-
Division Multiple
Access.

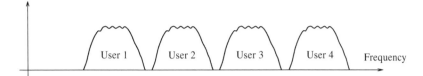

have the ability to maintain time-synchronism, in what is commonly referred to as *Time-Division Multiple Access* (TDMA). Note that FDMA allows completely uncoordinated transmissions in the time domain: no time-synchronization among the users is required. This advantage is not shared by TDMA where all transmitters and receivers must have access to a common clock.

The important feature of frequency-division and time-division multiaccess techniques is that, for all conceptual purposes, the various *users* are operating in separate noninterfering channels. To put it in the signal-space language of digital communications, those multiaccess techniques operate by ensuring that the signals transmitted by the various users are mutually *orthogonal*. Channel or receiver nonideal effects may require the insertion of guard times in TDMA (Figure 1.3) and spectral guard bands in FDMA (Figure 1.2) to avoid *cochannel interference*.

Why would it make sense to consider multiaccess techniques that do not adhere to the principle of dividing the channel into independent noninterfering subchannels? One reason is that noninterfering multiaccess strategies may waste channel resources when the number of potential users is much greater than the number of simultaneously active users at any given time. Think for example of wireless telephony; if each subscriber were assigned a fixed radio frequency channel, only a tiny fraction of the spectrum would be utilized at any given time. Analogously, in TDMA most of the time slots would be empty, at any given time.

How is it possible to assign the channel resources to the users in dynamic (in other words, on demand) rather than static fashion as above? At the expense of some increase in complexity, one possibility is to set up a separate *reservation* channel, where the users who want to use the channel notify the receiver, which, then, partitions the original channel using TDMA or FDMA among the active users only. This presupposes a separate feedback channel that notifies every user of the time or frequency slot where it is allowed to transmit. However, note that the reservation channel is a multiaccess channel and we still have to cope with the same issue as before, namely, how to partition the resources of that channel dynamically.

FIGURE 1.3.
Time-Division
Multiple Access
showing guard
times between
slots.

USER 1	USER 2	USER 3	USER 1	USER 2	USER 3	USER 1

1.3 RANDOM MULTIACCESS

Random multiaccess communication is one of the approaches to dynamic channel sharing. When a user has a message (usually referred to as a *packet* in this context) to transmit it goes ahead and transmits it as if it were the sole user of the channel. If indeed nobody else is transmitting simultaneously, then the message is received successfully. However, the users are uncoordinated and the possibility always exists that the message will interfere (in time and frequency) with another transmission. In such case, it is typically assumed in random multiaccess communication that the receiver cannot demodulate reliably several simultaneous messages. The only alternative left is to notify the transmitters that a *collision* has happened and, thus, their messages have to be retransmitted. Collisions would reoccur forever if upon notification of a collision, the transmitters involved were to retransmit immediately (or after a similar delay). To overcome this, users wait a random period of time before retransmitting. The main distinguishing feature among the existing random access communication systems is the algorithm used by the transmitters to determine the retransmission delay for each colliding packet.

The first random multiaccess communication system was the ALOHA system proposed for a radio channel in 1969 (Abramson [9]). Some coaxial-cable local area networks, typified by the widely used ETHERNET, employ a "polite" version of ALOHA, called Carrier-Sense Multiple Access (CSMA), where users listen to the channel before transmitting so as not to collide with an ongoing transmission (Kleinrock and Tobagi [211]). In general, random multiaccess communications are best suited for very bursty channels, in which it is not likely that more than one user will be transmitting simultaneously. The main theoretical advances in this area occurred in the 1970s through the mid 1980s.[3] *Polling* (Hayes and Sherman [139]) is another multiaccess strategy where simultaneous transmissions are avoided. The receiver asks every transmitter that shares a common channel (say, in round-robin fashion) whether it has anything to transmit.

1.4 CDMA

The channel-sharing approaches discussed so far are based on the philosophy of letting no more than one transmitter occupy a given time–frequency slot. Whenever this condition is violated in random-access

---

[3] See Bertsekas and Gallager [29] and Abramson [10].

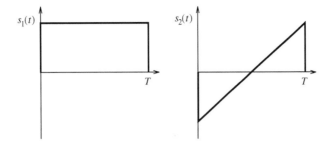

FIGURE 1.4.
Orthogonal
signals assigned
to two users.

communication, the receiver is unable to recover any of the colliding transmissions. As we remarked before, reception free from interchannel interference is a consequence of the use of orthogonal signaling. It is important to realize that this can be accomplished by signals that overlap both in time and in frequency. For example, consider the time-limited signals s_1 and s_2 in Figure 1.4. Those signals overlap both in the time and frequency domains (Figure 1.5), yet their *crosscorrelation* or inner product is zero:

$$\langle s_1, s_2 \rangle = \int_0^T s_1(t)s_2(t)\,dt = 0. \tag{1.1}$$

A simple two-user multiaccess communication system could be designed by letting users 1 and 2 modulate antipodally signals s_1 and s_2, respectively. This means that user i transmits $s_i(t)$ in order to send 1 and $-s_i(t)$ in order to send 0 (Figure 1.6) in successive time epochs of duration T. Let us assume that the system is *synchronous* in the sense that the transmission rate is the same for both users (equal to $1/T$ bits per second) and their bit epochs are perfectly aligned (Figure 1.6).

How can the information transmitted by both users be demodulated now that they overlap in both frequency and time? As we see in Figure 1.6, a hypothetical receiver that observed the sum of both signals (with identical amplitude) can easily demodulate the transmitted bit streams. The bit transmitted by user 1 is equal to 1 or 0 depending on whether the ramp received in the corresponding bit period has positive or negative absolute value; the bit transmitted by user 2 is equal to 1 or 0 depending on whether the ramp received in the corresponding bit period is increasing or decreasing. However, any receiver will actually observe the sum of both signals

FIGURE 1.5.
Fourier transforms
(magnitude,
$f > 0$) of
waveforms in
Figure 1.4.

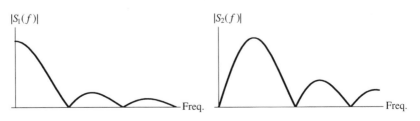

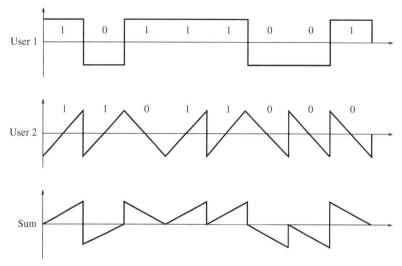

FIGURE 1.6.
Synchronous
antipodal
modulation of
orthogonal
signals.

embedded in additive *background noise*. A good strategy in that case (which minimizes bit-error-rate when the noise is white and Gaussian) is *matched filtering*, in which the received waveform is separately correlated in every bit period with both $s_1(t)$ and $s_2(t)$, followed by respective comparisons of the correlator outputs to zero thresholds. The matched filter for user 1 simply integrates the received signal in every bit period; if the result of such integration is positive/negative, then it outputs 1/0. The output of the correlator is indeed affected by the background noise (thus causing sporadic errors). However, thanks to the orthogonality condition (1.1) and the assumed synchronism, the output of the correlator for user 1 is not affected at all by the signal of user 2 regardless of the relative strengths of both signals. We can conclude that although the signals transmitted by both users overlap in both time and frequency, the bit-error-rate of this system will be the same as if the transmissions occurred in separate channels.

We have just seen a very simple example of a Code-Division Multiple-Access (CDMA) system. Users are assigned different "signature waveforms" or "codes"[4] (in an older terminology whence the term CDMA originated). Each transmitter sends its data stream by modulating its own signature waveform as in a single-user digital communication system. In the orthogonal two-user example considered above, the receiver need not concern itself with the fact that the signature waveforms overlap both in frequency and in time, because their orthogonality ensures that they will be transparent to the output of the other user's correlator.

---

[4] These are not to be confused with error-correcting codes that add redundancy to combat channel noise and distortion.

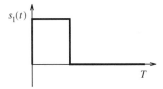

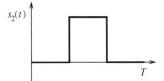

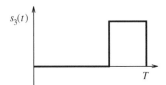

FIGURE 1.7.
Signature
waveforms in
TDMA.

The special case of orthogonal CDMA in which the signature waveforms do not overlap in the time domain corresponds to digital TDMA. For example, the three-user orthogonal CDMA system depicted in Figure 1.7 can be viewed as a TDMA system. Digital TDMA and FDMA systems can be seen as special cases of orthogonal CDMA where the signature waveforms are nonoverlapping in the time domain and the frequency domain, respectively.[5]

No signal can be both strictly time-limited and strictly band-limited. However, adopting a softer definition of bandwidth and/or duration (e.g., the percentage of energy outside the band $[-B, B]$ or outside the time interval $[0, T]$ not exceeding a given bound ϵ) we can pose the question: how many mutually orthogonal signals with (approximate) duration T and (approximate) bandwidth B can be constructed? Although no explicit answer in terms of T, B, and ϵ is known, bounds have been obtained that show that unless the product TB is small, the answer is essentially[6] $2TB$. An immediate consequence of this result is that a K-user orthogonal CDMA system employing antipodal modulation at the rate of R bits per second requires bandwidth approximately equal to

$$B = \frac{1}{2}RK. \tag{1.2}$$

---

[5] In practical TDMA systems the slots assigned to each user span more than one symbol, e.g., in *Global System for Mobile* (GSM) each slot consists of 148 bits, and guard-time/slot-duration $= 33/592$ (Redl et al. [358]).

[6] See Landau and Pollak [228], Petrich [332], and Dollard [69].

If the definition of duration or bandwidth is lax enough, then (1.2) is achievable by TDMA (cf. Problems 1.8, 1.9, 1.10). To achieve the minimum bandwidth in (1.2), the duration-bandwidth product of each signature waveform in an orthogonal CDMA system must be $K/2$. In contrast, achieving (1.2) with TDMA (or FDMA) requires a pulse with minimum duration-bandwidth product (which does not grow with the number of users). The need for signals with larger duration-bandwidth products in CDMA brings in benefits such as robustness against unknown channel distortion and antijamming capabilities. Waveforms with large duration-bandwidth products are called *spread-spectrum* signals. TDMA and FDMA can also use spread-spectrum signaling, but only at the expense of low *spectral efficiency*, defined as the aggregate data rate per unit bandwidth (bits per second per Hz).

The next important conceptual step is to realize that orthogonality of the signature waveforms is not imperative for CDMA. For example, the performance of the simple correlator considered above will be degraded in the presence of nonorthogonal interfering users, but we may be able to keep the degradation to tolerable levels provided there are not too many or too powerful interfering users. We have, in effect, dropped the requirement that the signature waveforms be orthogonal for the requirement that their mutual interference be sufficiently low. This still requires careful selection of the signature waveforms so that their crosscorrelations are fairly low (compared to the signature waveform energies $\|s_i\|^2 = \langle s_i, s_i \rangle$).[7] Removing the restriction of orthogonal signature waveforms has several major benefits that make CDMA an attractive multiaccess technique for many multiuser communication systems:

- The users can be *asynchronous*, that is, their time epochs need not be aligned, and yet "quasi-orthogonality" can be maintained by adequate design of spread-spectrum signature waveforms.
- The number of simultaneous users is no longer constrained to twice the duration-bandwidth product of the signature waveforms.
- Sharing of channel resources is inherently dynamic: reliability depends on the number of simultaneous users, rather than on the (usually much larger) number of potential users of the system. Thus, unlike orthogonal multiaccess, it is possible to trade off reception quality for increased capacity.

What factors determine the number of users that can be sustained in a CDMA system? As we mentioned, the bandwidth times the inverse of the

---

[7] In this sense, the signature waveforms in Figure 1.4 are not a particularly good choice, cf. Problem 1.2.

data rate determines the dimensionality of the signal space. Other factors that play a determining role include: the received signal-to-noise-ratios, signature waveform crosscorrelations, data redundancy, and the type of receiver used. The last factor brings us to the objective of *multiuser detection: the design and analysis of digital demodulation in the presence of multiaccess interference*. As we will see in detail, a consequence of the nonorthogonality of signature waveforms is that the simple correlator considered above is no longer optimal (even in the presence of white Gaussian background noise). For example, it suffers from the *near–far problem*: any interferer that is sufficiently powerful at the receiver causes arbitrarily high performance degradation. By suitable design of receivers that take into account the structure of the multiaccess interference it is possible to increase spectral efficiency, decrease output power, and robustify the system against imbalances in the received powers of various users.

The demodulators for nonorthogonal multiple access studied in this text are equally relevant when nonorthogonality occurs, not by design as in asynchronous CDMA, but through unintentional crosstalk. A classic example is a bundle of twisted pairs, which (unlike coaxial cables) radiate and capture energy. Digital transmissions carried by each pair are subject to contamination due to the superposition of the attenuated transmissions carried by neighboring pairs. A similar phenomenon occurs in adjacent tracks in magnetic recording. TDMA in mobile cellular telephony provides another illustration of cochannel interference. *Multipath* is a common channel impairment that causes the reception of the sum of several delayed replicas of the transmitted signal. In a TDMA system subject to multipath, signals that were originally in different time slots may actually overlap at the receiver. Another important source of cochannel interference in cellular TDMA is the simultaneous use of the same frequency band in different cells. Unless the receiver is robust against out-of-cell interference, careful allocation of frequency bands is required in those systems. The consequences of nonideal deviation from orthogonality are, naturally, dependent on the strength of the interfering sources.

To the foregoing incentives for contemplating the use of nonorthogonal multiaccess, we can add a basic lesson derived from *multiuser information theory*:[8] multiaccess channels whose capacity is achieved by orthogonal multiplexing techniques are the exception rather than the rule. For most situations, the number of users that can be accommodated by the channel is maximized by transmitting mutually interfering signals. In fact, optimal codes

---

[8] See, for example, El Gamal and Cover [101] and Gallager [100].

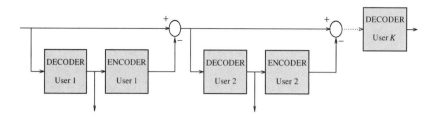

FIGURE 1.8.
Successive
decoding.

do not adhere to the structure imposed by nonorthogonal CDMA either, as the modulation of fixed signature waveforms is, in general, suboptimal. The burden of achieving optimal performance falls on the design of, not only suitable coding strategies, but on the receiver that has to decode the transmitted bit streams with arbitrary reliability from a noisy superposition of the transmitted signals. A receiver that achieves capacity under certain conditions (e.g., El Gamal and Cover [101]) uses the following *successive decoding* technique (Figure 1.8): a given user is decoded treating all the other users as noise; then the signal due to that user is remodulated with the decoded information and subtracted from the received waveform; and the process is repeated until all the transmitted information streams have been demodulated.

Among first-generation digital wireless telephony systems, (GSM Standard [82]) and (IS-54 Standard [441]) favor the simplicity and robustness of TDMA,[9] whereas the suitability of CDMA to bursty transmitters and the superiority of spread-spectrum signaling in harsh channel environments is favored in other commercial mobile cellular systems (IS-95 Standard [442]).[10]

In practical wireless multiaccess channels, it is very important to utilize precious channel bandwidth (and transmitted power) efficiently. Developed in the 1980s and without the benefit of a mature theory on multiuser communications, first-generation digital cellular systems do not come close to approaching channel capacity. As technology progresses, fundamental limits and advanced signal processing techniques are sure to become increasingly relevant for resource-efficient cost-effective design.

1.5 PROBLEMS

PROBLEM 1.1. To streamline the wiring of an airplane, an aircraft manufacturer decides to employ Time-Division Multiplexing in

---

[9] See, for example, Falconer et al. [85].

[10] In those commercial systems, FDMA is used in conjunction with either TDMA or CDMA. Hybrids of FDMA–TDMA–CDMA have also been proposed for commercial systems, e.g., Ruprecht et al. [366] and Blanz et al. [31].

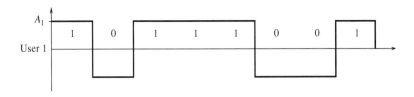

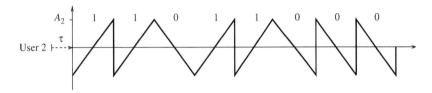

FIGURE 1.9.
Two-user
asynchronous
CDMA.

order to transmit twelve 10-kHz audio channels to each seat. Design such a system under the specification that no distortion of the reproduced waveforms is tolerated.

PROBLEM 1.2. Consider a two-user CDMA channel with the signature sequences in Figure 1.4. Users modulate those signature waveforms antipodally but do not maintain synchronism. Furthermore, both streams are received with different amplitude levels A_1 and A_2 (Figure 1.9).

(a) Find the minimum value of A_1/A_2 for which the correlator for user 1 gives error-free demodulation in the absence of background noise for all offsets $0 \le \tau \le T$.

(b) Find the maximum value of A_1/A_2 for which the correlator for user 2 gives error-free demodulation in the absence of background noise for all offsets $0 \le \tau \le T$.

PROBLEM 1.3. Let us say that two orthogonal signals are *strongly orthogonal* if they remain orthogonal after passing through an arbitrary linear time-invariant system. Show that a sufficient condition for s_1 and s_2 to be strongly orthogonal is that there exists a time T_0 such that, for all t,

$$s_1(T_0 - t) = s_1(T_0 + t),$$

$$s_2(T_0 - t) = -s_2(T_0 + t).$$

PROBLEM 1.4. If the number of users is a power of 2, a possible set of signature waveforms for orthogonal CDMA are the bipolar functions known as the Walsh functions. Every element in the set of 2^M Walsh functions can be represented by a vector $\mathbf{x}_M$ of 2^M

11

FIGURE 1.10.
Walsh functions of length 16.

components drawn from $\{+, -\}$ and defined recursively as

$$\mathbf{x}_0 = (+), \qquad (1.3)$$

$$\mathbf{x}_M = (\mathbf{x}_{M-1} \ \mathbf{x}_{M-1}) \quad \text{or} \quad (\mathbf{x}_{M-1} - \mathbf{x}_{M-1}). \qquad (1.4)$$

Figure 1.10 shows the set of Walsh functions for $M = 4$.

(a) Show that for all M, every pair of Walsh functions are orthogonal.

(b) Assume that the transmitted signals go through an ideal strictly band-limited low-pass filter of bandwidth equal to b/T, where T is the duration of the signature waveforms. Let the output of the filter due to the ith Walsh function be denoted by the signal $\tilde{s}_i(t)$. Choose a pair of nonstrongly orthogonal (Problem 1.3) Walsh functions of length 16 (Figure 1.10), and show a graph of the normalized crosscorrelation of their corresponding filtered versions,

$$\frac{|\langle \tilde{s}_i, \tilde{s}_j \rangle|}{\|\tilde{s}_i\| \|\tilde{s}_j\|},$$

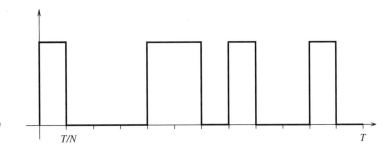

FIGURE 1.11.
On–off signature
waveform.

as a function of b. Do not take into account the effects of adjacent waveforms (intersymbol interference), that is, consider the transmission of one bit in isolation.

Note: The Hadamard matrices are a more general concept than the Walsh codes examined in this problem. A Hadamard matrix is a square matrix with elements in $\{-1, 1\}$ whose row vectors are pairwise orthogonal. A necessary condition for a Hadamard matrix of dimension $K > 2$ to exist is that K be divisible by 4. It is not known whether this condition is also sufficient. See also Problem 2.1.

PROBLEM 1.5. An on–off signature waveform of length N (Figure 1.11) can be written as

$$s(t) = \sum_{i=1}^{N} \alpha_i \, p\left(\frac{tN}{T} - i + 1\right),$$

where $\alpha_i \in \{0, A\}$, and $p(t) = 1$ if $0 \le t < 1$ and $p(t) = 0$ otherwise.

(a) Find the largest number of users that can be assigned on–off signature waveforms of length N in an orthogonal synchronous CDMA system. Assuming antipodal modulation, describe a receiver for such a system that demodulates all users error-free in the absence of background noise.

(b) Redo (a) for a CDMA system that need not be orthogonal but still requiring that all users be demodulated error-free in the absence of background noise regardless of their received (non-zero) amplitudes. Assume that the receiver knows the received amplitudes.

(c) Redo (b) if the receiver has no knowledge of the received amplitudes.

13

PROBLEM 1.6. Consider a communication channel where the transmitted signal is corrupted by additive independent noise:

$$r(t) = x(t) + n(t).$$

The transmitted signal $x(t)$ is power-limited to P and band-limited to B. The noise process $n(t)$ has power N. Assume that a binary stream can be "transmitted reliably"[11] through this channel if and only if its rate in bits per second is less than or equal to

$$C\left(\frac{P}{N}\right) = B \log_2\left(1 + \frac{P}{N}\right) \text{ bits/sec.} \qquad (1.5)$$

Suppose now that we consider a two-user channel such that

$$x(t) = x_1(t) + x_2(t),$$

where $x_1(t)$ and $x_2(t)$ are the outputs of two independent encoders, power-limited to P_1 and P_2, respectively, and whose binary inputs have rates R_1 and R_2, respectively.

(a) Show that if (R_1, R_2) does not belong to the pentagon

$$C = \left\{ 0 \le R_1 \le C\left(\frac{P_1}{N}\right), \ 0 \le R_2 \le C\left(\frac{P_2}{N}\right), \right.$$

$$\left. R_1 + R_2 \le C\left(\frac{P_1 + P_2}{N}\right) \right\},$$

then reliable communication is not possible. [*Hint:* If the encoders operate independently the total input power cannot exceed $P_1 + P_2$].

(b) Using successive decoding (Figure 1.8) show that reliable communication is possible at the rate pairs (R_1^a, R_2^a) and (R_1^b, R_2^b), where

$$R_1^a = C\left(\frac{P_1}{N}\right), \quad R_2^a = C\left(\frac{P_2}{P_1 + N}\right),$$

$$R_1^b = C\left(\frac{P_1}{P_2 + N}\right), \quad R_2^b = C\left(\frac{P_2}{N}\right).$$

(c) (Carleial [45]) and (Rimoldi and Urbanke [360]). Show that reliable communication is possible if and only if (R_1, R_2) belongs

---

[11] No prior knowledge of information theory is required for this problem; for reference purposes, we say that a data stream can be "transmitted reliably" through the channel if there exists an encoder (mapping from binary strings to $x(t)$) and a decoder (mapping from $r(t)$ to binary strings) such that it is possible to decode the transmitted stream "almost" error-free.

to **C**. [*Hint:* If $(\bar{R}_1, \bar{R}_2)$ is a convex combination of (R_1^a, R_2^a) and (R_1^b, R_2^b), then it must satisfy:

$$\bar{R}_2 = C\left(\frac{P_2}{N + \delta}\right),$$

$$\bar{R}_1 + \bar{R}_2 = C\left(\frac{P_1 + P_2}{N}\right)$$

for some $0 \le \delta \le P_1$. Consider an encoder for user 1 that multiplexes its incoming binary stream into two independent encoders with powers δ and $P_1 - \delta$, respectively. Use a successive decoder for a three-user channel.]

(d) Consider a general form of TDMA where users may have different priorities: users 1 and 2 remain silent in alternating slots of length $(1 - \alpha)$ time units and α time units, respectively. Find the set of rate pairs for which reliable communication is feasible when encoding is restricted to be TDMA (for all possible values of α). Show that TDMA is optimal (i.e., it achieves a point on the boundary of **C**) if and only if

$$\alpha = \frac{P_2}{P_1 + P_2}.$$

[*Hint:* The power constraints are averaged over all time, not just during the activity slots.]

PROBLEM 1.7. *Guard times in CDMA.* Let $\tilde{s}_1(t)$ denote the response of an ideal lowpass filter of bandwidth B to the rectangular signal $s_1(t)$ of duration T in Figure 1.4. Plot $0 < \alpha < 1$ as a function of BT so that

$$\int_0^\infty \tilde{s}_1(t)\tilde{s}_1(t - T - \alpha T)\,dt \le 0.95\|\tilde{s}_1\|^2.$$

(*Note:* The above crosscorrelation does not take into account interference between nonadjacent signals.)

PROBLEM 1.8. The root mean square (RMS) bandwidth, B (also called Gabor[12] bandwidth), of a signal, $s(t)$, time-limited to $[0, T]$ and with Fourier transform $S(f)$ is defined by

$$B^2 = \frac{\int_{-\infty}^{\infty} f^2 |S(f)|^2 df}{\int_{-\infty}^{\infty} |S(f)|^2 df} = \frac{1}{4\pi^2}\frac{\int_0^T \left(\frac{ds(t)}{dt}\right)^2 dt}{\int_0^T s^2(t)\,dt}.$$

---

[12] In honor of Dennis Gabor, 1971 Nobel laureate in Physics, who first defined this quantity.

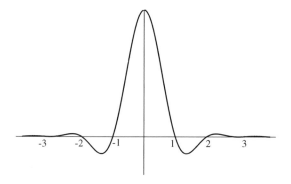

FIGURE 1.12.
Raised cosine
pulse with
$\alpha = 0.5$.

(a) Show that among those signals time-limited to $[0, T]$

$$s(t) = \sin \frac{\pi t}{T}$$

achieves minimum RMS bandwidth, equal to $1/(2T)$.

(b) Consider a K-user TDMA system operating at a per-user rate of R bits per second, with antipodally modulated strictly non-overlapping waveforms. Show that the required RMS band-width is equal to

$$B = \frac{1}{2} RK.$$

(c) (Nuttall and Amoroso [304]) Consider a K-user orthogonal CDMA system operating at a per-user rate of R bits per second and antipodal modulation. Denote the RMS bandwidth of the kth signature waveform by B_k. Show that

$$\min \max\{B_1, \ldots, B_K\} = \frac{1}{\sqrt{12}} R \sqrt{(K+1)\left(K + \frac{1}{2}\right)},$$

where the minimum is over the choice of signature waveforms. [*Hint:* Consider first the alternative criterion: $\min \frac{1}{K} \sum_{k=1}^{K} B_k^2$.]

PROBLEM 1.9. The *raised cosine pulse* with bandwidth $(1+\alpha)/2$ ($0 \le \alpha \le 1$) is the signal (Figure 1.12)

$$g_\alpha(t) = \text{sinc}(t) \frac{\cos \alpha \pi t}{1 - (2\alpha t)^2}$$

with Fourier transform (e.g., Proakis [345])

$$G_\alpha(f) = \begin{cases} 1, & 0 \le |f| \le \frac{1-\alpha}{2}; \\ \frac{1}{2}[1 + \cos(\frac{\pi}{2\alpha}(2|f| - 1 + \alpha))], & \frac{1-\alpha}{2} \le |f| \le \frac{1+\alpha}{2}; \\ 0, & |f| \ge \frac{1+\alpha}{2}. \end{cases}$$

(*Note:* It can be checked that at least 91% of the energy of the raised cosine pulse with bandwidth $(1+\alpha)/2$ resides in an interval of duration $2/(1 + \alpha)$), for all $0 \le \alpha \le 1$.)

(a) Consider a K-user multiaccess system where users transmit packets of M consecutive bits. User k transmits bit i in packet j by antipodal modulation of

$$g_\alpha\left(\frac{t}{T} - i - kM - jKM\right),$$

where $T = 1/(RK)$, and R is the per-user data rate in bits per second. Show that a receiver that samples the received waveform at multiples of T recovers the transmitted information error-free in the absence of any signals other than the superposition of the K transmitted signals. Show that the system is strictly band-limited to

$$B = \frac{(1 + \alpha)}{2} RK.$$

(b) Suppose that the sampling system used in (a) is subject to jitter and the received signal is sampled at times $nT + \epsilon T$. Give an equation that characterizes the minimum value of α necessary for the error-free condition in (a) to be satisfied. Assume that all users are received with equal power.

PROBLEM 1.10. Design a K-user antipodally modulated TDMA system with nonoverlapping signals and per-user rate equal to R bits per second, such that

$$B_{0.77} \leq \frac{1}{2} RK,$$

$$B_{0.86} \leq \frac{2}{3} RK,$$

where B_γ is the bandwidth within which a fraction γ of the energy is concentrated. [*Hint:* Consider the sinc function.]

PROBLEM 1.11. Redo Problem 1.10 under the specification

$$B_{0.95} \leq RK.$$

PROBLEM 1.12. (Liu et al. [249]) Consider a hybrid FDMA/CDMA system that has J nonoverlapping frequency bands in each of which users employ CDMA. Suppose that each of the K transmitters selects one of the J bands equiprobably and independently of other transmitters. Fix $\alpha > 0$. Show that the probability that more than $(1 + \alpha)K/J$ users select the jth band is upper bounded by

$$e^{-g(\alpha)K/J},$$

where

$$g(\alpha) = \frac{\alpha^2}{2}\left(1 - \frac{2\alpha}{3}\right).$$

Note: A standard large deviations result (Grimmett and Stirzaker [128]) says that if p is the probability that a biased coin shows heads, then

$$P[H_n \geq an] \leq e^{-nd(a\|p)}, \tag{1.6}$$

where H_n is the number of heads in n tosses and $p < a < 1$. In (1.6) we have employed the notation

$$d(a\|p) \stackrel{\text{def}}{=} a \log \frac{a}{p} + (1-a) \log \frac{1-a}{1-p}.$$

Furthermore, upon taking $\frac{1}{n}$ log of both sides of (1.6) the inequality becomes asymptotically tight.

CHAPTER TWO

CODE-DIVISION MULTIPLE-ACCESS CHANNELS

In this chapter we review several channel models that incorporate the major features of conceptual and practical interest in Code-Division Multiple Access. These channel models express the signal seen by the receiver as a function of the transmitted data and the channel noise. Sections 2.1 and 2.2 present the basic continuous-time CDMA channel model in its synchronous and asynchronous versions, respectively. Encompassing the majority of multiaccess systems with cochannel interference, this multiuser communication model is the main focus of interest in this text. Section 2.3 gives an overview of spread-spectrum signature waveforms, and Sections 2.4–2.8 discuss several ways in which the basic models can be modified to make them more realistic. Sections 2.9 and 2.10 introduce several discrete-time CDMA models used in succeeding chapters.

2.1 BASIC SYNCHRONOUS CDMA MODEL

Many of the essential ideas developed in succeeding chapters will originate from the analysis of the basic CDMA K-user channel model, consisting of the sum of antipodally modulated synchronous signature waveforms embedded in additive white Gaussian noise:

$$y(t) = \sum_{k=1}^{K} A_k b_k s_k(t) + \sigma n(t), \quad t \in [0, T]. \tag{2.1}$$

The notation introduced in (2.1) is defined as follows.

- T is the inverse of the data rate.
- $s_k(t)$ is the deterministic signature waveform assigned to the kth user, normalized so as to have unit energy

$$\|s_k\|^2 = \int_0^T s_k(t)\,dt = 1. \tag{2.2}$$

The signature waveforms are assumed to be zero outside the interval $[0, T]$, and therefore, there is no intersymbol interference.
- A_k is the received amplitude of the kth user's signal. A_k^2 is referred to as the energy of the kth user.
- $b_k \in \{-1, +1\}$ is the bit transmitted by the kth user.
- $n(t)$ is white Gaussian noise with unit power spectral density. It models thermal noise plus other noise sources unrelated to the transmitted signals. According to (2.1) the noise power in a frequency band with bandwidth B is $2\sigma^2 B$.[1]

As we will see, the performance of various demodulation strategies depends on the signal-to-noise ratios, A_k/σ, and on the similarity between the signature waveforms, quantified by their crosscorrelations defined as

$$\rho_{ij} = \langle s_i, s_j \rangle = \int_0^T s_i(t)s_j(t)\,dt. \tag{2.3}$$

Note that by the Cauchy–Schwarz inequality and (2.2):

$$|\rho_{ij}| = |\langle s_i, s_j \rangle| \le \|s_i\|\|s_j\| = 1.$$

The crosscorrelation matrix

$$\mathbf{R} = \{\rho_{ij}\} \tag{2.4}$$

has diagonal elements equal to 1 and is symmetric nonnegative definite, because for any K-vector $\mathbf{a} = (a_1, \cdots, a_K)^T$

$$\mathbf{a}^T \mathbf{R} \mathbf{a} = \left\| \sum_{k=1}^K a_k s_k \right\|^2 \ge 0.$$

Therefore the crosscorrelation matrix $\mathbf{R}$ is positive definite if and only if the signature waveforms $\{s_1, \ldots, s_K\}$ are linearly independent.

It is pedagogically sound to work with the minimal number of parameters that bring out the essential features of a problem. For this reason, we will frequently pay particular attention to the basic CDMA channel model in the

---

[1] In the literature, the noise one-sided spectral level $2\sigma^2$ is frequently denoted by N_0.

special case of two users:

$$y(t) = A_1 b_1 s_1(t) + A_2 b_2 s_2(t) + \sigma n(t). \tag{2.5}$$

In the two-user synchronous case we will find it convenient to drop the subindices from the crosscorrelation

$$\rho = \int_0^T s_1(t) s_2(t) \, dt.$$

2.2 BASIC ASYNCHRONOUS CDMA MODEL

In the synchronous model considered in Section 2.1 bit epochs are aligned at the receiver. This requires closed-loop timing control or providing the transmitters with access to a common clock (such as the Global Positioning System). In cellular systems, the design of the reverse link (mobile-to-base station) is considerably simplified if the users need not be synchronized.

As we saw in Chapter 1, symbol-synchronism is not necessary for CDMA to operate, and it is possible to let the users transmit completely asynchronously. We need to introduce offsets that model the lack of alignment of the bit epochs at the receiver: $\tau_k \in [0, T), k = 1, \ldots, K$ (Figure 2.1). The symbol-epoch offsets are defined with respect to an arbitrary origin (we will often take $\tau_1 = 0$). Note that we still require that the data rate $1/T$ be identical for all users. In practice, even if the data rates are nominally the same, small differences translate into (slowly) time-varying offsets. Whenever performance is to be averaged with respect to the relative offsets, it makes sense to assume that they are independent, uniformly distributed on $[0, T]$.

The synchronous model (2.1) is a one-shot model, as in that case it is sufficient to restrict attention to the received waveform in an interval of

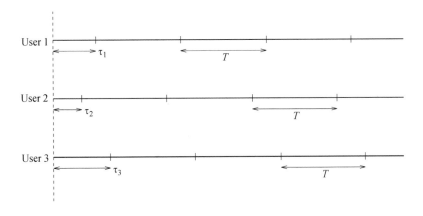

FIGURE 2.1.
Offsets modeling
asynchronism

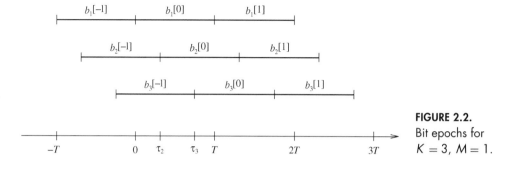

FIGURE 2.2.
Bit epochs for
$K = 3$, $M = 1$.

length T, the bit duration. In the asynchronous case (cf. Figure 2.1) we must take into account the fact that the users send a stream of bits:

$$b_k[-M], \ldots, b_k[0], \ldots, b_k[M],$$

where we have assumed without essential loss of generality that the length of the frames or packets transmitted by each user is equal (to $2M + 1$). This assumption can be readily dropped at the expense of more cumbersome notation. Generalizing (2.1) to the asynchronous case, the CDMA channel model now becomes

$$y(t) = \sum_{k=1}^{K} \sum_{i=-M}^{M} A_k b_k[i] s_k(t - iT - \tau_k) + \sigma n(t). \qquad (2.6)$$

In the context of this model, users initiate and terminate their transmissions within T time units of each other, which presupposes some form of block-synchronism, if not symbol-synchronism. This assumption allows us to focus on offsets modulo-T and does not impact the generality of the analysis because of the typically large values of M. Figure 2.2 shows the symbol epochs for three asynchronous users in a case where $M = 1$.

The synchronous channel (2.1) corresponds to the special case of (2.6) where all the offsets are identical,

$$\tau_1 = \cdots = \tau_K, \qquad (2.7)$$

in which case, we may as well consider the one-shot version of (2.6), if, as we assumed in Section 2.1, there is no intersymbol interference.

It is worth pointing out another special case of (2.6): the case in which all the received amplitudes and all the signature waveforms are equal:

$$A_1 = \cdots = A_K = A \qquad (2.8)$$

and

$$s_1 = \cdots = s_K = s, \qquad (2.9)$$

and in which the offsets satisfy

$$\tau_k = \frac{(k-1)T}{K}.\tag{2.10}$$

In this particular case, (2.6) becomes

$$y(t) = \sum_{k=1}^{K} \sum_{i=-M}^{M} A b_k[i] s(t - iT - (k-1)T/K) + \sigma n(t)$$

$$= \sum_j A b[j] s(t - jT/K) + \sigma n(t),\tag{2.11}$$

where we have denoted

$$b[iK + k - 1] = b_k[i].$$

The channel in (2.11) is, in fact, the standard single-user white Gaussian channel with signals subject to intersymbol interference. In this model, which occupies a prominent role in communication theory, every bit overlaps with a fixed finite number of neighboring bits. Since the waveform s in (2.11) has duration T, every bit $b[j]$ in (2.11) overlaps with $2K - 2$ bits.

As an example, consider the intersymbol interference scenario depicted in Figure 2.3: every symbol overlaps with three preceding symbols and three succeeding symbols. This is equivalent to an asynchronous CDMA channel with four "users" (each free from intersymbol interference) where the signature waveforms are common to all users and the offsets are $\tau_1 = 0$, $\tau_2 = T/4, \tau_3 = T/2$, and $\tau_4 = 3T/4$. Each "user" carries every fourth bit of the original data stream, so the individual data rate for each "user" is $1/4$ of the original data rate. As we will see, many of the demodulation strategies used for channel (2.11) find counterparts in the more general channel (2.6). Moreover, oftentimes we will find that it is conceptually advantageous to use the synchronous CDMA channel as the starting point for the design and analysis of demodulators for asynchronous CDMA.

Synchronous or asynchronous CDMA systems with intersymbol interference are also encountered in applications. Intersymbol interference may be present due to channel distortion (Section 2.6) or introduced intentionally (for example, to increase the signature time-bandwidth product) in which case it is also known as partial-response signaling. The models in (2.1) and (2.6) remain valid in the presence of intersymbol interference, but the signature waveforms are longer than T and the one-shot approach used in Section 2.1 is then suboptimal.

TDMA can also be seen as a special case of the asynchronous CDMA model in (2.6): the same signature waveform is assigned for all users (2.9)

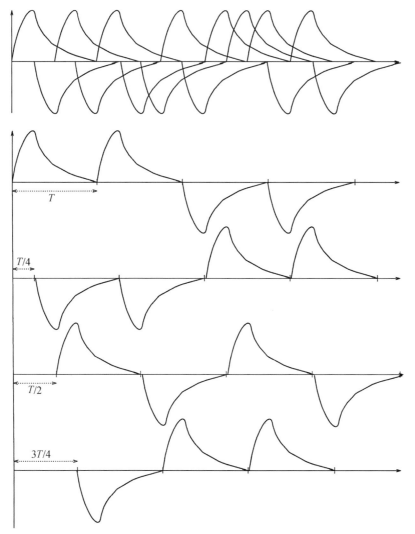

FIGURE 2.3.
Intersymbol interference as an asynchronous CDMA channel.

and the consecutive offsets are T/K apart (2.10). Except for the possibly unequal received powers in TDMA, this special case is conceptually similar to the intersymbol interference model (2.11). Ideal orthogonal TDMA satisfies

$$\int s\left(t - \frac{iT}{K}\right) s\left(t - \frac{jT}{K}\right) dt = 0, \qquad (2.12)$$

for $i \neq j$. For example, (2.12) is satisfied if $s(t)$ has duration T/K. Due to channel dispersion, (2.12) may fail to hold in nonideal TDMA (particularly for $|i - j| = 1$), thereby resulting in cochannel interference. TDMA systems where users transmit more than one bit on each time slot (cf. Footnote 5 of Chapter 1) require a more general model than (2.6).

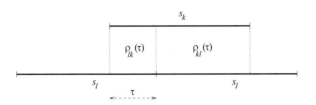

FIGURE 2.4.
Definition of asynchronous crosscorrelations ($k < l$).

We have already noticed that the synchronous model in (2.1) can be put as a special case of the asynchronous model in (2.6), in the sense that any asynchronous demodulation technique (and its analysis) must work for any arbitrary set of offsets including $\tau_1 = \cdots = \tau_K = 0$. Yet, the synchronous channel (2.1) is a canonical model in the sense that we can view the asynchronous model (2.6) as a special case. Think of each bit in (2.6), $\{b_k[i], k = 1, \ldots, K, i = -M, \ldots, M\}$, as coming from a different "user" in a synchronous channel whose bit interval is $[-MT, MT + 2T]$. In this view, the number of fictitious users is equal to $(2M + 1)K$. For example, the asynchronous 3-user case depicted in Figure 2.2 can be seen as a 9-user synchronous channel. Sometimes in the upcoming development, we will find that this view fails to fully exploit the structure of the problem; often, however, it will facilitate the analysis of asynchronous multiuser detectors.

As we mentioned in Section 2.1, performance depends on the signature waveforms through their crosscorrelations. However, in the asynchronous channel the "synchronous" crosscorrelations

$$\rho_{kl} = \int_0^T s_k(t)s_l(t)\, dt$$

are no longer sufficient to determine performance. In asynchronous CDMA, we must define two crosscorrelations between every pair of signature waveforms that depend on the offset between the signals. If $k < l$, then we denote

$$\rho_{kl}(\tau) = \int_\tau^T s_k(t)s_l(t - \tau)\, dt, \tag{2.13a}$$

$$\rho_{lk}(\tau) = \int_0^\tau s_k(t)s_l(t + T - \tau)\, dt, \tag{2.13b}$$

where $\tau \in [0, T]$. Notice that the first subindex of $\rho_{kl}(\tau)$ denotes the "left" signal in the correlation (Figure 2.4). The length of the integration interval is τ or $T - \tau$ according as the sign of $l - k$. For example, it can be checked that the crosscorrelations of the (unit-energy) signature waveforms in Figure 1.4 are given by

$$\rho_{12}(\tau) = -\rho_{21}(\tau) = \frac{\sqrt{3}}{T^2}\tau(\tau - T).$$

25

If the users are labeled so that their offsets are increasing, then for a given set of offsets, the value of τ of interest in (2.13) is equal to $\tau_l - \tau_k$. With a slight abuse of notation, we will denote

$$\rho_{kl} = \rho_{kl}(\tau_l - \tau_k),$$

where in contrast to the "synchronous" crosscorrelations, ρ_{kl} is, in general, asymmetric.[2]

In the foregoing special cases of single-user intersymbol interference and TDMA, (2.9) and (2.10), the crosscorrelations are

$$\rho_{kl} = \begin{cases} R\left(\frac{l-k}{K}T\right), & l > k, \\ R\left(T - \frac{l-k}{K}T\right), & l < k, \end{cases}$$

where the autocorrelation function of s is denoted by

$$R(\tau) = \int s(t)s(t-\tau)\,dt.$$

2.3 SIGNATURE WAVEFORMS

2.3.1 DIRECT-SEQUENCE SPREAD SPECTRUM

For many results on multiuser detection, it is not necessary to place any specific structure on the signature waveforms. Yet, it is instructive to examine the structure of the signature waveforms employed in many CDMA systems, namely, *direct-sequence spread spectrum*. As we saw in Chapter 1, spread-spectrum signaling formats feature large duration-bandwidth products. Direct-sequence refers to a specific approach to construct spread-spectrum waveforms, characterized by:

1. chip waveform p_{T_c} such that

$$\int_{-\infty}^{\infty} p_{T_c}(t)p_{T_c}(t - nT_c)\,dt = 0, \quad n = 1, 2, \ldots, \tag{2.14}$$

2. number of chips per bit N, and
3. binary sequence of length N: $(c_1, \ldots, c_N)$.

If the binary sequence is used to modulate the chip waveform antipodally

---

[2] The sequence-design literature refers to the asynchronous crosscorrelations ρ_{jk}, ρ_{kj} as *partial crosscorrelations* and to $\rho_{jk} + \rho_{kj}$ and $\rho_{jk} - \rho_{kj}$ as the *even and odd periodic crosscorrelations* respectively.

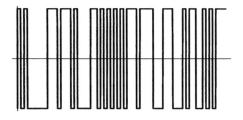 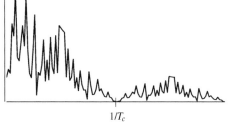

$1/T_c$

FIGURE 2.5.
Direct-sequence
spread-spectrum
signature
waveform with
$N = 63$ in the
time and
frequency
domains
(rectangular chip
waveform).

we obtain the direct-sequence spread-spectrum waveform with duration NT_c:

$$s(t) = A \sum_{i=1}^{N} (-1)^{c_i} p_{T_c}(t - (i - 1)T_c). \qquad (2.15)$$

It is important to realize that the binary sequence $(c_1, \ldots, c_N)$ does not carry information: it is assumed to be known by the intended receiver. An example of a direct-sequence spread-spectrum waveform is shown in Figure 2.5. In that case, the chip waveform is the rectangular pulse

$$p_{T_c}(t) = \begin{cases} 1, & \text{if } 0 \leq t < T_c, \\ 0, & \text{otherwise.} \end{cases}$$

The magnitude of the Fourier transform (in linear scale) of the direct-sequence waveform shown in Figure 2.5 shows considerable frequency content beyond the spectral null at $1/T_c$. For the sake of spectral efficiency, some CDMA systems choose a smooth chip waveform with very little energy beyond the chip rate. For example, Figure 2.6 shows the same binary sequence used in Figure 2.5, but with the chip waveform

FIGURE 2.6.
Direct-sequence
spread-spectrum
signature
waveform with
$N = 63$ in the
time and
frequency
domains (sinc
chip waveform).

$$p_{T_c}(t) = \text{sinc}\left(\frac{2t}{T_c} - 1\right),$$

and with the overall signature waveform truncated to the interval $[0, NT_c]$.

 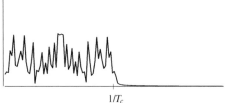

$1/T_c$

27

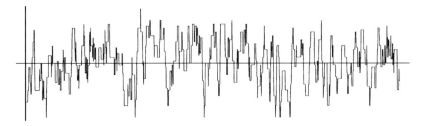

FIGURE 2.7.
Noiseless sum of six modulated direct-sequence waveforms with $N = 63$.

With this choice, the spectrum of the direct-sequence waveform falls abruptly beyond $1/T_c$ (Figure 2.6).

Figure 2.7 shows the noiseless superposition over two bit periods (of user 1) of six asynchronous antipodally modulated direct-sequence waveforms ($N = 63$ with rectangular chip waveforms). The received amplitudes of the six users are within 5 dB of each other. The realization shown in Figure 2.7 would be typical of a broadband noise process. Yet, from the viewpoint of a receiver that knows the signature waveforms, received amplitudes, and timing of each user, the waveform shown in Figure 2.7 only has 17 bits of uncertainty.[3]

2.3.2 SPREADING FACTOR

We now summarize some of the effects of the choice of the number of chips per symbol N, also known as the *spreading factor, spreading gain,* or *processing gain*:

- For fixed duration of the signature waveform, its bandwidth is proportional to N.
- For a given signal-to-noise ratio, the single-user bit-error-rate in a white Gaussian noise channel is independent of N (Chapter 3).
- In an orthogonal synchronous direct-sequence spread-spectrum CDMA system, the number of users that can be supported is less than or equal to N (Problem 2.1).
- For a given maximum crosscorrelation level, the number of synchronous or asynchronous signature waveforms grows with N (Problem 2.2). The variance of the crosscorrelation between two signature waveforms with randomly and independently chosen codes is $1/N$ (Section 2.3.5).
- Large values of N contribute to the *privacy* of the system, as they hinder unintended receivers wishing to unveil signature waveforms so as to eavesdrop on the transmitted information.

---

[3] The figure shows two full periods of one user and three (partial) periods of the other five users.

- Because the energy spectral level of the signature waveform is inversely proportional to N, large values of N contribute to reduce the interference caused on coexisting narrowband transmissions. For the same reason, *noise camouflage*[4] from an interceptor that monitors the radio spectrum is enhanced with increasing values of N.
- The performance degradation caused by a narrowband jammer decreases with N. In the absence of background noise, a narrowband jammer with power up to $10 \log_{10} N$ dB above the spread-spectrum signal induces no errors (Problem 2.17).
- The reliability with which the time of arrival of the signal can be determined increases with N (cf. (2.16)). This facilitates timing synchronization at the receiver and the ability to combat multipath.

2.3.3 SIGNATURE SEQUENCES

Ordinarily, both the number N of chips per bit and the chip waveform are common to all users in a direct-sequence spread-spectrum CDMA system. What distinguishes the different signature waveforms is the assignment of the binary "sequence" or "code" $(c_1, \ldots, c_N)$. A wealth of combinatorial techniques exists for constructing *pseudonoise* (PN) signature sequences, which, for given N and K, achieve low crosscorrelations for all possible offsets. Foremost among those techniques are the Gold, Kasami, and bent sequences (Simon et al. [410]).

The normalized crosscorrelation between two synchronous direct-sequence waveforms with signature sequences $(c_{j1}, \ldots, c_{jN})$ and $(c_{k1}, \ldots, c_{kN})$ is given by

$$\rho_{jk} = -1 + \frac{2}{N} \sum_{i=1}^{N} 1\{c_{ji} = c_{ki}\}.$$

Pseudonoise sequences can be (but need not be) easily generated by linear feedback shift registers such as the one shown in Figure 2.8. There are a total of 2^r different states if r is the number of binary registers. Therefore, the output of the pseudonoise shift-register generator is periodic, with a period determined by the taps and the initial state. The taps are chosen so as to prevent the state from ever becoming all-zero because in such an event it would remain there forever. Discounting this forbidden state, we see that the periodicity of the output sequence can never be larger than $2^r - 1$. Sequences (or equivalently the linear feedback shift registers that

---

[4] Usually referred to as *low probability of intercept* in the spread-spectrum literature.

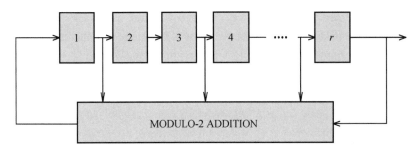

FIGURE 2.8.
Generation of
pseudonoise
sequence with a
feedback shift
register.

produce them) that achieve this bound are called *maximal-length*. Their
use is widespread because they achieve the most noiselike sequence with
a given amount of shift register complexity. For example, if we let $r = 4$
and the new value of register 1 is equal to its previous value plus the value
of register 4 (in modulo-2 arithmetic), then the shift register puts out a
periodic repetition of the 15-bit sequence 000111101011001. Since the
periodicity is exponential in the number of registers, it is very simple to
generate pseudonoise sequences with very long periods. Maximal-length
shift-register sequences have pseudorandom properties such as:

a) There are 2^{r-1} ones and $2^{r-1} - 1$ zeros.
b) The periodic autocorrelation function of the generated sequence c_1,
   $c_2, \ldots$ is

$$R(\kappa) = \frac{1}{N} \sum_{i=1}^{N} (-1)^{c_i + c_{i+\kappa}} = \begin{cases} 1, & \kappa = 0, N, 2N, \ldots, \\ -1/N, & \text{otherwise,} \end{cases} \quad (2.16)$$

with $N = 2^r - 1$. This means that if we compare any two strings of length
N within the sequence, they are either the same string or the number of
agreements is equal to the number of discrepancies minus 1.

2.3.4   LONG SEQUENCES

An alternative way to view a direct-sequence spread-spectrum
waveform is as the product of a signal modulated at the data rate and a
periodic repetition of a pseudonoise waveform:

$$Ag(t) \sum_{i=-M}^{M} b[i]p(t - iT), \quad (2.17)$$

with

$$p(t) = \begin{cases} 1, & \text{if } 0 \le t < T, \\ 0, & \text{otherwise,} \end{cases}$$

30

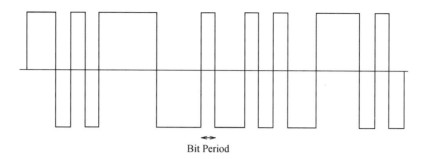

Bit Period

PN Sequence Period

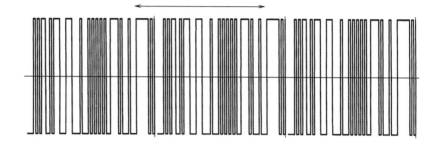

FIGURE 2.9.
Direct-sequence
spread spectrum
system with
$N_0 = 63$ and
$N = 7$.

and

$$g(t) = \sum_{j=-\infty}^{\infty} \sum_{n=1}^{N_0} (-1)^{c_n} p_{T_c}(t - (n + jN_0 - 1)T_c),$$

where (2.15) coincides with this model if $T = NT_c$ and $N_0 = N$. In some practical CDMA systems (e.g., IS-95 Standard [442]), the period of the pseudonoise waveform is larger than the symbol period (Figure 2.9),

$$L = N_0/N > 1.$$

In those systems, the foregoing conclusions about the choice of N apply verbatim, except in the case of the privacy feature where N_0 replaces N. In particular, the processing gain is determined by the number of chips per bit rather than the periodicity of the pseudonoise sequence. If $L > 1$, then the CDMA channel no longer fits the basic model in Sections 2.1 and 2.2. It is necessary to replace the kth user stream

$$\sum_{i=-M}^{M} A_k b_k[i] s_k(t - iT)$$

by

$$\sum_{i=-M}^{M} A_k b_k[i] s_{k,(i)_L}(t - iT),$$

where $(i)_L$ denotes i modulo L, and the concatenation of $s_{k,1}, \ldots, s_{k,L}$ gives one period of the PN sequence assigned to user k.

As we saw, maximal-length shift-register sequences of very long periodicity are easy to generate; thus, it is easy to implement transmitters where L, the ratio of the periodicity of the pseudonoise waveform to the spreading gain, is very large.[5] In that case, no attempt is made to design signature waveforms with low crosscorrelations, and the large values of L enable the approximation of signature codes as purely random for the purposes of analysis. As we mentioned, except for privacy, the benefits of large spreading gains are not enhanced by choosing $L > 1$. Large values of L hinder the synchronization task of an eavesdropper. At any rate, encryption of the information streams is a more robust way to obtain privacy, and the use of some multiuser demodulation strategies becomes more cumbersome when $L > 1$, as, in that case, the crosscorrelations vary at the data rate.

2.3.5   RANDOM SEQUENCES

Averaging performance of multiuser detectors with respect to the choice of direct-sequence spread-spectrum signatures is relevant to systems with large L. Furthermore, such average analyses give a baseline of comparison (as a function of K and N) for systems with $L = 1$ and signature sequences designed to achieve low crosscorrelations.

In the random signature model, we have

$$s_k(t) = \frac{1}{\sqrt{N}} \sum_{i=1}^{N} d_{ki} \, p_{T_c}(t - iT_c + T_c), \qquad (2.18)$$

where $\{d_{ki} \in \{-1, +1\}, k = 1, \ldots, K; i = 1, \ldots, N\}$ are independent and equally likely and the deterministic chip waveform p_{T_c} is assumed to have unit energy and autocorrelation function

$$R_p(\tau) = \int_{-\infty}^{\infty} p_{T_c}(t) p_{T_c}(t - \tau) \, dt. \qquad (2.19)$$

Note that (2.14) implies

$$R_p(nT_c) = 0, \quad n \neq 0. \qquad (2.20)$$

The asynchronous crosscorrelation function (2.13a) can be expressed in terms of signature codes and chip autocorrelation function as

$$\rho_{kl}(\tau) = \int s_k(t) s_l(t - \tau) \, dt$$

$$= \frac{1}{N} \sum_{i=1}^{N} \sum_{j=1}^{N} d_{ki} d_{lj} R_p(\tau + (j - i)T_c). \qquad (2.21)$$

---

[5] In IS-95 Standard [442], $N = 128$, $N_0 = 2^{42} - 1$, $L \approx 2^{35}$.

Clearly, since $\{d_{ki} \in \{-1, +1\}, k = 1, \ldots, K; i = 1, \ldots, N\}$ are independent and equally likely,

$$E[\rho_{kl}(\tau)] = 0, \qquad (2.22)$$

for $k \neq l$. The second moment of the crosscorrelation function is

$$E\left[\rho_{kl}^2(\tau)\right] = \frac{1}{N^2} \sum_{i=1}^{N} \sum_{j=1}^{N} \sum_{m=1}^{N} \sum_{n=1}^{N} E[d_{ki} d_{lj} d_{km} d_{ln}]$$

$$\times R_p(\tau + (j - i)T_c) R_p(\tau + (n - m)T_c)$$

$$= \frac{1}{N^2} \sum_{i=1}^{N} \sum_{j=1}^{N} R_p^2(\tau + (j - i)T_c). \qquad (2.23)$$

Taking into account (2.20), we can readily particularize (2.23) at $\tau = 0$ to obtain the second moment of the synchronous crosscorrelation $\rho_{kl} = \rho_{kl}(0)$:

$$E\left[\rho_{kl}^2\right] = \frac{1}{N^2} \sum_{i=1}^{N} R_p^2(0)$$

$$= \frac{1}{N}. \qquad (2.24)$$

In the asynchronous case, it is useful to further average (2.23) with respect to the uniformly distributed delay $\tau \in [0, T]$:

$$\frac{1}{T} \int_0^T E\left[\rho_{kl}^2(\tau)\right] dt = \frac{1}{N^2 T} \sum_{i=1}^{N} \sum_{j=1}^{N} \int_0^T R_p^2(\tau + (j - i)T_c) \, d\tau$$

$$= \frac{1}{2N^2 T} \sum_{i=1}^{N} \sum_{j=1}^{N} \int_{-T}^{T} R_p^2(\tau + (j - i)T_c) \, d\tau. \qquad (2.25)$$

The evaluation of (2.25) is considerably simplified if $R_p(\tau) = 0$ for $|\tau| \geq T_c$, that is, when the chip waveform has duration T_c. Then, (2.25) becomes

$$\frac{1}{T} \int_0^T E\left[\rho_{kl}^2(\tau)\right] dt = \frac{1}{2N^2 T} \sum_{i=1}^{N} \sum_{j=1}^{N} \int_{-T_c}^{T_c} R_p^2(\tau) \, d\tau$$

$$= \frac{1}{NT_c} \int_0^{T_c} R_p^2(\tau) \, d\tau. \qquad (2.26)$$

In the special case of rectangular chip waveforms

$$R_p(\tau) = 1 - \frac{\tau}{T_c}, \quad 0 \leq \tau \leq T_c,$$

and (2.26) particularizes to

$$\frac{1}{T} \int_0^T E\left[\rho_{kl}^2(\tau)\right] dt = \frac{1}{N} \int_0^1 (1 - x)^2 \, dx$$

$$= \frac{1}{3N}. \qquad (2.27)$$

33

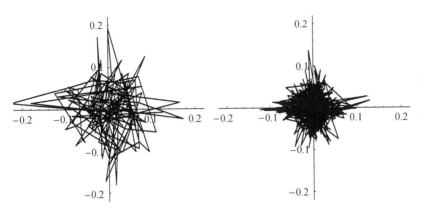

FIGURE 2.10.
Locus of crosscorrelations (ρ_{12}, ρ_{21}) for $N = 128$ (left) and $N = 512$ (right) with rectangular chips.

Obviously, by symmetry, the "left" crosscorrelation (2.13b) has the same moments when the delay is uniformly distributed.

Figure 2.10 shows a realization of the locus of pairs of partial crosscorrelations

$$(\rho_{12}(\tau), \rho_{21}(\tau)) \quad \text{for all } \tau \in [0, T]$$

for direct-sequence spread-spectrum waveforms with randomly selected signature sequences and rectangular chip waveforms. According to (2.27), the standard deviation of the crosscorrelations in the case $N = 512$ is half of that in $N = 128$, which is validated in Figure 2.10.

For future reference, we pause to give a key asymptotic ($K \to \infty$) result that we will use to analyze performance when the choice of direct-sequence spread-spectrum codes is random. This result gives the asymptotic distribution of the eigenvalues of the crosscorrelation matrix **R**:

Proposition 2.1 *(Bai and Yin [19]) Suppose that K users employ direct-sequence spread-spectrum waveforms with N chips per symbol. Let*

$$\lim_{K \to \infty} \frac{K}{N} = \beta \in (0, \infty).$$

*Suppose that the choice of signature sequences is completely random: the sequences assigned to each user are independent, and all binary sequences are equally likely. Then, the percentage of the K eigenvalues of **R** that lie below x converges (as $K \to \infty$) to the cumulative distribution function of the probability density function*

$$f_\beta(x) = [1 - \beta^{-1}]^+ \delta(x) + \frac{\sqrt{[x - a]^+ [b - x]^+}}{2\pi\beta x}, \qquad (2.28)$$

where

$$[z]^+ = \max\{0, z\},$$

and

$$a = (1 - \sqrt{\beta})^2,$$
$$b = (1 + \sqrt{\beta})^2. \tag{2.29}$$

Moreover, if $\beta \leq 1$, the smallest eigenvalue converges almost surely to a.

2.3.6 OTHER SPREAD-SPECTRUM FORMATS

Direct-sequence is not the only spread-spectrum signaling format suitable for CDMA. The direct-sequence format (2.15) can be generalized in several directions:

1. The "codes" need not be binary valued.
2. While preserving their orthogonality, the N chip waveforms need not be delayed versions of each other.

Such a generalization is reflected in the following spread-spectrum model:

$$s(t) = A \sum_{i=1}^{N} \beta_i \psi_i(t), \tag{2.30}$$

where $(\beta_1, \ldots, \beta_N)$ is the spreading code and the N deterministic signals ψ_i satisfy the orthogonality conditions

$$\langle \psi_i, \psi_l \rangle = 0,$$

if $i \neq l$. Model (2.30) particularizes to direct sequence and to multicarrier CDMA:

DIRECT SEQUENCE

In direct-sequence spread-spectrum, the chip waveforms are delayed versions of each other:

$$\psi_i(t) = p_{T_c}(t - (i - 1)T_c), \tag{2.31}$$

where recall from (2.14) that the chip waveform p_{T_c} is orthogonal to any version of itself delayed an integer multiple of T_c. In the special case of direct-sequence spread-spectrum we have considered until now, $\beta_j \in \{-1, 1\}$. (See Problem 2.16 for the more general nonbinary setting.)

35

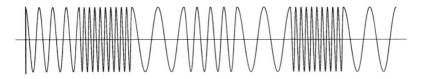

FIGURE 2.11.
Frequency-
hopping
spread-spectrum
signature
waveform, $N = 7$.

MULTICARRIER CDMA

The dual of (2.31) in the frequency domain is obtained by considering chip waveforms whose Fourier transform satisfies

$$\Psi_i(f) = p_{1/T}\left(f - \frac{i-1}{T}\right), \tag{2.32}$$

which implies that

$$\psi_i(t) = e^{j2\pi(i-1)t/T} P_{1/T}(t),$$

where $P_{1/T}$ is the inverse Fourier transform of $p_{1/T}$. In this model, β_j is usually complex-valued.

FREQUENCY HOPPING

We could even lift the restriction that chip waveforms are linearly modulated by the code coefficients, in which case we get

$$s(t) = A \sum_{i=1}^{N} \psi_i(t; \beta_i), \quad \beta_i \in \mathcal{A}, \tag{2.33}$$

where

$$\langle \psi_i(\cdot; \beta), \psi_l(\cdot; \gamma) \rangle = 0,$$

if $i \neq l$.

A special case of (2.33) is that of frequency hopping:

$$\psi_i(t; \beta_i) = \begin{cases} \cos(2\pi\beta_i t), & \text{if } iT_c - T_c \leq t \leq iT_c; \\ 0, & \text{otherwise.} \end{cases}$$

In *frequency-hopping* spread-spectrum the chips are modulated in frequency, rather than in phase as in direct-sequence. Each of the N chips corresponds to a sinusoid whose frequency is chosen pseudorandomly from a given set of frequencies. Figure 2.11 shows a frequency-hopping spread-spectrum signature waveform, where $N = 7$ and there are three different frequencies ($|\mathcal{A}| = 3$). In a related, but different, signaling strategy called *slow frequency hopping*, each symbol is modulated by a single narrowband signal whose carrier frequency is periodically changed, every few symbols.

2.4 DATA STREAMS

A basic assumption we will use for the design and analysis of multiuser detectors is that all possible data streams

$$\{b_k[-M], \ldots, b_k[M], \ k = 1, \ldots, K\}$$

are equiprobable. This means that bits are independent, equally likely to be $+1$ or -1.

In practice, the raw data may be subject to error-control coding, in which case the assumption of independent bits no longer holds. Other types of data dependencies may be introduced in order to shape the spectrum of the transmitted signals or to satisfy constraints on the digital streams imposed by communication protocols.

In general, it is suboptimal to separate the demodulation and error-control decoding operations. Nevertheless, conceptually and practically, it makes sense to study the demodulation operation separately from error-control decoding. The bit-error-rate for the raw independent data, although not the ultimate performance measure of interest in an error-control coded system, provides an excellent way to compare different demodulation strategies. In practice, complexity considerations usually rule out optimal decoders that work with the received waveform directly. Instead, error-control decoders work either with demodulated data ("hard-decisions") or with real-valued scalars ("soft-decisions") which give an assessment of the reliability of those decisions. Hard-decision demodulators can be modified to that effect.

The separation of demodulation and decoding is akin in spirit to the separation of encoding and modulation described in Chapter 1. Having each user modulate a signature waveform imposes a structure on the transmitted signal that incurs a loss of capacity. Yet, this structure facilitates (particularly when the signature waveforms are spread-spectrum) the challenging task of the receiver.

Differential encoding of the data streams is another feature present in many systems of interest. In differential encoding, the kth user sends

$$\sum_{i=-M}^{M} A_k e_k[i] s_k(t - iT),$$

where the transmitted stream is not the data stream (with or without redundancy) itself but its transitions:

$$e_k[i] = e_k[i-1] b_k[i].$$

The differential encoder is initialized with a bit $e_k[-M-1]$ known to the receiver, which makes decisions on the consecutive products

$$e_k[i-1]e_k[i] = b_k[i].$$

The purpose of differential encoding is to facilitate demodulation in the presence of phase uncertainties. For example, at the expense of some performance loss, differential encoding enables differentially coherent demodulation whereby the signal received in the previous interval is used to provide a carrier-phase reference in lieu of an internally generated sinusoid in phase with the received waveform.

Some systems offer *multirate* capability, whereby the information rates transmitted by various users need not coincide. An obvious way in which a user can multiply its data rate by a factor of k is by serial-to-parallel conversion of its data stream into k data streams, each of which modulates a different signature waveform (Problem 2.20). Alternatively, the spreading gain can be chosen inversely proportional to the data rate while keeping the chip rate (and, hence, bandwidth) of the transmitted signal invariant to the data rate.

2.5 MODULATION

2.5.1 CARRIER MODULATION

The basic CDMA model in Sections 2.1 and 2.2 is a baseband model. It is easy to incorporate frequency translation by multiplying the signature waveforms by sinusoids. Since different users' carriers (even if they have the same frequency) are not phase-synchronized, carrier modulation has the effect of multiplying each crosscorrelation by a random factor. The function $\bar{s}_k$ denotes the unit-energy modulated version of s_k using a carrier with frequency f_c and phase ϕ_k:

$$\bar{s}_k(t) = \frac{s_k(t)\cos(2\pi f_c t + \phi_k)}{\|s_k(t)\cos(2\pi f_c t + \phi_k)\|}.$$

We can write the energy and inner products of the modulated signature waveforms as

$$\|s_k(t)\cos(2\pi f_c t + \phi_k)\|^2$$

$$= \frac{1}{2}\int_0^T s_k^2(t)\,dt + \frac{1}{2}\int_0^T s_k^2(t)\cos(4\pi f_c t + 2\phi_k)\,dt, \qquad (2.34)$$

$$\langle s_k(t)\cos(2\pi f_c t + \phi_k), s_j(t)\cos(2\pi f_c t + \phi_j)\rangle$$

$$= \frac{1}{2}\cos(\phi_k - \phi_j)\langle s_k, s_j\rangle + \frac{1}{2}\int_0^T s_k(t)s_j(t)\cos(4\pi f_c t + \phi_k + \phi_j)\,dt.$$

$$(2.35)$$

38

If, as customary, the bandwidth of the signature waveforms is less than f_c, then the integrals of signals at $2f_c$ in (2.34) and (2.35) can be neglected. Thus,

$$\bar{\rho}_{kj} = \langle \bar{s}_k, \bar{s}_j \rangle = \rho_{kj} \cos(\phi_k - \phi_j). \tag{2.36}$$

Because the phases ϕ_k are independent and uniformly distributed on $[0, 2\pi]$, the probability density function of $\bar{\rho}_{kj}$ is given by

$$f_{\bar{\rho}_{kj}}(r) = \frac{1}{\pi \sqrt{\rho_{kj}^2 - r^2}}, \quad r \in (-\rho_{kj}, \rho_{kj}). \tag{2.37}$$

A similar result is obtained for the asynchronous crosscorrelations (2.13). Conditioned on ρ_{kj}, the moments of $\bar{\rho}_{kj}$ are

$$E[\bar{\rho}_{kj}] = 0,$$

$$E[\bar{\rho}_{kj}^2] = \frac{\rho_{kj}^2}{2}. \tag{2.38}$$

If the asynchronous crosscorrelations are averaged with respect to random sequences, delays, and phases, then their variance is equal to $1/(6N)$, in the case of rectangular chip waveforms (2.27).

As usual, carrier-modulated signals are conveniently represented by complex-valued baseband signals. If $a(t)$ and $\phi(t)$ are real-valued baseband signals, we can express the carrier-modulated signal[6] as

$$a(t) \cos(2\pi f_c t + \phi(t)) = \Re\{a(t)e^{j2\pi f_c t + \phi(t)}\} \tag{2.39}$$

$$= \Re\{a(t)e^{j2\pi f_c t}\}$$

$$= \Re\{a(t)\} \cos(2\pi f_c t)$$

$$- \Im\{a(t)\} \sin(2\pi f_c t), \tag{2.40}$$

where we have introduced the complex-valued signal

$$a(t) = a(t)e^{j\phi(t)} = a(t)\cos\phi(t) + ja(t)\sin\phi(t).$$

As $e^{j2\pi f_c t}$ does not carry any information, it is useful to view the carrier-modulated signal (2.39) as the complex-valued baseband signal $a(t)$. The real and imaginary parts of $a(t)$ are known as the *quadrature* components of $a(t) \cos(2\pi f_c t + \phi(t))$.

We can now generalize the basic synchronous CDMA model to

$$y(t) = \sum_{k=1}^{K} A_k b_k s_k(t) + \sigma n(t), \tag{2.41}$$

---

[6] The real and imaginary parts of a complex number are denoted by $a = \Re\{a\} + j\Im\{a\}$.

where $y(t)$, A_k, $s_k(t)$, and $n(t)$ are complex-valued,[7] and the real and imaginary components of $n(t)$ are independent and white Gaussian, each with spectral level equal to 1. The complex-valued signature waveforms $s_k(t)$ are normalized to

$$\|s_k\|^2 = \int_0^T \Re^2\{s_k(t)\}\, dt + \int_0^T \Im^2\{s_k(t)\}\, dt = 1.$$

An analogous complex-valued counterpart can be formulated for the asynchronous model. The complex-valued model in (2.41) is particularly advisable when the transmitted signals undergo fading and/or the demodulation is noncoherent.

2.5.2   NONANTIPODAL MODULATION

Inasmuch as the data bits are $\{-1, +1\}$, channels (2.1) and (2.6) assume a simple type of data modulation: binary antipodal.[8] This assumption is common in the analysis of CDMA systems and constitutes the most energy-efficient binary modulation strategy (Problem 2.9). At the expense of some notational complexity and negligible conceptual difficulty it is possible to generalize many of the results that we will present to any linear modulation format, by letting the data symbols belong to an arbitrary finite alphabet. This type of modulation is known as *Pulse Amplitude Modulation* (PAM).

In practice, a relatively simple way to improve bandwidth efficiency is to use quadrature modulation systems in which parallel data streams modulate two phases of the same carrier separated by 90°. Allowable phase transitions are controlled to further reduce bandwidth in modulation systems such as $\pi/4$-QPSK (Feher [89]). Different signature waveforms may be assigned to each quadrature channel. Note that by doubling K, (2.41) is general enough to encompass quadrature modulation systems.

When nonlinear modulation is used, the kth user transmits

$$\sum_{i=-M}^{M} A_k s_k(t - iT; b_k[i]),$$

where $b_k[i]$ need not be binary-valued. The special case

$$s_k(t; b) = b s_k(t)$$

---

[7] Complex quantities are denoted in sans serif typeface.

[8] Binary antipodal modulation of carrier-modulated signals is known as binary phase-shift keying (BPSK) or phase-reversal keying (PRK).

corresponds to linear modulation. Nonlinear modulation techniques are employed in some important multiuser communication systems. For example, frequency-hopping spread-spectrum waveforms are not modulated antipodally; instead, m-ary frequency shift keying and noncoherent demodulation are the usual choices. The GSM TDMA-based mobile digital telephony system (GSM Standard [82]) employs Gaussian minimum shift keying (GMSK): the information stream $\{b_i\}$ is first antipodally modulated,

$$z(t) = \sum_i b[i] f(t - iT),\qquad(2.42)$$

and then exponentially modulated,

$$s(t) = A \cos(2\pi f_c t + \pi h z(t)).$$

The antipodally modulated pulse in (2.42) is not time-limited and is chosen to control the spillover of frequency content beyond $1/T$:

$$f(t) = \frac{1}{\sqrt{2\pi}T} \int_{-\infty}^{t} \int_{\sqrt{2}\alpha(\tau-T)}^{\sqrt{2}\alpha\tau} e^{-\lambda^2/2}\, d\lambda\, d\tau.$$

Another example of nonlinear modulation is the 2^J-ary modulation format employed in the uplink of the IS-95 CDMA-based mobile digital telephony system (IS-95 Standard [442]).[9] This system can be best explained as the basic antipodally modulated CDMA model (2.6) where the stream of modulating bits $b_k[i]$ is not the information stream itself but the result of encoding blocks of J consecutive information bits into blocks of 2^J channel bits. That encoding is carried out by using the block of J information bits as the address of a Walsh function of length 2^J (Problem 1.4). If the $\ell \in \{1, \ldots, 2^J\}$ Walsh function is denoted by

$$h[\ell, 1], \cdots h[\ell, 2^J],$$

then J bits of information can be transmitted with the signal

$$\sum_{i=1}^{2^J} h[\ell, i] s_k(t - iT),\qquad(2.43)$$

where s_k has duration T and $h[\ell, i] \in \{-1, 1\}$. Thus, in this nonlinear-modulation scheme, the data rate is $J2^{-J}/T$. Note that we can view Walsh nonlinear modulation as conventional antipodal linear modulation coupled with a block error-control code that constrains each block of consecutive modulating J bits to take only J different values.

---

[9] $J = 6$ in the IS-95 system.

In fact, a K-user channel employing m-ary nonlinear modulation is always equivalent to a Km-user linearly modulated channel where the data of different "users" are dependent:

$$\sum_{k=1}^{K} A_k s_k(t; b_k) = \sum_{n=1}^{mK} \bar{A}_n \bar{b}_n \bar{s}_n(t), \qquad (2.44)$$

with

$$\bar{s}_{j+(k-1)m}(t) = s_k(t; j), \qquad (2.45)$$

$$\bar{b}_{j+(k-1)m} = 1\{b_k = j\}, \qquad (2.46)$$

$$\bar{A}_{j+(k-1)m} = A_k. \qquad (2.47)$$

2.6 FADING

In the basic CDMA models (2.1) and (2.6), the amplitudes A_k and signature waveforms s_k are those obtained at the receiver. Often the received amplitudes and signature waveforms do not coincide with those sent by the transmitters because of channel attenuation and distortion.

Fading refers to time-varying channel conditions. Any system with mobile transmitters and/or receivers is subject to fading. Even if the receivers and transmitters are not mobile, fading may be present in systems such as shortwave ionospheric communication. *Frequency-flat fading* affects the received amplitudes but does not introduce signature waveform distortion. *Frequency-selective fading* affects the received signals in both strength and shape.

2.6.1 FREQUENCY-FLAT FADING

When propagation conditions change, due for example to mobility, the received amplitudes vary with time. This feature is easily incorporated in the CDMA synchronous and asynchronous models:

$$y(t) = \sum_{k=1}^{K} \sum_{i=-M}^{M} A_k[i] b_k[i] s_k(t - iT) + \sigma n(t), \qquad (2.48)$$

$$y(t) = \sum_{k=1}^{K} \sum_{i=-M}^{M} A_k[i] b_k[i] s_k(t - iT - \tau_k) + \sigma n(t). \qquad (2.49)$$

Whether or not the receiver is able to track the time-varying coefficients $A_k[i]$, its performance will depend on the statistical properties of those

FIGURE 2.12.
Probability density
functions: (a)
Rayleigh; (b) Rice
$(d = 2)$; (c)
Nakagami
$(d = 3)$; (d)
log–normal
$(\sigma_P = 10)$.

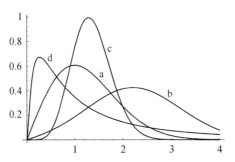

random processes. In most systems, it is safe to assume that the random processes $\{A_k[i]\}$ are independent from user to user. Furthermore, it is usually assumed that $\{A_k[i]\}$ are wide-sense stationary processes. Regarding the first-order statistics of the received amplitude (probability density function of $|A_k[i]|$), it is convenient to write it as the product of a deterministic component times a random component:

$$|A_k[i]| = A_k R[i].$$

Typical probability density functions of $R[i] \geq 0$ include:

Rayleigh: $\quad f_R(r) = r e^{-r^2/2},$ (2.50)

Rice: $\quad f_R(r) = r e^{-\frac{r^2+d^2}{2}} I_0(rd),$ (2.51)

Nakagami: $\quad f_R(r) = c(d) r^{2d-1} e^{-dr^2/2},$ (2.52)

log–normal: $\quad f_R(r) = \dfrac{20 \log_{10} e}{r\sigma_P \sqrt{2\pi}} \exp\left(-\dfrac{(20 \log_{10} r)^2}{2\sigma_P^2}\right).$ (2.53)

Figure 2.12 shows the Rayleigh, Rice,[10] Nakagami, and log–normal probability density functions. The second moments of the Rayleigh (2.50), Rice (2.51), and Nakagami (2.52) probability density functions are equal to 2. The second moment of the log–normal probability density function (2.53) is $1.02686^{\sigma_P^2}$ (Problem 2.31).

Rayleigh fading arises when the real and imaginary parts of $A_k[i]$ are independent zero-mean Gaussian random processes, in which case the phase of $A_k[i]$ is uniformly distributed on $[0, 2\pi]$. If the real and imaginary parts of $A_k[i]$ are independent Gaussian random processes with nonzero mean A_k, then $|A_k[i]| = A_k R[i]$ with $R[i]$ having a Rice distribution with parameter $d = |A_k|/A_k$. Rice fading arises when there is a direct line of sight between transmitter and receiver. The additional degree of freedom in the

---

[10] I_0 is the zero-order Bessel function $I_0(x) = \frac{1}{2\pi} \int_0^{2\pi} e^{x \cos(\alpha+\beta)} d\alpha.$

Nakagami distribution relative to the Rayleigh distribution enables a better fit to experimental measurements in urban channels.

Coarse fading variations due to shadowing by obstacles in the line of sight between receiver and transmitter are usually modeled by log–normal fading, for which the received power in dB is a Gaussian random variable. In typical urban cells, the differences in power received at the base station from different users may be as high as 90 dB: this is the *near–far problem*. It is challenging to implement robust signal processing, time-acquisition, and demodulation for such enormous dynamic ranges. *Power control* systems counter those imbalances by dynamically adjusting transmitted power, based on feedback information from the receiver. As we will illustrate in future chapters, for most modulation/demodulation systems, equal received power is the optimal power allocation for a fixed sum of received powers. Consequently, power control not only alleviates the effects of fading but it contributes to a savings in power. Power control proves particularly vital for systems that neglect the structure of the multi-access interference at the receiver (Chapter 3). Practical power controllers do not have unlimited capabilities for fast and accurate response, particularly in systems such as mobile-satellite communications. Thus, it is important to analyze the performance of multiuser detectors subject to power imbalances.

Several models are used in the literature for the autocorrelation of the fading amplitudes:

$$\Phi_k[n] = E\left[A_k[i]A_k^*[i+n]\right].\tag{2.54}$$

For example, in narrowband communication the Clarke–Jakes autocorrelation is frequently used:

$$\Phi_k[n] = \gamma_k J_0(\zeta_k n)$$

$$\stackrel{\text{def}}{=} \frac{\gamma_k}{2\pi} \int_0^{2\pi} \cos(\zeta_k n \sin\alpha)\, d\alpha.\tag{2.55}$$

Autoregressive moving-average models are also popular choices for the fading coefficients. According to those models, the complex fading coefficients $A_k[i]$ are the response of finite-dimensional linear time-invariant discrete-time systems to white processes.

The lower the spectral content of the fading amplitudes, the more accurately they can be estimated, and the longer the average *fade* duration becomes. A fade is a period where the magnitude of the received amplitude is below a threshold of acceptable performance.

2.6.2 FREQUENCY-SELECTIVE FADING

In many multiuser channels, not only do the received amplitudes vary with time but so do the received signature waveforms due to channel distortion. The additive multiple-access channel is invalidated by nonlinearities, in the presence of which the received waveform no longer comprises the noisy superposition of the various users' waveforms. Fortunately, channel distortion is often accurately modeled by a linear transformation, and we limit ourselves to that case.

The signature waveform of the kth user undergoes a linear time-varying transformation fully characterized by the complex-valued (baseband) impulse response:

$$h_k(t, \tau), \qquad (2.56)$$

which denotes the response of the system at time t due to a delta function at time τ, $\delta(t - \tau)$. In the special case of a time-invariant system, the dependence of $h_k(t, \tau)$ on its arguments is only through $t - \tau$. The effect of frequency-selective fading on the basic CDMA model is that the signature waveform seen at the receiver is not $s_k(t)$ but the convolution

$$\int_0^t h_k(t, \lambda)s_k(\lambda)\, d\lambda.$$

Time-varying linear distortion is particularly prevalent in mobile communication systems, and we will focus our discussion of frequency-selective fading in that scenario. Mobile transmitters see different channels – hence the explicit dependence on k in (2.56). To understand the basic features of time-varying linear distortion present in mobile channels of practical interest we determine the impulse response in several highly simplified settings:

Free space + Stationary. In the simplest setting, the transmitter and receiver are in free space and stationary. There is only one (line-of-sight) path between the two antennas and the signal is received undistorted. Indeed, the channel has the following time-invariant impulse response:

$$h(t, \tau) = A\delta\left(t - \tau - \frac{d_0}{c}\right), \qquad (2.57)$$

where d_0 is the distance between transmitter and receiver and c is the speed of light.

Free space + Motion. Suppose now that the transmitter travels in free space toward the receiver at speed v. This means that the distance is given by

$$d(t) = d_0 - vt.$$

45

The impulse response is

$$h(t, \tau) = A_t \delta \left(t - \tau - \frac{d_0}{c} + \frac{v}{c} t \right)$$

$$= A_t \delta \left(\left(1 + \frac{v}{c} \right) t - \tau - \frac{d_0}{c} \right). \qquad (2.58)$$

In free space, the power loss with distance is inversely proportional to the square of the distance: $|A_t| \propto 1/d(t)$.

We now have a time-varying system whose effect on a sinusoid is

$$A_t \int_{-\infty}^{\infty} h(t, \lambda) e^{j2\pi f_0 \lambda} d\lambda = A_t e^{-j2\pi \frac{d_0}{c}} e^{j2\pi f_0 (1 + \frac{v}{c}) t}. \qquad (2.59)$$

Therefore, in addition to a time-varying amplitude due to increased signal strength as the transmitter moves toward the receiver, the sinusoid undergoes an increase in frequency: the *Doppler*[11] shift. For example, a mobile operating at 900 MHz and approaching the base station at 100 km/h undergoes a positive Doppler shift of 83 Hz. The effect of (2.58) on input signals other than sinusoids is not a simple translation in the frequency domain (such as that obtained by multiplication with a sinusoid) because, as we can see from (2.59), the Doppler shift is proportional to the input frequency.

Remote reflection + Stationary. The transmitter and receiver are stationary, but they are not in free space. Due to the presence of *remote reflectors* there are L different paths between transmitter and receiver. Since the paths have different lengths, the impulse response has the *specular multipath* form:

$$h(t, \tau) = \sum_{j=1}^{L} A_j \delta \left(t - \tau - \frac{d_j}{c} \right). \qquad (2.60)$$

The time-invariance of (2.60) is due to the assumed stationarity of transmitter, receiver, and remote reflectors.

In contrast to free space where the power loss is inversely proportional to the distance squared, propagation over a flat reflecting surface results in the power loss being inversely proportional to the fourth power of the distance; in practice, the path-loss exponent depends heavily on the nature of the channel and generally ranges between 2 and 5 (e.g., Stuber [433]).

---

[11] Christian J. Doppler (1803–1853) described this phenomenon for the first time.

The *delay spread* is the time it takes for light to travel a distance equal to the longest path minus the shortest path:

$$\frac{\max_j d_j - \min_j d_j}{c}.$$

For example, a path difference of 30 m corresponds to a delay spread of 0.1 μs (typical of indoor picocellular communication), whereas a path difference of 6 km corresponds to a delay spread of 20 μs (typical of mountainous terrain). The inverse of the delay spread is called the *coherence bandwidth* of the channel. Coherence bandwiths of 3 MHz (resp. 100 kHz) are typical of indoor (resp. outdoor) wireless channels (Jung et al. [182]). Signals whose bandwidth is much smaller than the coherence bandwidth are not distorted appreciably by multipath (which explains the term *frequency-flat* fading used in the previous subsection.) In contrast, when the coherence bandwidth is comparable or smaller than the symbol rate, multipath introduces not only distortion but intersymbol interference. Note that unlike Doppler effects, multipath distortion is not due to mobility; however, when either the receiver or the transmitter moves, the delays $\frac{d_j}{c}$ and complex weights A_j in (2.60) change with time. Accordingly, to take full advantage of the transmitted power the receiver must track the time-varying channel impulse response. For example, if the parameter variations are not too rapid relative to the symbol rate, an adaptive algorithm known as the *rake* can be used to track $A_j(t)$, $d_j(t)$, $j = 1, \ldots, L$ and combine (coherently or noncoherently) the delayed replicas of the transmitted signal.

Even though multipath complicates the task of taking advantage of all the transmitted energy at the receiver, it provides a potential source of beneficial *diversity*. To explain this concept, it is useful to compare a channel with L paths (2.60) with a channel without multipath given by (2.60) and $L = 1$. If A_j, $j = 1, \ldots, L$ are independent, Gaussian, complex random variables, the strength of each path is Rayleigh-distributed. This implies that with nonnegligible probability the path strength will be very small (cf. Figure 2.12). Due to the assumed independence of the path strengths, we can expect performance to be worst if $L = 1$, since in that case the probability that the received energy falls below an acceptable threshold is maximized.

The large time-bandwidth product of spread-spectrum signals allows the designer to choose signature waveforms with good autocorrelation properties in the sense that their autocorrelation function resembles a delta function:

$$\int s_k(t)s_k(t + \lambda)\, dt \approx \delta(\lambda). \tag{2.61}$$

This is important for two reasons: (a) robustness and (b) identification. (a) A delayed replica of a signature waveform is very different from the signature waveform itself and results in small interference. (b) If the response of the channel to a signal s_k that satisfies (2.61) is passed through a filter with impulse response $s_k(-t)$, then the output of that matched filter approximates the impulse response of the channel (Problem 2.27) (assuming the channel time variations are small relative to the duration of s_k). For direct-sequence spread-spectrum signals, the autocorrelation cannot be made negligible for arguments smaller than the chip duration, so multipath delays smaller than the chip duration are not resolvable. The number of resolvable paths is essentially given by the bandwidth of the spread-spectrum signal times the delay spread.

Remote scattering + Stationary. Specular reflections of electromagnetic waves are produced by sharp boundaries between two different media. Far more commonly, waves are *scattered* (i.e., they are reflected by multiple particles in a random disorganized fashion), in which case the phases of the scattered waves superpose randomly. No longer a superposition of a finite number of delta functions, the resulting time-invariant impulse response $h(t, 0) = h(\lambda + t, \lambda)$ is typically modeled as a (complex-valued) Gaussian process. The *scattering/multipath intensity profile* is defined as

$$\varphi(t) = E[|h(t, 0)|^2].$$

The *rms delay spread*, σ_φ, is defined as the standard deviation of the probability density function

$$\frac{\varphi}{\int_0^\infty \varphi(t)\, dt}.$$

Since $\varphi(t)$ is usually nonzero, but insignificant, at arguments much larger than the rms delay spread, a more pragmatic definition of *coherence bandwidth* than the one given in the foregoing paragraph on remote reflection is that coherence bandwidth is the reciprocal of a multiple[12] of σ_φ. In intuitive terms, spectral components of the transmitted signal that differ by no more than the coherence bandwidth are subject to similar channel attenuation/phase rotation.

---

[12] Usually ranging between 5 and 50, depending on the application (Sklar [413]).

Local scattering/reflection + Motion. Even in the absence of remote scatterers/reflectors, objects near the mobile produce multiple transmitter/receiver paths. Although the length of those paths is essentially the same, some path lengths increase while others decrease with motion, depending on the location of the scattering objects relative to the motion vector. In the case of a finite number of local reflectors we obtain

$$h(t, \tau) = \sum_{j=1}^{L} B_j \delta\left(t - \tau - \frac{d_0}{c} + \frac{v_j}{c}t\right). \qquad (2.62)$$

The response of such a linear system to a sinusoid can be readily computed using (2.59). Therefore, we observe that the effect of motion and local reflection on a delta function in the frequency domain is dual to the effect of fixed multipath on a delta function in the time domain.

With a continuum of local scatterers, the spreading function is no longer composed of a finite sum of delta functions. A simple idealized example of a continuum of local scatterers is the case where the energy received by the mobile is independent of the angle of arrival. In that case, a pure sinusoid transmitted at frequency f_c is received as a signal with power spectral density (Problem 2.33)

$$S(f) = \frac{k}{\sqrt{f_m^2 - (f - f_c)^2}}, \qquad f \in (f_c - f_m, f_c + f_m),$$

where $f_m = vf_0/c$ and v is the speed of the mobile. This provides a simple example of a linear time-varying system whose response to a sinusoid is a signal with a continuous spectrum. The *Doppler spread* is equal to the bandwidth increase of the transmitted signal, equal to $2f_m$ in the special case of a pure tone.

Remote scattering + Local scattering + Motion. The combination of the foregoing effects is described by the statistical properties of the impulse response. To express the key statistical properties of the impulse response it is convenient to work with its Fourier transform. Since the transfer function (at frequency f) in the familiar time-invariant case is the response of the system to $e^{j2\pi ft}$ divided by $e^{j2\pi ft}$, we can define the "time-varying transfer function" in the same way:

$$H(t, \theta) = e^{-j2\pi\theta t} \int_{-\infty}^{\infty} h(t, \tau)e^{j2\pi\theta\tau} \, d\tau$$

$$= \int_{-\infty}^{\infty} h(t, t - \lambda)e^{-j2\pi\theta\lambda} \, d\lambda. \qquad (2.63)$$

Often, $h(t, t - \lambda)$ is modeled as a wide-sense stationary process for all λ, in which case $H(t, \theta)$ is also wide-sense stationary for all θ. Furthermore, in *uncorrelated scattering* $h(t_1, \lambda_1)$ and $h(t_2, \lambda_2)$ are uncorrelated whenever $t_1 - \lambda_1 \neq t_2 - \lambda_2$ so we can write

$$E[h(t_1, \tau_1)h^*(t_2, \tau_2)] = \Xi(t_1 - t_2, t_1 - \tau_1)\delta(t_1 - \tau_1 - t_2 + \tau_2), \quad (2.64)$$

where the scattering intensity profile defined before can be obtained as a special case of the correlation function Ξ:

$$\Xi(0, t) = \varphi(t).$$

Using (2.64), we can verify that the following correlation function depends only on the difference between the corresponding arguments:

$$E[H(t_1, \theta_1)H^*(t_2, \theta_2)]$$

$$= \int_{-\infty}^{\infty} \int_{-\infty}^{\infty} E[h(t_1, t_1 - v_1)h^*(t_2, t_2 - v_2)]e^{-j2\pi(v_1\theta_1 - v_2\theta_2)} \, dv_1 \, dv_2$$

$$= \int_{-\infty}^{\infty} \Xi(t_1 - t_2, v_1)e^{-j2\pi v_1(\theta_1 - \theta_2)} \, dv_1$$

$$\stackrel{\text{def}}{=} \Phi(t_1 - t_2, \theta_1 - \theta_2). \quad (2.65)$$

Upon taking the two-dimensional Fourier transform of the correlation function in (2.65) we get the *scattering function*:

$$\Psi(f, \tau) = \int_{-\infty}^{\infty} \int_{-\infty}^{\infty} \Phi(t, \theta)e^{-j2\pi ft}e^{j2\pi\tau\theta} \, dt \, d\theta.$$

The range of (positive) values of τ (resp. f) for which the scattering function is nonnegligible is the multipath spread (resp. Doppler spread). As a rough rule of thumb, the Doppler spread is inversely proportional to the time it takes for the mobile to traverse a wavelength of the carrier frequency.

2.6.3 HOMOGENEOUS FADING

We conclude this section with a simple model of an extreme type of frequency-selective fading that originates from the consideration of independent Rayleigh frequency-flat fading at the chip level. Consider the complex-valued version of the spread-spectrum signal model in (2.30) and assume that each dimension is affected by a different random coefficient:

$$s(t) = \sum_{i=1}^{N} |C_i|e^{j\theta_i}\beta_i\psi_i(t). \quad (2.66)$$

Further, let us assume that $\{C_i = |C_i|e^{j\theta_i}, i = 1, \ldots N\}$ are independent, zero-mean, complex Gaussian variables. In the conventional direct-sequence model (2.31) where the basis functions ψ_i are delayed versions of each other, the homogeneous fading model applies to scenarios with very fast fading. In multicarrier CDMA (2.32), homogeneous fading models are used when the coherence bandwidth is smaller than the data rate.

Notice that in a CDMA system where the spreading coefficients $(\beta_1, \ldots, \beta_N)$ have identical magnitude (such as in the usual bipolar direct-sequence case), all the information contained in the spreading codes gets washed out by this kind of fading. In effect, the "signature sequences" become random and totally controlled by the fading. Moreover, depending on the application, the values of $\{C_i \ i = 1, \ldots, N\}$ may not be easy to track at the receiver, in which case the time-domain structure of the signals can no longer provide distinguishability among users, unless multilevel spreading coefficients are employed.

2.7 ANTENNA ARRAYS

Antenna arrays consist of several antennas arranged together in space. They are employed in multiuser wireless communications for two primary purposes:

Diversity: The use of multiple antennas to provide diversity dates back to the early days of single-user radio communication. The rationale is that if the array elements are sufficiently separated, fading does not affect the signal received by each element equally, and performance (and, in particular, outage probability) is improved.

User separation: It is possible to exploit the fact that users are physically separated by providing the receiver with an antenna whose radiation pattern[13] is not omnidirectional. Likewise, directive antennas can concentrate transmitted power in the direction of the intended receiver.

At a given location in space, the received signal depends not only on time but also on the angle of arrival:[14]

$$y(t, \alpha), \quad t \in (-\infty, +\infty), \quad \alpha \in [0, 2\pi). \qquad (2.67)$$

---

[13] The *radiation pattern* (also known as directivity pattern or simply antenna pattern) is the distribution of the power radiated (or received) by the antenna as a function of the angle.

[14] It is customary to restrict attention to two-dimensional patterns since, in most applications, users are roughly located on a plane.

Conceptually, we could view the angle (or "space") variable similarly to the time variable, and we could generalize the basic CDMA model accordingly. However, the signal in (2.67) is not observable at the output of the antenna. Rather, we can think of the antenna as a "correlator" in the space domain:

$$\frac{1}{2\pi} \int_0^{2\pi} y(t, \alpha) R(\alpha) \, d\alpha, \qquad (2.68)$$

where $R(\alpha)$ is the radiation pattern. Because the radiation pattern of an antenna depends on the frequency of the received signal, (2.68) gives an overly simplified view, which is nevertheless useful when dealing with the usual case of radio-frequency signals whose bandwidth is small relative to their center frequency.

The simplest one-element antenna has an omnidirectional pattern: $R(\alpha)$ equal to a constant. Linear arrays of equispaced omnidirectional antennas are the most common approach to beamforming. The number of antennas in the array determines the achievable angle discrimination, in much the same way that the achievable sharpness of frequency discrimination of a finite-impulse response digital filter is determined by the number of filter taps. Once the number of elements in the array and their distances are fixed, the resulting antenna pattern depends on how the element outputs are combined. Let us consider the simplest case of a two-element array (Figure 2.13) separated by d. A planar sinusoidal field at frequency f_c is incident upon the array at angle α. The field received by element 2 is delayed relative to the field received by element 1 by the time it takes for light to travel the distance $d \cos \alpha$. Thus, the respective responses are

$$x_1(t) = e^{j2\pi f_c t + j\theta},$$

$$x_2(t) = e^{j2\pi f_c(t - \frac{d}{c}\cos\alpha) + j\theta}.$$

We now combine both responses as

$$\begin{aligned} z(t) &= x_1(t) + g e^{j\varphi} x_2(t) \\ &= e^{j2\pi f_c t + j\theta}(1 + g e^{j\varphi - j2\pi f_c \frac{d}{c}\cos\alpha}) \\ &= e^{j2\pi f_c t + j\theta'}\left(1 + g^2 + 2g\cos\left(\varphi - 2\pi f_c \frac{d}{c}\cos\alpha\right)\right)^{1/2} \qquad (2.69) \\ &= e^{j2\pi f_c t + j\theta'}(1 + g^2 + 2g\cos(\varphi - \pi\cos\alpha))^{1/2}, \qquad (2.70) \end{aligned}$$

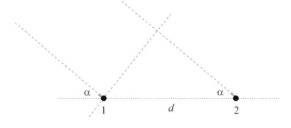

FIGURE 2.13. Two-element array.

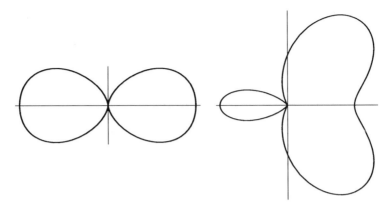

FIGURE 2.14.
Two-element array radiation patterns with equal gains and phase offsets of $\varphi = \pi$ (left) and $\varphi = \pi/2$ (right).

where g and φ are design parameters. In (2.70) we have chosen d to be one half of the wavelength of the sinusoid field at frequency f_c:

$$d = \frac{\lambda_c}{2} = \frac{c}{2 f_c}.$$

This choice is the most convenient one as smaller values of the inter-element distance d do not let $2\pi f_c \frac{d}{c} \cos\alpha$ range over the full interval $[-\pi, \pi)$. Figure 2.14 shows polar plots of the magnitude of the two-element array radiation pattern:

$$|R(\alpha)|^2 = 1 + g^2 + 2g \cos(\varphi - \pi \cos\alpha).$$

In general, an array of D elements spaced at half a wavelength whose outputs are combined with the *steering vector* weights

$$(g_1 e^{j\varphi_1}, \ldots, g_D e^{j\varphi_D})$$

has the radiation pattern

$$R(\alpha) = \sum_{n=1}^{D} g_n e^{j\varphi_n - j\pi(n-1)\cos\alpha}. \tag{2.71}$$

We now proceed to generalize the complex-valued scalar CDMA model in (2.41) to the scenario of array observations. To that end, we assume that the carrier frequency is common to all users and that it is much higher than the bandwidth of the transmitted signals, so that the antenna has approximately the same pattern for all received frequencies. If user k transmits the signal

$$A_k b_k s_k(t) e^{j2\pi f_c t}$$

during the interval $t \in [0, T]$, the response of the nth array element will be denoted by

$$c_{nk} A_k b_k s_k(t) e^{j2\pi f_c t}.$$

We can think of the vector

$$\mathbf{c}_k = \begin{bmatrix} c_{1k} \\ \vdots \\ c_{Dk} \end{bmatrix}$$

as the *spatial signature* of the kth user. If the energy transmitted by user k arrives exclusively at direction α_k, then we can be more explicit about the coefficient c_{nk}; according to the foregoing development,

$$c_{nk} = e^{-j2\pi f_c(n-1)\frac{d}{c}\cos\alpha_k}$$

$$= e^{-j\pi(n-1)\cos\alpha_k}, \tag{2.72}$$

where (2.72) assumes half-wavelength spacing.

In the presence of multipath, the signal transmitted by each user arrives at the receiver not from one direction, but from a continuum, in which case the form of the coefficients in (2.72) is not applicable. The *angle spread* gives the range of angle values for which significant energy is received. Typical values are $360°$ in indoor picocell systems, $20°$ in urban cells, and $1°$ in a flat rural environment (Paulraj and Papadias [326]).

Applying superposition, the baseband vector-valued, continuous-time synchronous CDMA array response model is

$$y_1(t) = \sum_{k=1}^{K} c_{1k} A_k b_k s_k(t) + \sigma n_1(t),$$

$$\vdots \tag{2.73}$$

$$y_D(t) = \sum_{k=1}^{K} c_{Dk} A_k b_k s_k(t) + \sigma n_D(t),$$

where the Gaussian noises $\{n_i\}$ are usually assumed to be independent. If the transmitted waveforms are spread-spectrum waveforms with spreading factor N and they are synchronized, the vector signal obtained at the output of the array as the superposition of the K transmitted signals belongs to an $N \times D$-dimensional vector space. Whether the dimensions or degrees of freedom come from the time domain only or from both time and space makes little conceptual difference. Practically, there is indeed a marked difference because the system designer has control over the similarity of signals in the time domain (albeit, not total control due to unknown channel distortion) but no control over similarity in the space domain. Indeed, users may be arbitrarily close to each other, so the crosscorrelation between the spatial signatures $(c_{1k}, \ldots c_{Dk})$ and $(c_{1j}, \ldots c_{Dj})$ may be arbitrarily close to unity. Furthermore, in mobile applications no prior knowledge of the spatial signatures can be assumed.

Pure *Space-Division Multiple Access* corresponds to the special case of (2.73) in which users are not differentiated in the time domain, that is, $s_k(t) = s(t)$ for all k.

In addition to antenna arrays, there are other mechanisms that provide diversity: error-control coding, multipath, cross-polarization, frequency diversity, etc. We can generalize (2.73) to an arbitrary diversity system by defining $s_{dk}(t)$ as the signature of the kth user in the dth branch. Then

$$y_1(t) = \sum_{k=1}^{K} A_{1k} b_k s_{1k}(t) + \sigma n_1(t),$$

$$\vdots \tag{2.74}$$

$$y_D(t) = \sum_{k=1}^{K} A_{Dk} b_k s_{Dk}(t) + \sigma n_D(t).$$

This diversity model encompasses the array response model in the special case in which

$$s_{dk}(t) = c_{dk} s_k(t).$$

2.8 BACKGROUND NOISE

The additive noise $n(t)$ in (2.1) and (2.6) models any part of the received signal not due to the transmitters in the multiuser communication system. It incorporates receiver thermal noise, background electromagnetic noise, as well as man-made interference. Typical scenarios where $n(t)$ deviates from the white Gaussian noise assumption include narrow band interference, out-of-cell interference, and impulsive noise.

Narrowband interference may originate from intentional jamming or from the coexistence of different communication systems in a frequency band. For example, one of the advantage of spread-spectrum techniques is that they can be overlaid on top of an existing narrowband channel while causing an inappreciable rise in the background noise level of the narrowband channel. Although we do not deal with narrowband interference rejection techniques in subsequent chapters, the sections on bibliographical notes contain references to works where multiuser detection techniques have proven useful for narrowband interference suppression.

Even though any interferer (in-cell or out-of-cell) has the same structure as the multiser signals that are incorporated in the basic models (2.1) and (2.6), frequently it makes sense for the sake of reduced complexity to regard the aggregate signal due to weak out-of-cell interferers as white and Gaussian.

Accurately modeling the non-Gaussian nature of the background noise in channels with impulsive noise may lead to important performance gains in multiuser channels. Typical non-Gaussian noise models involve probability density functions that are the mixture of two Gaussian distributions with different variances.

2.9   DISCRETE-TIME SYNCHRONOUS MODELS

Multiuser detectors commonly have a front-end whose objective is to obtain a discrete-time process from the received continuous-time waveform $y(t)$. Continuous-to-discrete-time conversion can be realized by conventional sampling, or more generally, by correlation of $y(t)$ with deterministic signals. Two types of deterministic signals are of principal interest: the signature waveforms and orthonormal signals.

2.9.1   MATCHED FILTER OUTPUTS

One way of converting the received waveform into a discrete-time process is to pass it through a bank of matched filters (Figure 2.15), each

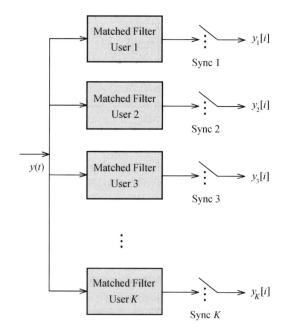

FIGURE 2.15.
Discrete-time
K-dimensional
vector of matched
filter outputs.

matched to the signature waveform of a different user. In the synchronous case, the output of the bank of matched filters is

$$y_1 = \int_0^T y(t)s_1(t)\,dt,$$

$$\vdots \qquad\qquad (2.75)$$

$$y_K = \int_0^T y(t)s_K(t)\,dt.$$

Using (2.1) and (2.3) we can express the output of the kth matched filter as

$$y_k = A_k b_k + \sum_{j \neq k} A_j b_j \rho_{jk} + n_k, \qquad (2.76)$$

where

$$n_k = \sigma \int_0^T n(t)s_k(t)\,dt \qquad (2.77)$$

is a Gaussian random variable with zero mean and variance equal to σ^2 (from Proposition 3.5 and the fact that s_k is normalized to have unit energy).

It is convenient to express (2.76) in vector form:

$$\mathbf{y} = \mathbf{RAb} + \mathbf{n}, \qquad (2.78)$$

where $\mathbf{R}$ is the normalized crosscorrelation matrix,

$$\mathbf{y} = [y_1, \ldots, y_K]^T,$$

$$\mathbf{b} = [b_1, \ldots, b_K]^T,$$

$$\mathbf{A} = \text{diag}\{A_1, \ldots, A_K\},$$

and $\mathbf{n}$ is a zero-mean Gaussian random vector with covariance matrix equal to

$$E[\mathbf{nn}^T] = \sigma^2 \mathbf{R}. \qquad (2.79)$$

In Chapter 4, we will show that no information relevant to demodulation is lost by the bank of matched filters; in other words, $y(t)$ can be replaced by $\mathbf{y}$ without loss of optimality.

The unnormalized crosscorrelation matrix whose (j, k) element is $\langle A_j s_j, A_k s_k \rangle$ is denoted as

$$\mathbf{H} = \mathbf{ARA}. \qquad (2.80)$$

To analyze any detector whose front-end consists of a bank of matched filters, the original channel model can be replaced by the linear Gaussian

K-dimensional model in (2.78). Recall that in the synchronous case it is sufficient to restrict attention to a one-shot model; thus, we have omitted the dependence of $\mathbf{y}$, $\mathbf{b}$, and $\mathbf{n}$ on the symbol index.

We can easily generalize (2.78) so as to encompass the complex-valued model in (2.41):

$$y(t) = \sum_{k=1}^{K} A_k b_k s_k(t) + \sigma n(t). \qquad (2.81)$$

The outputs of the matched filters are

$$y_k = \langle y, s_k \rangle$$

$$= \int_0^T y(t) s_k^*(t)\, dt$$

$$= A_k b_k + \sum_{j \neq k} A_j b_j \rho_{kj} + n_k, \qquad (2.82)$$

where

$$\rho_{kj} = \int_0^T s_k^*(t) s_j(t)\, dt. \qquad (2.83)$$

Then, the same vector model as in (2.78),

$$\mathbf{y} = \mathbf{RAb} + \mathbf{n}, \qquad (2.84)$$

can be used to represent (2.82) with a Hermitian matrix $\mathbf{R}$, a complex diagonal matrix $\mathbf{A}$, and a complex-valued Gaussian vector $\mathbf{n}$ with independent real and imaginary components and covariance matrix equal to $2\sigma^2 \mathbf{R}$.

Although (2.78) will emerge as the most convenient discrete-time model in future chapters, occasionally we will have the opportunity to employ other models that contain the same information as the bank of matched filter outputs. For example, consider the following result:

Proposition 2.2 *(Cholesky factorization) For every positive definite matrix* $\mathbf{R}$ *there exists a unique lower triangular matrix* $\mathbf{F}$ *(i.e.,* $F_{ik} = 0$ *for* $i < k$*) with positive diagonal elements such that*

$$\mathbf{R} = \mathbf{F}^T \mathbf{F}.$$

For brevity, we shall denote the inverse of the transpose of a matrix by

$$\left(\mathbf{F}^T\right)^{-1} \overset{\text{def}}{=} \mathbf{F}^{-T}. \qquad (2.85)$$

If the matched filter outputs $\mathbf{y}$ are processed by the matrix $\mathbf{F}^{-T}$, we obtain the equivalent discrete-time model

$$\bar{\mathbf{y}} = \mathbf{F}^{-T}\mathbf{y} \tag{2.86}$$

$$= \mathbf{F}^{-T}\mathbf{F}^{T}\mathbf{F}\mathbf{A}\mathbf{b} + \mathbf{F}^{-T}\mathbf{n}$$

$$= \mathbf{F}\mathbf{A}\mathbf{b} + \bar{\mathbf{n}}, \tag{2.87}$$

where $\bar{y}_k$ contains contributions from users $1, \ldots, k$ but not from users $k+1, \ldots, K$. The covariance matrix of $\bar{\mathbf{n}}$ is

$$E[\bar{\mathbf{n}}\bar{\mathbf{n}}^{T}] = \sigma^2\mathbf{F}^{-T}\mathbf{R}\mathbf{F}^{-1} = \sigma^2\mathbf{I}, \tag{2.88}$$

where $\mathbf{I}$ is the identity matrix. Reflecting the independence of the noise components, (2.87) is called a *whitened matched filter* model. Because there is a one-to-one correspondence between $\bar{\mathbf{y}}$ and $\mathbf{y}$, both models contain the same information about the data.

The discrete-time model in (2.78) was derived from a scalar continuous-time model. It is conceptually straightforward to generalize (2.97) to the case in which the original observations are D dimensional as in (2.73). For example, in a synchronous system we can make L equal to the spreading gain times the dimensionality of the array. Alternatively, it is often advantageous to distinguish between the roles played by the spatial signatures and the time-domain signatures of the users. If the D antenna output waveforms in (2.73) are each processed by a bank of K matched filters, we obtain a $D \times K$ matrix $\mathbf{Y}$ with entries

$$Y_{dj} = \langle y_d, s_j \rangle = \sum_{k=1}^{K} c_{dk} A_k b_k \rho_{jk} + n_{dj} \tag{2.89}$$

and

$$E[n_{dj} n_{al}^*] = 2\sigma^2 \rho_{jl} \delta_{da}. \tag{2.90}$$

We can rewrite (2.89) in matrix form as

$$\mathbf{Y} = \mathbf{CABR}^{T} + \mathbf{N}, \tag{2.91}$$

where $\mathbf{C}$ denotes the $D \times K$ matrix of spatial signatures,

$$\mathbf{B} = \text{diag}\{b_1, \ldots, b_K\},$$

and $\mathbf{N}$ is a $D \times K$ zero-mean Gaussian matrix with correlations given in (2.90).

In order to obtain a discrete-time model for the general diversity model in (2.74),

$$y_1(t) = \sum_{k=1}^{K} A_{1k} b_k s_{1k}(t) + \sigma n_1(t),$$

$$\vdots$$

$$y_D(t) = \sum_{k=1}^{K} A_{Dk} b_k s_{Dk}(t) + \sigma n_D(t), \qquad (2.92)$$

we will only consider the matched filter outputs

$$y_{dk} = \langle y_d, s_{dk} \rangle$$
$$= A_{dk} b_k + \sum_{l \neq k} A_{dl} b_l \rho_{lk}(d) + \sigma n_{dk} \qquad (2.93)$$

where the crosscorrelations of the signals in the dth branch are denoted by

$$\rho_{kj}(d) = \int_0^T s_{dk}^*(t) s_{dj}(t) \, dt. \qquad (2.94)$$

Instead of adopting the matrix notation in (2.89) it is now more convenient to represent (2.93) in vector form:

$$\mathbf{y}_d = \mathbf{R}(d) \mathbf{A}(d) \mathbf{b} + \mathbf{n}_d, \quad d = 1, \dots, D, \qquad (2.95)$$

where $\mathbf{R}(d)$ is the matrix defined in (2.94), $\mathbf{n}_d$ is complex Gaussian with covariance matrix $2\sigma^2 \mathbf{R}(d)$, and

$$\mathbf{A}(d) = \mathrm{diag}\{A_{d1}, \dots, A_{dK}\}.$$

2.9.2 ORTHONORMAL PROJECTIONS

The dimensionality of the vectors in (2.78) and (2.87) is equal to the number of users. In some situations (such as when the signature waveforms of some interferers are not known a priori) other models (with possibly different dimensionality) are useful. Let $\{\psi_1, \dots, \psi_L\}$ be a set of L orthonormal signals defined on $[0, T]$. The *signature vector*, $\mathbf{s}_k$, of the kth user is the L-dimensional representation of s_k on the basis $\{\psi_1, \dots, \psi_L\}$, that is, the l component of the column vector $\mathbf{s}_k$ is

$$s_{kl} = \int_0^T s_k(t) \psi_l(t) \, dt.$$

Furthermore, we define the components of the vector $\mathbf{r}$ as

$$r_l = \int_0^T y(t)\psi_l(t)\,dt.$$

We can then write the column vector

$$\mathbf{r} = \sum_{k=1}^K A_k b_k \mathbf{s}_k + \sigma\mathbf{m} \tag{2.96}$$

$$= \mathbf{SAb} + \sigma\mathbf{m}, \tag{2.97}$$

where $\mathbf{m}$ is an L-dimensional Gaussian vector with independent unit-variance components, and we have introduced the $L \times K$ matrix of signature vectors

$$\mathbf{S} = [\,\mathbf{s}_1 \quad | \quad \cdots \quad | \quad \mathbf{s}_K\,]. \tag{2.98}$$

Since bits of different users are uncorrelated, the covariance matrix of (2.97) is equal to

$$E[\mathbf{rr}^T] = \sigma^2\mathbf{I} + \sum_{k=1}^K A_k^2 \mathbf{s}_k \mathbf{s}_k^T \tag{2.99}$$

$$= \sigma^2\mathbf{I} + \mathbf{SA}^2\mathbf{S}^T. \tag{2.100}$$

The finite-dimensional model in (2.97) holds regardless of whether the L orthonormal signals $\{\psi_1, \ldots, \psi_L\}$ span the signature waveforms $\{s_1, \ldots, s_K\}$. An example of a set of orthonormal signals that span the signature waveforms is a direct-sequence spread-spectrum system where L is equal to the number of chips per symbol and the orthonormal signals are delayed chip waveforms:

$$\psi_i(t) = p_{T_c}(t - (i-1)T_c).$$

If the signature waveforms are spanned by $\{\psi_1, \ldots, \psi_L\}$, then the $K \times K$ crosscorrelation matrix becomes simply (Problem 2.43)

$$\mathbf{R} = \mathbf{S}^T\mathbf{S}. \tag{2.101}$$

Moreover,

$$\|\mathbf{s}_k\| = 1, \tag{2.102}$$

and all the information contained in $\mathbf{y}$ is contained in $\mathbf{r}$ because the matched filter outputs can be expressed as linear combinations of the components of $\mathbf{r}$:

$$\mathbf{y} = \mathbf{S}^T\mathbf{r}. \tag{2.103}$$

Note that in the case of Space-Division Multiple Access, (2.89) reduces to

$$\langle y_d, s \rangle = \sum_{k=1}^{K} c_{dk} A_k b_k + n_d,$$

which is identical to the model in (2.97) with the spatial signature matrix $\mathbf{C}$ taking the role of $\mathbf{S}$.

2.10 DISCRETE-TIME ASYNCHRONOUS MODELS

For the asynchronous channel, notation is simplified if users are labeled chronologically (i.e., by their time of arrival), which is equivalent to assuming without loss of generality that

$$\tau_1 \leq \tau_2 \cdots \leq \tau_K.$$

Then, taking into account (2.6) and (2.13), the matched filter outputs can be expressed as

$$y_k[i] = A_k b_k[i] + \sum_{j<k} A_j b_j[i+1]\rho_{kj} + \sum_{j<k} A_j b_j[i]\rho_{jk}$$

$$+ \sum_{j>k} A_j b_j[i]\rho_{kj} + \sum_{j>k} A_j b_j[i-1]\rho_{jk} + n_k[i], \quad (2.104)$$

where

$$n_k[i] = \sigma \int_{\tau_k+iT}^{\tau_k+iT+T} n(t)s_k(t-iT-\tau_k)\,dt. \quad (2.105)$$

We can write (2.104) in matrix form:

$$\mathbf{y}[i] = \mathbf{R}^T[1]\mathbf{A}\mathbf{b}[i+1] + \mathbf{R}[0]\mathbf{A}\mathbf{b}[i] + \mathbf{R}[1]\mathbf{A}\mathbf{b}[i-1] + \mathbf{n}[i], \quad (2.106)$$

where the zero-mean Gaussian process $\mathbf{n}[i]$ has autocorrelation matrix

$$E[\mathbf{n}[i]\mathbf{n}^T[j]] = \begin{cases} \sigma^2 \mathbf{R}^T[1], & \text{if } j = i+1; \\ \sigma^2 \mathbf{R}[0], & \text{if } j = i; \\ \sigma^2 \mathbf{R}[1], & \text{if } j = i-1; \\ \mathbf{0}, & \text{otherwise,} \end{cases} \quad (2.107)$$

and the matrices $\mathbf{R}[0]$ and $\mathbf{R}[1]$ are defined by

$$R_{jk}[0] = \begin{cases} 1, & \text{if } j = k; \\ \rho_{jk}, & \text{if } j < k; \\ \rho_{kj}, & \text{if } j > k; \end{cases} \quad (2.108)$$

$$R_{jk}[1] = \begin{cases} 0, & \text{if } j \geq k; \\ \rho_{kj}, & \text{if } j < k. \end{cases} \quad (2.109)$$

FIGURE 2.16.
K-dimensional channel of matched filter outputs for asynchronous CDMA channel.

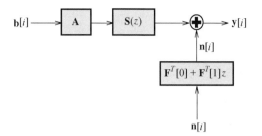

For example, in the three-user case,

$$R[0] = \begin{bmatrix} 1 & \rho_{12} & \rho_{13} \\ \rho_{12} & 1 & \rho_{23} \\ \rho_{13} & \rho_{23} & 1 \end{bmatrix}, \qquad (2.110)$$

$$R[1] = \begin{bmatrix} 0 & \rho_{21} & \rho_{31} \\ 0 & 0 & \rho_{32} \\ 0 & 0 & 0 \end{bmatrix}. \qquad (2.111)$$

The vector discrete-time model in (2.106) can be represented in the *z*-transform domain as shown in Figure 2.16 where

$$S(z) = R^T[1]z + R[0] + R[1]z^{-1} \qquad (2.112)$$

and $\bar{n}[i]$ is independent Gaussian with covariance matrix $\sigma^2 I$.

Note that if the signature waveforms have duration larger than T, then the model has to be generalized to incorporate crosscorrelation matrices $R[2], \ldots, R[L]$, where L is the length of the intersymbol interference.

The choice of the $K \times K$ matrices $F[0]$ and $F[1]$ in Figure 2.16 is governed by the following generalization of Proposition 2.2:

Proposition 2.3 *The matrix in (2.112) can be expressed as*

$$S(z) = [F[0] + F[1]z]^T [F[0] + F[1]z^{-1}], \qquad (2.113)$$

where $F[0]$ *is lower triangular and* $F[1]$ *is upper triangular with zero diagonal, such that*

$$R[0] = F^T[0]F[0] + F^T[1]F[1], \qquad (2.114)$$

$$R[1] = F^T[0]F[1], \qquad (2.115)$$

$$\det F[0] = \exp\left(\frac{1}{2}\int_0^1 \log(\det S(e^{j2\pi f}))\,df\right). \qquad (2.116)$$

Furthermore, if $\det S(e^{-j\omega}) > 0$ *for all* $\omega \in [-\pi, \pi]$, *then*

$$\left[F[0] + F[1]z^{-1}\right]^{-1}$$

is causal and stable.

In parallel with (2.87), if the vector sequence of matched filter outputs $\mathbf{y}[i]$ is fed to the filter $\left[\mathbf{F}^T[0] + \mathbf{F}^T[1]z\right]^{-1}$, the output sequence is given by

$$\bar{\mathbf{y}}[i] = \mathbf{F}[0]\mathbf{Ab}[i] + \mathbf{F}[1]\mathbf{Ab}[i-1] + \bar{\mathbf{n}}[i], \qquad (2.117)$$

where, as we mentioned above, $\bar{\mathbf{n}}[i]$ is independent Gaussian with covariance matrix $\sigma^2\mathbf{I}$.

As in the synchronous case, alternative finite-dimensional models can be used with a set of orthonormal waveforms that span all the signature waveforms and their delayed versions. In a direct-sequence spread-spectrum system this can be accomplished by chip matched filters sampled at the chip rate times the number of users. Nevertheless, for (approximately) band-limited chip waveforms it is sufficient to sample at the Nyquist rate.

2.11 BIBLIOGRAPHICAL NOTES

Spread-spectrum signaling was introduced at the time of World War II for military communications with antijamming capabilities. The multiple-access potential of signals with large duration-bandwidth products was noted by Claude Shannon, the father of information theory, in 1949 (Price [343]). Compendia of the history, theory, and practice of direct-sequence and frequency-hopping spread-spectrum communications can be found in Scholtz [397], Simon et al. [410], Pickholtz et al. [333], Dixon [68], and Glisic and Vucetic [119]. A chronology of the evolution of CDMA radio communication systems is given in Scholtz [398]. CDMA for satellite communications is reviewed in Gaudenzi et al. [104]. Examples of optical and acoustical CDMA communication systems appear in Tamura et al. [439] and Catipovic et al. [48], respectively.

Reference Kaiser et al. [192] coined the term "chip" to refer to the elementary direct-sequence spread-spectrum waveform. An early reference on the analysis of the crosscorrelation properties of spread-spectrum signals with random signatures is Wolf and Elspas [539]. Formula (2.27) is obtained in Stiglitz [426]. Multicarrier CDMA is reviewed in Prasad and Hara [342] and Prasad [341]. The Walsh functions were introduced in Walsh [519]. Synchronous CDMA based on Walsh functions was advocated as an alternative to TDMA for satellite multiaccess in Golay [122]. Other orthogonal families have been proposed for synchronous CDMA in Chang [49], Suehiro and Hatori [435], and Lindsey [243].

The implementation and properties of maximal-length shift register sequences are compiled in Golomb [124]. The design of pseudonoise sequences with good autocorrelation and crosscorrelation properties has been an active research field for the past thirty years (Kasami [200]; Gold [123]; Olsen et al. [307]; Sarwate and Pursley [381]; Simon et al. [410]; Kohno [213]). An up-to-date tutorial introduction to the elements of pseudonoise sequences is given in Helleseth and Kumar [143]. The design of CDMA signature waveforms given a desired number of users K and available dimensionality N is an interesting problem in the "oversaturated" case: $K > N$. Optimal designs to maximize the Shannon capacity of the basic synchronous CDMA channel (Verdú [493, 497]) are found in Rupf and Massey [364] (see also Cheng and Verdú [55]; Cheng and Verdú [56]; and Parsavand and Varanasi [321]); preservation of the minimum distance achieved in the orthogonal case is the goal of Ross and Taylor [362]; and tree-structured designs that facilitate multiuser detection are proposed in Learned et al. [230]. Partial response CDMA systems (where the signature duration is greater than the intersymbol time) are proposed in Wornell [542], Ruprecht et al. [366], and Blanz et al. [31].

The use of multiuser detection to combat multiaccess interference in TDMA (due to channel distortion and out-of-cell interference) is studied in Wales [518], Ranta et al. [350], Ranta et al. [351], Kadaba et al. [190], and Caire et al. [41]. Multiuser detection has also been explored in hybrid TDMA/CDMA systems in Honig and Madhow [154], Jung et al. [185], Blanz et al. [31], and Kammerlander [197]. Multiuser detection for multirate CDMA systems is proposed in Ojanpera et al. [306], Hottinen [162], and Hottinen and Pehkonen [164].

Statistical models for fading channels are extensively treated in Proakis [345], Clarke [59], Jakes [174], Loo [251], Stuber [433], Parsons [322], and Sklar [413], for example. The autocorrelation function in (2.54) was obtained in Clarke [59] and popularized in Jakes [174]. Reviews of techniques for the mitigation of frequency-flat and frequency-selective fading are given in Stein [421] and Sklar [412]. A tutorial on the multipath resistance of spread-spectrum formats is found in Turin [456]. The rake receiver to adaptively equalize multipath dates back to the 1950s (Price and Green [344]); it was originally developed for a 45.5 bits-per-second, $N = 220$, direct-sequence spread-spectrum shortwave teletype link known as the *Noise Modulation and Correlation system* (Ward [524]). Many other channel estimation techniques for spread-spectrum have been explored, e.g., Turin [456], Sanada et al. [377], Mowbray et al. [290], and Steiner and Jung [424]. The generalization of the rake to take into account the different spatial signatures of the various paths has been considered in a number of recent works reviewed in Paulraj and Papadias [326].

Proposition 2.1 due to Bai and Yin [19] has its roots in the classic semicircle law of Wigner [531]. For the homogeneous fading model, the exact nonasymptotic distribution of the eigenvalues of the unnormalized crosscorrelation matrix **H** can be found in Telatar [440].

Johnson and Dudgeon [180] is a general reference on antenna arrays from a signal processing viewpoint. There are numerous adaptive signal processing methods to obtain the spatial signatures; foremost among them are the "multiple signal classification (MUSIC)" method of Schmidt [391] and Bienvenue and Kopp [30] and the "estimation of signal parameters via rotational invariance techniques (ESPRIT)" method of Paulraj et al. [327]. Models for the aggregate effect of out-of-cell interference are given in Rappaport and Milstein [356], Viterbi et al. [511], and Agashe and Woerner [13].

The K-dimensional discrete-time model in (2.106) was introduced in Verdú [486]. Based on classical results, Proposition 2.3 can be found in Duel-Hallen [72, 74]. Whitened matched filters were introduced in Forney [96] in the context of single-user channels subject to intersymbol interference.

Oversampling beyond the Nyquist rate (Problem 2.39) has been shown to bring about a number of benefits for the implementation of linear receivers (Gitlin and Weinstein [118]) and for the identification of unknown channels (Tong et al. [448]). Tutorials on narrowband interference rejection can be found in Milstein [278], Rusch and Poor [368], and Laster and Reed [229].

2.12 PROBLEMS

PROBLEM 2.1. Let $M(N)$ denote the maximum number of orthogonal, synchronous, direct-sequence spread-spectrum signals with spreading factor N. Show

(a) $M(N) = 1$ if N is odd.
(b) $M(N) = N$ if N is a power of 2.
(c) $M(N) = 2$ if N is even and it is not divisible by 4 (e.g., Sec. 5.7 Peterson [331]).

Note: A general expression for $M(N)$ is a longstanding open problem.

PROBLEM 2.2. *Welch's bound* (Welch [529]) Consider a synchronous CDMA system with K users employing direct-sequence spread-spectrum signals of length N.

66

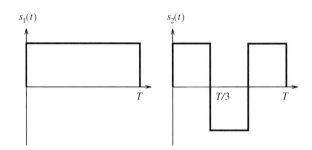

FIGURE 2.17.
Direct sequence
signature
waveforms with
$N = 3$.

(a) Show the following lower bound on the sum of the squares of the crosscorrelation matrix:

$$\sum_{i=1}^{K} \sum_{j=1}^{K} \rho_{ij}^2 \geq \frac{K^2}{N}. \qquad (2.118)$$

Note that this result is trivial if $K \leq N$. [*Hint:* The signature waveforms need not be bipolar for this result to hold.]

(b) Derive an upper bound on the maximum number of synchronous CDMA users for which direct-sequence spread-spectrum signature waveforms of length N can be found with absolute crosscorrelations not exceeding $|\rho_{max}|$. Assume $|\rho_{max}| \leq 1/\sqrt{N}$.

PROBLEM 2.3. Consider the signature waveforms in Figure 2.17.

(a) Find the crosscorrelation functions $\rho_{12}(\tau)$, $\rho_{21}(\tau)$ for $\tau \in [0, T]$.
(b) Find the probability density function of $\rho_{12}(\tau)$ if τ is uniformly distributed and the signals in Figure 2.17 are modulated by equal-frequency carriers with independent uniformly distributed phases.

PROBLEM 2.4. Define the *mean-square crosscorrelation* between s_1 and s_2 as

$$\frac{1}{T} \int_0^T \left(\rho_{12}^2(\tau) + \rho_{21}^2(\tau) \right) d\tau.$$

Find the mean-square crosscorrelation of

(a) $s_1(t) = s_2(t) = 1/\sqrt{T}$,
(b) the signals in Figure 2.17.

PROBLEM 2.5. Show that in a maximal-length shift-register sequence, the number of ones is equal to the number of zeros plus 1.

PROBLEM 2.6. Show that the matrix whose elements are

$$\{\rho_{kj}^2, k = 1, \ldots K, j = 1, \ldots K\}$$

is nonnegative definite.

PROBLEM 2.7. Suppose that all users in an asynchronous direct-sequence CDMA system are assigned the same signature waveform. Express the periodic autocorrelation function of the signature waveform in terms of the functions $\rho_{12}(\tau)$ and $\rho_{21}(\tau)$.

PROBLEM 2.8. Find the power spectral density of the wide-sense stationary process:

$$z(t) = \sum_{n=-\infty}^{\infty} b[n]s(t - nT - \tau),$$

where $b[n] \in \{-1, 1\}$ is an independent equiprobable sequence, independent of τ (which is uniformly distributed on $[0, T]$), and $s(t)$ is the direct-sequence spread-spectrum signal

$$s(t) = \sum_{i=1}^{N} (-1)^{c_i} p_{T_c}(t - (i-1)T_c). \tag{2.119}$$

Express your answer in terms of the energy spectral density of the chip waveform (magnitude square of its Fourier transform) and of the signature $(c_1, \ldots, c_N)$.

PROBLEM 2.9. Consider a single-user communication system that transmits the signal $z_1(t)$ to send "1" and the signal $z_0(t)$ to send "0". Assume that z_i is zero outside the interval $[0, T_i]$ and its energy is constrained to E_i.

(a) Describe the set of pairs (z_1, z_0) that maximize $\|z_1 - z_0\|$ under those constraints.

(b) Describe the subset of the optimal set found in (a) that maximizes the L_1-distance:

$$\int_0^{\max\{T_0, T_1\}} |z_1(t) - z_0(t)| \, dt.$$

PROBLEM 2.10. A telephone user speaks for periods of length $T_A[i]$, interspersed with silent periods of length $T_S[i]$. Let us assume that $\{T_A[i]\}$ and $\{T_S[i]\}$ are independent processes with means m_A and m_S respectively. Find the distribution of the number of users who are simultaneously speaking out of a total population of K active users. *Note:* Nonorthogonal CDMA is suited to take advantage of silent periods: if no (or reduced) energy is transmitted during silent periods, the effective number of simultaneous users is reduced on average ($2m_S \approx 3m_A$ is typical).

PROBLEM 2.11. (A. McKellips) In a single-user *full-response continuous phase modulation* the transmitted signal is given by

$$S(t; b[0], \dots b[2M]) = \cos\left(2\pi f_c t + \sum_{i=0}^{2M} b[i]\Phi(t - iT)\right),$$

(2.120)

where f_c is the carrier frequency, the data are $\{-1, +1\}$, and $\Phi(\cdot)$ is a continuous monotone increasing deterministic function that satisfies

$$\Phi(t) = 0, t < 0, \quad \Phi(t) = \frac{\pi}{2}, t \geq T.$$

Define the deterministic sequence $q[n]$ that has period equal to 4 and

$$q[0] = 1; q[1] = -1; q[2] = -1; q[3] = 1,$$

and let

$$d[i] \stackrel{\text{def}}{=} q[i + 1]\prod_{j=0}^{i} b[j].$$

Note that there is a one-to-one correspondence between the strings $b[0], \dots, b[2M]$ and $d[0], \dots, d[2M]$; thus, if the $\{b[i]\}$ are independent and equiprobable, so are the $\{d[i]\}$. Show that the single-user signal in (2.120) can be expressed as a two-user CDMA signal:

$$S(t; b[0], \dots b[2M]) = \sum_{i=0}^{M}\{d[2i]s_1(t - 2iT)$$

$$+ d[2i + 1]s_2(t - 2iT - T)\},$$

where each "user's" signature waveform has duration $2T$:

$$s_1(t) = \sin(2\pi f t)\left[\sin \Phi(t)p(t) + \cos \Phi(t - T)p(t - T)\right],$$

$$s_2(t) = \cos(2\pi f t)\left[\sin \Phi(t)p(t) + \cos \Phi(t - T)p(t - T)\right],$$

and $p(t)$ is a unit-amplitude rectangular pulse of length T.

Hint: For all $L, x_1, \dots x_L$:

$$\cos\left(\sum_{i=1}^{L} x_i\right) = \sum_{v\in\{0,1\}^{L-1}} q[w(v)]cs_{p[w(v)]}(x_1)\prod_{i=2}^{L} cs_{v_{i-1}}(x_i),$$

$$w(v) \stackrel{\triangle}{=} \sum_{i=0}^{M-1} v_i,$$

$$cs_j(x) \stackrel{\triangle}{=} \begin{cases} \cos(x), & j = 0, \\ \sin(x), & j = 1, \end{cases}$$

$$p[n] \stackrel{\triangle}{=} \begin{cases} 0, & n \text{ even}, \\ 1, & n \text{ odd}. \end{cases}$$

PROBLEM 2.12. Suppose that a transmitted signal s is subject to multipath and the receiver obtains

$$\tilde{s}(t) = s(t) + \sum_{j=1}^{2} \alpha_j s(t - d_j),$$

where $d_j \geq T/15$ and

$$\alpha_1^2 + \alpha_2^2 = \nu.$$

(a) Find the minimum value (over d_j and α_j) of $\langle s, \tilde{s} \rangle$ as a function of ν if s is the maximal-length shift register sequence of length 15,

$$[- - - + + + + - + - + + - - +],$$

with unit energy and duration T ($s(t) = 0, t \notin [0, T]$).

(b) Find the minimum value of $\langle s, \tilde{s} \rangle$ if s is a constant in $[0, T]$:

$$s(t) = \begin{cases} \frac{1}{\sqrt{T}}, & \text{if } t \in [0, T]; \\ 0, & \text{otherwise.} \end{cases}$$

PROBLEM 2.13. Consider a simple two-path channel with impulse response:

$$h(t) = \delta(t) - A\delta(t - d).$$

Find the magnitude spectrum of this channel as a function of A and d. Superpose a plot of the magnitude spectrum of the channel on top of the spectrum of the direct-sequence spread-spectrum function in Figure 2.5 if (a) $(A, d) = (1, T_c/2)$, (b) $(A, d) = (-1, T_c)$, (c) $(A, d) = (1, 10T_c)$.

PROBLEM 2.14. Prove or disprove: the crosscorrelation matrix defined in (2.13) is nonnegative definite for all $0 \leq \tau < T$.

PROBLEM 2.15. Show that in the random signature sequence model with asynchronous users and uniformly distributed delays, the crosscorrelations ρ_{12} and ρ_{21} are dependent and uncorrelated.

PROBLEM 2.16. A generalized form of direct-sequence spread-spectrum uses nonbinary codes to multiply the chip waveforms, that is (cf. (2.15)), the signature waveforms are

$$s(t) = A \sum_{i=1}^{N} \beta_i p_{T_c}(t - (i - 1)T_c), \qquad (2.121)$$

where $(\beta_1, \ldots, \beta_N)$ is a vector with real-valued components and A is such that $\|s\| = 1$. The *Kabatyanskii–Levenshtein–Mazo* bound

(Mazo [270]) states that if the maximum absolute crosscorrelation between any pair of signature waveforms is forbidden to exceed

$$|\langle s_i, s_j \rangle| \leq \rho_{\max},$$

then the maximum number of synchronous real-valued direct-sequence spread-spectrum waveforms with N chips per symbol (2.121) is bounded by

$$K_{\max} \leq \min_j \frac{(1 - \rho_{\max}^2)N(N + 2) \cdots (N + 2j)}{((2j + 1) - (N + 2j)\rho_{\max}^2)(1 \cdot 3 \cdot 5 \cdots (2j - 1))},$$

(2.122)

where the minimum is over all positive integers j for which the denominator is positive.

Particularize (2.122) to the case

$$\rho_{\max}^2 = \frac{1}{M}.$$

PROBLEM 2.17. Suppose that a maximal-length shift-register direct-sequence spread-spectrum waveform $s(t)$ with N chips per bit and chip waveform p_{T_c} is modulated antipodally and transmitted with carrier frequency f_c:

$$A \sum_{i=-M}^{M} b[i]s(t - iT)\cos(2\pi f_c t).$$

If the signal is received embedded in a jamming signal given by

$$A_J \cos(2\pi f_c t + \phi),$$

find the maximum value of A_J/A for which a correlator gives error-free demodulation for all ϕ in the absence of any other noise source.

PROBLEM 2.18. Redo Problem 1.5.b for $N = 2, 3, 4$ if the users are asynchronous and the receiver knows the relative delays between the users.

PROBLEM 2.19. Consider a synchronous CDMA system with three users and baseband signature waveforms s_1, s_2, s_3 modulated by respective carriers at frequency f_c and independent uniformly distributed phases ϕ_1, ϕ_2, ϕ_3.

(a) Find a closed-form expression for the probability that the absolute crosscorrelation between the modulated signals satisfies

$$|\bar{\rho}_{12}| \leq \alpha |\rho_{12}|$$

for $0 < \alpha < 1$.

(b) Describe the joint distribution of the random variables $\bar{\rho}_{12}$, $\bar{\rho}_{13}$, and $\bar{\rho}_{23}$. (Note: $\bar{\rho}_{12}$, $\bar{\rho}_{13}$, and $\bar{\rho}_{23}$ are not jointly continuous; you may give a "probability density function" that includes delta functions)

PROBLEM 2.20. In *Multicode* CDMA (I et al. [168]), a user requiring m times the basic data rate serial-to-parallel converts its data into m independent streams that modulate m synchronous signature waveforms. Consider a three-user asynchronous system where user k transmits at k times the data rate of user 1. Give the structure of the 6×6 crosscorrelation matrices $\mathbf{R}[0]$ and $\mathbf{R}[1]$ that describe this channel.

PROBLEM 2.21. A synchronous direct-sequence spread-spectrum CDMA system uses signature waveforms randomly chosen so that the components of the codes $(c_{k1}, \ldots, c_{kN})$ are independently chosen and equally likely to be 0 or 1.

(a) Show that the crosscorrelations have the even binomial distribution:

$$P\left[\rho_{kl} = 1 - \frac{2i}{N}\right] = \binom{N}{i} 2^{-N}, \quad i = 0, 1, \ldots, N. \quad (2.123)$$

(b) Show that the $K(K-1)/2$ crosscorrelations are pairwise independent but not jointly independent.
(c) Show that for all $j = 1, \ldots, K$, the random variables $\{\rho_{jk}, k = 1, \ldots, K\}$ are independent.
(d) Show that as $N \to \infty$, the sequence $\sqrt{N}\rho_{jk}$ converges in distribution to a zero-mean, unit-variance, Gaussian random variable.
(e) Denote

$$\beta = \frac{K}{N}.$$

Show that

$$\lim_{K\to\infty} \sum_{\substack{j=1 \\ j\neq k}}^{K} \rho_{jk}^2 = \beta,$$

in the sense of mean-square convergence. [*Note:* Convergence almost surely also holds.]

PROBLEM 2.22. (E. Telatar) In this problem we consider a direct-sequence spread-spectrum CDMA system where chips are modulated by real-valued quantities as in Problem 2.16:

$$s(t) = A \sum_{i=1}^{N} \beta_i p_{T_c}(t - (i-1)T_c). \qquad (2.124)$$

Suppose that the signature codes $(\beta_1, \ldots, \beta_N)$ are assigned randomly and independently from user to user. Unlike Problem 2.21 where the signatures were chosen from the vertices of a hypercube, here $(\beta_1, \ldots, \beta_N)$ is chosen uniformly from the surface of the N-dimensional unit sphere. Show that the normalized crosscorrelation $-1 \le \rho_{jk} \le 1$ between different synchronous waveforms has the following probability density function for $N \ge 3$:

$$f_{\rho_{jk}}(x) = \frac{1}{c(N)} \frac{(N-2)(N-4)(N-6) \cdots}{(N-3)(N-5)(N-7) \cdots} \left(1 - x^2\right)^{(N-3)/2},$$
$$(2.125)$$

where $c(N)$ is equal to 2 if N is odd and equal to π if N is even.

PROBLEM 2.23. Assume that users i and j are assigned direct-sequence spread-spectrum waveforms with N chips per bit, and rectangular chip waveforms and that signatures are chosen independently of each other and with equal probability on the set of all possible bipolar signatures.

(a) Find the mean and variance of $\rho_{ij}(lT_c + \lambda)$ as a function of $l = 0, \ldots, N-1, \lambda \in [0, T_c)$.
(b) (Burr [39]) Find the expected mean-square crosscorrelation (Problem 2.4).

PROBLEM 2.24. Consider the synchronous complex-valued model in (2.48),

$$y(t) = \sum_{k=1}^{K} A_k[i]b_k s_k(t) + \sigma n(t), \quad t \in [0, T], \qquad (2.126)$$

and define the complex matched-filter outputs

$$y_k = \int_0^T y(t)s_k^*(t) \, dt.$$

Find the covariance matrix of $[y_1, \ldots, y_K]$ assuming that $A_k[i]$ has zero mean and variance A_k^2.

PROBLEM 2.25. Give a complex-valued representation of the frequency-hopping signature waveform in Figure 2.11.

PROBLEM 2.26. Consider a K-user TDMA system that uses a rectangular signal of duration T_0 with guard times equal to γT_0.

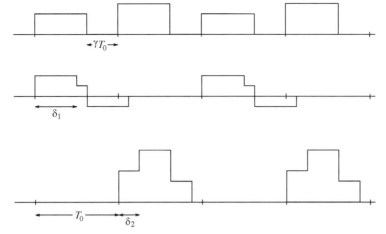

FIGURE 2.18.
TDMA signals
subject to
multipath. $K = 2$,
$\alpha_1 = -0.5$,
$\alpha_2 = 2/3$,
$\gamma = 3/8$.

The transmission of the ith user is subject to two-path reception
modeled by the impulse response (Figure 2.18):

$$h_i(t) = \delta(t) + \alpha_i \delta(t - \delta_i T_0).$$

(a) Find $\delta_{\max}$, the largest value of δ_i that ensures the absence of
single-user intersymbol interference for the ith user.

(b) Find ρ_{12} and ρ_{21} for the received signature waveforms as a
function of $0 \leq \delta_1 \leq \delta_{\max}$ and $0 \leq \delta_2 \leq \delta_{\max}$ if $\gamma = 3/8$,
$\alpha_1 = -0.5$, and $\alpha_2 = 2/3$.

PROBLEM 2.27. Suppose that a signal $s(t)$ goes through the cascade of a linear time-invariant system with impulse response $h(t)$
and a linear time-invariant system with impulse response $s(-t)$.
Show that if the autocorrelation function of $s(t)$ satisfies

$$\int s(t)s(t + \lambda)\, dt = \delta(\lambda), \qquad (2.127)$$

then the output of the cascade is $h(t)$. (Naturally, only an approximation to (2.127) can be realized.)

PROBLEM 2.28. Orthogonal Frequency Division Multiplexing
(OFDM) is the following special case of the basic synchronous
CDMA model (2.41) where the signature waveforms are given by

$$s_k(t) = \frac{1}{\sqrt{T}} e^{j2\pi(k-1-\frac{K-1}{2})t/T}, \qquad t \in [0, T].$$

(a) Verify that this is a form of orthogonal CDMA, that is,

$$\langle s_k, s_l \rangle = 0.$$

(b) Suppose that there is no noise and all users are transmitted with the same amplitude. Show that a K-point discrete Fourier transform of

$$[y(0), y(T/K), \ldots, y((K-1)T/K)],$$

where

$$y(t) = \sum_{k=1}^{K} Ab_k s_k(t)$$

recovers the transmitted information vector.

(c) Find the power spectral density of the transmitted signal (e.g., Stuber [432]).

PROBLEM 2.29. A base station in a cellular TDMA system receives the undistorted transmission of the K signals in its cell plus the K signals from another cell operating at the same frequency. Find the crosscorrelations in the corresponding asynchronous $2K$-user CDMA model under the following assumptions:

1. Transmitters within a cell are asynchronous with transmitters in another cell.
2. No guard times are employed.
3. All transmitters use phase-shift-keying with the same frequency and independent carrier phases.
4. One bit is transmitted per slot.

[*Hint:* The $2K$ users may be numbered depending on the relative offset between the cells.]

PROBLEM 2.30. Suppose that the received amplitude of a transmitted signal is equal to

$$A[i] = A_0 R[i],$$

where A_0 is deterministic and $R[i]$ is a stationary ergodic Rayleigh-distributed random process whose first-order probability density function is

$$f_{R[i]}(r) = r e^{-r^2/2}.$$

An "outage" is declared at symbol [i] if $A[i] \le \alpha A_0$. Show that the percentage of time the system is in outage is equal to

$$1 - e^{-\alpha^2/2}.$$

Do you have enough information to compute the average duration of outage periods?

PROBLEM 2.31. Show that if R has the log–normal probability density function

$$f_R(r) = \frac{20 \log_{10} e}{r \sigma_P \sqrt{2\pi}} \exp\left(-\frac{(20 \log_{10} r)^2}{2\sigma_P^2}\right), \tag{2.128}$$

then

$$E[R^2] = 10^{\sigma_P^2(\ln 10)/200}.$$

PROBLEM 2.32. Express the cumulative distribution function of the log–normal density function in (2.128) in terms of the Q-function defined in (3.29).

PROBLEM 2.33. A receiver moves in a straight line toward the transmitter with velocity v. Assume that due to local scattering the energy received by the mobile antenna is isotropic on a two-dimensional plane (i.e., it is independent of the angle of arrival θ).

(a) If the transmitter sends a pure tone at frequency f_c, show that the frequency received at angle θ is equal to

$$f(\theta) = f_m \cos \theta + f_c,$$

where

$$f_m = \frac{v}{c} f_c$$

and c is the speed of light.

(b) Show that a pure tone at frequency f_c produces a received signal with power spectral density given by

$$S(f) = \frac{\alpha}{\sqrt{f_m^2 - (f - f_c)^2}}, \quad f \in (f_c - f_m, f_c + f_m).$$

(c) Show that the impulse response is

$$h(t, \tau) = \frac{\beta}{\sqrt{\left(\frac{vt}{c}\right)^2 - \left(t - \tau - \frac{d_0}{c}\right)^2}}, \quad t \in \left(\frac{\tau + \frac{d_0}{c}}{1 + \frac{v}{c}}, \frac{\tau + \frac{d_0}{c}}{1 - \frac{v}{c}}\right).$$

PROBLEM 2.34. Consider a *diversity* scheme by means of which the receiver obtains D independent measurements:

$$y_1(t) = AR_1 s(t) + \sigma n_1(t),$$

$$\vdots$$

$$y_D(t) = AR_D s(t) + \sigma n_D(t),$$

where the signal s has unit energy, $n_1, \ldots n_D$ are independent noises with unit spectral density, and $R_1, \ldots, R_D$ are independent with Rayleigh density function

$$f_R(r) = re^{-r^2/2}.$$

(a) In the strategy known as *selection combining*, the receiver selects the branch with the largest energy:

$$z(t) = AR_i s(t) + \sigma n_i(t),$$

where

$$i = \operatorname{argmax} R_j^2.$$

Find a closed-form expression for the cumulative distribution function of the signal-to-noise ratio of z, that is, of

$$\frac{A^2}{\sigma^2} \max \left\{ R_1^2, \ldots R_D^2 \right\}.$$

(b) In *maximal-ratio combining*, the receiver computes

$$z(t) = \sum_{i=1}^{D} R_i y_i(t).$$

Find a closed-form expression for the moment-generating function of the (amplitude) signal-to-noise ratio of z, that is,

$$E\left[e^{-sA \sum_{i=1}^{D} R_i^2 / (\sigma\sqrt{D})} \right].$$

(c) In *equal-gain combining*, the receiver computes

$$z(t) = \sum_{i=1}^{D} y_i(t).$$

Find a closed-form expression for the moment-generating function of the (amplitude) signal-to-noise ratio of z, that is,

$$E\left[e^{-sA \sum_{i=1}^{D} R_i / (\sigma\sqrt{D})} \right].$$

Note: The Laplace transform of $re^{-r^2/2}$ is

$$1 - \sqrt{2\pi} s e^{s^2/2} Q(s),$$

where Q is defined in (3.29).

PROBLEM 2.35. Assuming that all six element outputs are combined with equal gain and phase, plot the radiation pattern of the equilateral triangular antenna array in Figure 2.19.

PROBLEM 2.36. Design a linear array with the minimum number of elements that has nulls at angles $0°$, $120°$, and $240°$.

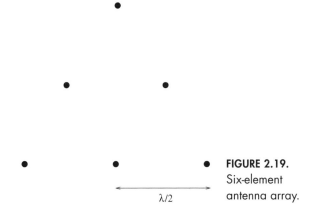

FIGURE 2.19.
Six-element
antenna array.

$\lambda/2$

PROBLEM 2.37. Suppose that all D elements in a linear (half-wavelength spaced) array are combined with equal magnitude.

(a) Find the magnitude of the radiation pattern obtained when all elements are combined with the same phase, that is,

$$\varphi_1 = \cdots = \varphi_D = 0.$$

(b) A scanning radar that contains no moving parts can be designed by letting

$$\varphi_n = n\beta.$$

Find the direction of maximum radiation as a function of β.

PROBLEM 2.38. Two base stations are $2d$ apart. A mobile transmitter located at distance $x > d$ from base station 1 adjusts its power so that it is received with power P at base station 2. Find the power with which the mobile is received at base station 1. Assume a path-loss exponent of 4 and consider both $x < 2d$ and $x > 2d$.

PROBLEM 2.39. (Tong et al. [448]) Suppose that the output of the single-user channel

$$y(t) = \sum_i b[i]s(t - iT) + \sigma n(t)$$

is low-pass filtered with an ideal filter with bandwidth B, and denote the responses of the filter to $y(t)$, $s(t)$, and $n(t)$ by $x(t)$, $v(t)$, and $\tilde{n}(t)$, respectively. Define the m-vector

$$\mathbf{x}[i] = (x(t_0 + iT + \Delta), \ldots, x(t_0 + iT + m\Delta))^T,$$

the d-vector

$$\mathbf{b}[i] = (b[i], \ldots, b[i + d - 1])^T,$$

and the m-vector

$$\tilde{\mathbf{n}}[i] = (\tilde{n}(t_0 + iT + \Delta), \ldots, \tilde{n}(t_0 + iT + m\Delta))^T .$$

(a) Show that $x(t)$ is a wide-sense *cyclostationary* process, that is, its autocorrelation function satisfies

$$R_X(t_1, t_2) = E[x(t_1)x(t_2)] = R_X(t_1 + T, t_2 + T).$$

(b) Express the m-vector $\mathbf{x}[i]$ as

$$\mathbf{x}[i] = \mathbf{V}\mathbf{b}[i] + \mathbf{n}[i], \tag{2.129}$$

by finding the $m \times d$ matrix $\mathbf{V}$ as a function of $v(t)$.

(c) Find the autocorrelation function

$$R_b[i, j] = E[\mathbf{b}[i]\mathbf{b}[j]].$$

(d) Find the autocorrelation function

$$R_{\tilde{n}}[i, j] = E[\tilde{\mathbf{n}}[i]\tilde{\mathbf{n}}[j]].$$

(e) Show that the discrete-time m-dimensional process $\mathbf{x}[i]$ is wide-sense stationary regardless of the values of B, T, t_0, m, and Δ.

PROBLEM 2.40. (Van Hesswyk et al. [144]) K users employ direct-sequence spread-spectrum waveforms:

$$s_k(t) = \sum_{i=1}^{N} z_{ik} p_{T_c}(t - (i-1)T_c), \tag{2.130}$$

where the last chip is blanked: $z_{Nk} = 0$. Assume that the users are *quasisynchronous*: $0 \le \tau_k \le T_c$. The receiver computes the observables:

$$r_i = \int y(t) p_{T_c}(t - (i-1)T_c) \, dt.$$

Denote

$$\mathbf{s}_k = [z_{1k} \quad z_{2k} \quad z_{3k} \quad \cdots \quad z_{N-1\,k} \quad 0],$$

$$\bar{\mathbf{s}}_k = [0 \quad z_{1k} \quad z_{2k} \quad \cdots \quad z_{N-2\,k} \quad z_{N-1\,k}].$$

Show that the observed vector obeys the following synchronous discrete-time model (cf. (2.96)):

$$\mathbf{r} = \sum_{k=1}^{K} A_k b_k [\alpha_k(\tau_k)\mathbf{s}_k + \bar{\alpha}_k(\tau_k)\bar{\mathbf{s}}_k] + \sigma \mathbf{m}$$

and determine $\alpha_k(\tau_k)$ and $\bar{\alpha}_k(\tau_k)$ as functions of p_{T_c}. [*Note:* If delays are bounded by $q T_c$, then the foregoing method can be generalized by appending q blank chips.]

PROBLEM 2.41. A set of K signature N-vectors that satisfies

$$\mathbf{SS}^T = \frac{K}{N}\mathbf{I} \qquad (2.131)$$

is called a Welch-bound-equality set because (2.131) is a necessary and sufficient condition for (2.118) to be satisfied with equality (Massey and Mittelholzer [267]).

(a) Prove that $K \geq N$ is necessary for $\mathbf{S}$ to satisfy (2.131).
(b) Show that in a bipolar, N-chip, direct-sequence spread-spectrum system with $K = 2^{N-1}$ where no two users are assigned the same waveform (or its antipodal), the signature waveform set is Welch-bound-equality.
(c) Let m be an integer and consider N orthonormal vectors, each of which is assigned to m users (i.e., $K = mN$). Show that the resulting $N \times mN$ matrix $\mathbf{S}$ is Welch-bound-equality.

PROBLEM 2.42. Consider the hypothetical case in which the background noise is absent ($\sigma = 0$) in the model (2.97). Let the receiver observe

$$\mathbf{r}[i] = \mathbf{SAb}[i], \quad i = 0,\ldots.$$

Assume that the signature waveforms are linearly independent.

(a) Give a scheme to obtain $\mathbf{b}[i]$ from $\mathbf{r}[i]$ and $\mathbf{SA}$. (Note that $\mathbf{SA}$ need not be a square matrix.)
(b) Assume now that $\mathbf{S}$ and $\mathbf{A}$ are unknown to the receiver. Suppose that the indices i_j are such that the received L-vectors

$$\{\mathbf{r}[i_1], \ldots, \mathbf{r}[i_{2^K}]\}$$

are all different.

Give a scheme to obtain the matrix $\mathbf{SA}$ (up to a sign uncertainty in each row) based on a list of the $2^{2K-1} - 2^{K-1}$ differences:

$$\{\mathbf{r}[i_n] - \mathbf{r}[i_m], \ m = n+1, \ldots 2^K, \ n = 1, \ldots 2^K - 1\}.$$

PROBLEM 2.43.

(a) Show that if $\{s_1, \ldots, s_K\}$ are spanned by the orthonormal basis $\{\psi_1, \ldots, \psi_L\}$, then

$$\mathbf{R} = \mathbf{S}^T\mathbf{S}, \qquad (2.132)$$

where $\mathbf{S}$ is defined in (2.98).
(b) Give an example where (2.132) is satisfied and $L < K$.

PROBLEM 2.44. Consider a direct-sequence spread-spectrum system with N chips per bit and rectangular chip waveforms. Continuous-to-discrete-time conversion is realized by a chip matched filter that is sampled at the chip rate but with a timing offset $0 \leq \tau \leq T_c$:

$$s_l = \int_{lT_c - T_c + \tau}^{lT_c + \tau} p_{T_c}(t - (l-1)T_c - \tau)s(t)\, dt,$$

where $s(t)$ is given by (2.15).

(a) Express the signature vector $\mathbf{s}$ at the output of the chip matched filter as a function of τ and the binary signature sequence $(c_1, \ldots, c_N)$.

(b) Find the expected energy of the signature vector, $\|\mathbf{s}\|^2$, as a function of τ assuming that the signature waveform is randomly chosen (all choices being equiprobable).

PROBLEM 2.45. Consider the synchronous crosscorrelation matrix $\mathbf{R}$ when all K signature waveforms are equicorrelated:

$$\rho_{jk} = \rho$$

for $j \neq k$. Show that $\mathbf{R}$ has an eigenvalue of multiplicity $K - 1$ equal to $1 - \rho$ and an eigenvalue equal to $1 + \rho K - \rho$.

PROBLEM 2.46. Let $\bar{\mathbf{y}}$ be the output of a bank of $K + 1$ matched filters with linearly independent unit energy $\{s_1, \ldots, s_{K+1}\}$, when the input is

$$y(t) = \sum_{k=1}^{K} A_k b_k s_k(t) + \sigma n(t).$$

Show how to obtain $\{\sigma, A_1, \ldots A_K\}$ from the $(K+1) \times (K+1)$ covariance matrix $E[\bar{\mathbf{y}}\bar{\mathbf{y}}^T]$ and $\bar{\mathbf{R}}$, the crosscorrelation matrix of $\{s_1, \ldots, s_{K+1}\}$.

PROBLEM 2.47.

(a) Verify that if $\gamma^2 > 0$ is an eigenvalue of

$$\mathbf{H} = \mathbf{A}\mathbf{R}\mathbf{A},$$

then $\sigma^2 + \gamma^2$ is an eigenvalue of $E[\mathbf{r}\mathbf{r}^T]$, the covariance matrix of the observations in the L-dimensional discrete-time model given in (2.99).

(b) Verify that if the signature vectors are linearly independent, then σ^2 is an eigenvalue of $E[\mathbf{r}\mathbf{r}^T]$ with multiplicity $L - K$.

(c) Find the eigenvectors and eigenvalues of $E[\mathbf{r}\mathbf{r}^T]$ if $K = 2$.

PROBLEM 2.48. We construct KN-dimensional unit-energy vectors $s_1, \ldots, s_K$ by drawing all NK components independently equally likely to be $1/\sqrt{N}$ or $-1/\sqrt{N}$. We will let N and K grow without bound so that

$$\lim_{N \to \infty} \frac{N}{K} = \alpha \in (0, +\infty).$$

Show that the $N \times N$ matrix

$$\frac{N}{K} \sum_{k=1}^{K} s_k s_k^T$$

has an asymptotic eigenvalue distribution equal to the cumulative distribution function of the density function

$$f_\alpha(x) = [1 - \alpha^{-1}]^+ \delta(x) + \frac{\sqrt{[x - a]^+ [b - x]^+}}{2\pi \alpha x}, \qquad (2.133)$$

where

$$a = \left(1 - \sqrt{\alpha}\right)^2,$$

$$b = \left(1 + \sqrt{\alpha}\right)^2. \qquad (2.134)$$

PROBLEM 2.49. Show that

$$R[-1] = R^T[1], \ R[0], \ R[1],$$

defined in (2.108) and (2.109), is a nonnegative-definite sequence in the sense that for any choice of finitely many K-vectors $x_1, \ldots, x_n$

$$\sum_{j,k} x_j^T R[j - k] x_k \geq 0,$$

assuming $R[m] = 0$, with $|m| > 1$.

PROBLEM 2.50. Solve Equations (2.114) and (2.115) to obtain $F[0]$ and $F[1]$ in Proposition 2.3 for the two-user asynchronous channel. [*Hint:* Express the solution in terms of $\rho_{12} + \rho_{21}$ and $\rho_{12} - \rho_{21}$ and use:

$$\sin(\alpha + \phi) = \cos \alpha \sin \phi + \cos \phi \sin \alpha.]$$

PROBLEM 2.51. Show that in the Cholesky factorization of Proposition 2.2,

$$F_{11}^2 (R^{-1})_{11} = 1.$$

PROBLEM 2.52. Find the Cholesky lower triangular factor of

$$R = \begin{bmatrix} 1 & \rho_{12} & \rho_{13} \\ \rho_{12} & 1 & \rho_{23} \\ \rho_{13} & \rho_{23} & 1 \end{bmatrix}.$$

PROBLEM 2.53. In the statement of Proposition 2.2, the positive-definite crosscorrelation matrix is expressed as the product of an upper triangular matrix $\mathbf{F}^T$ times a lower triangular matrix $\mathbf{F}$. Show that Proposition 2.2 is equivalent to: For every positive definite matrix $\mathbf{R}$ there exists a unique lower triangular matrix $\mathbf{L}$ with positive diagonal elements such that

$$\mathbf{R} = \mathbf{L}\mathbf{L}^T.$$

PROBLEM 2.54.

(a) Show by induction that the jth column of the lower triangular matrix $\mathbf{F}$ in the decomposition of the crosscorrelation matrix $\mathbf{R} = \mathbf{F}^T\mathbf{F}$ depends only on $\{s_j, \ldots, s_K\}$.
(b) Show that the jth column of $\mathbf{F}^{-1}$ depends only on $\{s_j, \ldots, s_K\}$.

PROBLEM 2.55. Denote the Cholesky decompositions

$$\mathbf{R} = \mathbf{F}^T\mathbf{F},$$

$$\mathbf{R}[k, K] = \mathbf{F}[k, K]^T\mathbf{F}[k, K], \tag{2.135}$$

where $\mathbf{R}[k, K]$ is the $(K-k+1) \times (K-k+1)$ submatrix of $\mathbf{R}$ that results from eliminating rows/columns $1, \ldots, k-1$. Show that for $1 \leq i \leq K - k + 1, 1 \leq j \leq K - k + 1$,

$$F_{ij}[k, K] = F_{k+i-1\,k+j-1}, \tag{2.136}$$

and, in particular,

$$F_{kk} = \frac{1}{(\mathbf{R}[k, K])_{11}^{-1}}. \tag{2.137}$$

PROBLEM 2.56. Let us consider a synchronous CDMA channel where all users transmit training sequences (known at the receiver) and where the crosscorrelation matrix $\mathbf{R}$ is nonsingular. An estimate of the $D \times K$ matrix of spatial signatures $\mathbf{C}$ is (cf. (2.91))

$$\hat{\mathbf{C}}[J] = \frac{1}{J} \sum_{j=1}^{J} \mathbf{Y}[i]\mathbf{R}^{-T}\mathbf{B}[i]\mathbf{A}^{-1}.$$

(a) Show that the $\{dk\}$ element of $\hat{\mathbf{C}}[J]$ satisfies $E[\hat{c}_{dk}[J]] = c_{dk}$.
(b) Find $\mathrm{var}(\hat{c}_{dk}[J])$.

PROBLEM 2.57. (Muñoz-Medina and Fernández-Rubio [293]) Let Ξ_k be the covariance matrix of the D-vector (cf. (2.89))

$$\begin{bmatrix} \langle y_1, s_k \rangle \\ \vdots \\ \langle y_D, s_k \rangle \end{bmatrix}.$$

Instead of assuming that training sequences are available as in Problem 2.56, let us assume that the covariance matrices Ξ_k, $k = 1, \ldots, K$ are known. (In practice, estimates can be obtained by computing sample autocorrelations.) Define the $D \times D$ matrix

$$\mathbf{M}_k = \sum_{j=1}^{K} q_{kj} \Xi_j$$

with $\mathbf{Q}$ equal to the inverse of the matrix with coefficients $\{\rho_{ik}^2\}$.

(a) Show that

$$\mathbf{M}_k - A_k^2 \mathbf{c}_k \mathbf{c}_k^*$$

is a multiple of the identity matrix.

(b) Show that the spatial signature of the kth user, $\mathbf{c}_k$, is the largest-eigenvalue eigenvector of $\mathbf{M}_k$.

PROBLEM 2.58. Consider a digital subscriber line subject to crosstalk such that the sampled received process is

$$y_i = \sum_j \left[x_j h_{i-j} + \sum_{k=2}^{K} z_{kj} g_{i-j} \right] + n_i, \qquad (2.138)$$

where $\{x_j\}$ is the desired data sequence and $\{z_{kj}\}$ is the data sequence of the kth interfering digital subscriber line. All data sequences are independent, taking values in $\{-1, +1\}$. The process $\{n_i\}$ is an independent zero-mean Gaussian with variance equal to σ^2. Assuming that $h_i = g_i = 0$ for $i < 0$ and $i > L$, generalize (2.106) to encompass (2.138) including the specification of the matrices $\mathbf{R}[i]$ therein.

PROBLEM 2.59. In Code-Time-Division Multiple Access (Ruprecht et al. [366]), users are assigned the same signature waveform, $s(t)$, and their mutual offsets are controlled so that $\tau_k - \tau_{k-1} = T/K$, that is, the channel is equivalent to the single-user intersymbol interference channel (Figure 2.3) with the exception that the amplitudes are periodically time-varying. Express the matrices $\mathbf{R}[0]$ and $\mathbf{R}[1]$ in terms of the autocorrelation function of the assigned waveform $s(t)$.

CHAPTER THREE

SINGLE-USER
MATCHED FILTER

In this chapter we analyze the simplest strategy to demodulate CDMA signals: the single-user matched filter. This is the demodulator that was first adopted in the implementation of CDMA receivers. In the multiuser detection literature, it is frequently referred to as the *conventional detector*. In this chapter we also introduce the major performance measures used to compare multiuser detectors. We start by discussing some basic concepts in hypothesis testing used in this and future chapters. Then, we derive the matched filter as the optimal solution in the single-user channel. The performance of the coherent matched filter in multiuser channels is examined both with and without fading. Differentially coherent and noncoherent demodulation are also studied.

3.1 HYPOTHESIS TESTING

3.1.1 OPTIMAL DECISIONS

A certain observed random quantity has a distribution known to belong to a finite set. Each of the possible distributions constitutes a different hypothesis. We must make a guess as to which is the "true" distribution (hypothesis) on the basis of the value taken by the observed quantity. Data demodulation is a hypothesis testing problem: the observed quantity is a noise-corrupted version of a transmitted signal, and there are as many hypotheses as different values for the transmitted data. For example, in the basic synchronous K-user CDMA channel model (2.1) there are 2^K hypotheses and the observed quantity is a waveform on the interval $[0, T]$.

In hypothesis testing, the goal is to design and analyze *decision rules*, that is, mappings from the space of observations to the set of hypotheses.

Decision rules are specified by partitioning the observation space into *decision regions*, each of which corresponds to a different hypothesis. As we mentioned in Section 2.4, of special interest to us is the case when all values taken by the transmitted data are equiprobable. The following simple result gives the optimal decision rule for equiprobable hypotheses.

Proposition 3.1 *Consider m equiprobable hypotheses under which an observed random vector Z has density functions:*[1]

$$H_1 : Z \sim f_{Z|1},$$

$$\vdots$$

$$H_m : Z \sim f_{Z|m}.$$

Then the following decision regions minimize error probability:

$$\Omega_i = \{z : f_{Z|i}(z) = \max_{j=1,\dots,m} f_{Z|j}(z)\} - \bigcup_{j=1}^{i-1} \Omega_j, \quad i = 1, \dots, m. \quad (3.1)$$

To justify Proposition 3.1, we can simply write the probability of error, P, as

$$P = 1 - \frac{1}{m} \sum_{i=1}^{m} P[Z \in \Omega_i | i]$$

$$= 1 - \frac{1}{m} \sum_{i=1}^{m} \int_{\Omega_i} f_{Z|i}(z)\,dz$$

$$\geq 1 - \frac{1}{m} \int \max_{j=1,\dots,m} f_{Z|j}(z)\,dz, \quad (3.2)$$

where the last integral is over the whole observation space. Choosing Ω_i as in (3.1), the lower bound in (3.2) is satisfied with equality. The decision regions given in (3.1) are not the only optimal choices. The nonuniqueness of the optimum solution arises because there are points in the observation space at which the maximum density is achieved by several densities simultaneously. As far as error probability is concerned, it is immaterial how those elements are assigned to the decision regions of the maximizing hypotheses; in (3.1) they were arbitrarily assigned to the maximizing hypothesis with the lowest index. It is easy to generalize Proposition 3.1 to encompass the general case where the hypotheses are not equiprobable. The unconditional probabilities, $P[H_i]$, of the hypotheses (before the data are observed) are

---

[1] The notation $\sim$ stands for "is distributed according to."

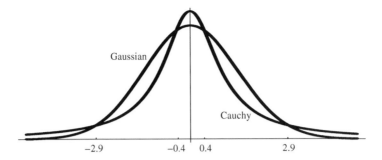

FIGURE 3.1.
Cauchy vs.
Gaussian binary
hypothesis test.

usually called *a priori* probabilities. Conditioned on a particular realization, z, of the observation, the *a posteriori* probabilities can be computed using the Bayes formula:

$$P[H_i|z] = \frac{f_{Z|i}(z)P[H_i]}{\sum_{j=1}^{m} f_{Z|j}(z)P[H_j]}.$$

In general, the minimum-error-probability decision rule is the so-called *maximum a posteriori* (MAP) rule, which selects the hypothesis with the highest $P[H_i|z]$. The decisions given in (3.1) are also known as *maximum-likelihood* (ML) decisions. Maximum likelihood decisions are sensible choices whenever it is not possible to rely on knowledge of the a priori probabilities.

The maximum-likelihood solution given in Proposition 3.1 is illustrated in Figure 3.1, which corresponds to a binary hypothesis testing problem between a standard Cauchy distribution and a zero-mean Gaussian with variance equal to 2:

$$H_1 : Z \sim f_{Z|1}(z) = \frac{1/\pi}{1 + z^2},$$

$$H_2 : Z \sim f_{Z|2}(z) = \frac{1}{2\sqrt{\pi}} e^{-z^2/4}.$$

Applying the result of Proposition 3.1, we see from Figure 3.1 that a rule that minimizes the error probability if both hypotheses are equally likely is to declare hypothesis H_2 if the observed value lies in $(-2.92, -0.42) \cup (0.42, 2.92)$ and to declare hypothesis H_1 otherwise. (See also Problem 3.10.)

Another example of the application of Proposition 3.1 is an m-hypothesis testing problem where the observation is a Gaussian L-vector with independent components (variance equal to σ^2 under each of the hypotheses). The distributions under the various hypotheses are distinguished by their means: the mean of the jth component under hypothesis H_i is a_{ij}. Therefore, the probability density function corresponding to H_i is given by

$$f_{Z|i}(z_1, \ldots, z_L) = \frac{1}{(2\pi)^{L/2}\sigma^L} \exp\left(-\frac{1}{2\sigma^2} \sum_{j=1}^{L} (z_j - a_{ij})^2\right), \qquad (3.3)$$

87

and the minimum probability of error decision regions for equiprobable hypotheses are

$$\Omega_i = \{(z_1, \ldots, z_L) : \sum_{j=1}^{L}(z_j - a_{ij})^2 = \min_{k=1,\ldots,m} \sum_{j=1}^{L}(z_j - a_{kj})^2\} - \bigcup_{j=1}^{i-1}\Omega_j,$$

(3.4)

which means that we select the hypothesis whose mean vector is closest to the observed vector in Euclidean distance.

3.1.2 CONTINUOUS-TIME SIGNALS IN WHITE GAUSSIAN NOISE

In some of the specific hypothesis testing problems we encounter in data demodulation, the observed quantity is not a vector as required by Proposition 3.1. For example, the CDMA receiver for either (2.1) or (2.6) observes a real-valued function defined on a finite time interval. Sometimes, we will place some structure on the receiver so that decisions are based on functions of the received waveform (called *observables* or *decision statistics*) which are either scalars or vectors. In those cases, we can use Proposition 3.1. Whenever we do not want to assume any prespecified receiver structure, Proposition 3.1 is insufficient because we have not defined the probability density function of real-valued functions. In order to sidestep a rigorous definition of such a concept, for which we would need measure-theoretic prerequisites we are not assuming, we will invoke the following counterpart to Proposition 3.1.

Proposition 3.2 *Let $x_1, \ldots, x_m$ be finite-energy deterministic functions defined on an interval of the real line $\mathcal{I}$. Let $n(t)$ be white Gaussian noise with unit power spectral density. Consider m equiprobable hypotheses:*

$$H_1 : y(t) = x_1(t) + \sigma n(t), \ t \in \mathcal{I},$$

$$\vdots$$

$$H_m : y(t) = x_m(t) + \sigma n(t), \ t \in \mathcal{I}.$$

Then, the following decision regions minimize error probability:

$$\Omega_i = \{y = \{y(t), t \in \mathcal{I}\} : f[y|x_i] = \max_{j=1,\ldots,m} f[y|x_j]\} - \bigcup_{j=1}^{i-1}\Omega_j, \quad (3.5)$$

where

$$f[y|x_i] = \exp\left(-\frac{1}{2\sigma^2}\int_{\mathcal{I}}[y(t) - x_i(t)]^2\, dt\right). \quad (3.6)$$

According to Proposition 3.2, the *likelihood* functions $f[y|x_i]$ defined in (3.6) play the role of the (unnormalized) conditional probability density functions in Proposition 3.1. Let us use the structure of the likelihood functions in (3.6) to simplify the decision regions in (3.5). The comparison of two likelihood functions,

$$f[y|x_i] \lessgtr f[y|x_j],$$

is equivalent (via (3.6)) to each of the following:

$$-\frac{1}{2\sigma^2} \int_{\mathcal{I}} [y(t) - x_i(t)]^2 \, dt \;\lessgtr\; -\frac{1}{2\sigma^2} \int_{\mathcal{I}} [y(t) - x_j(t)]^2 \, dt, \qquad (3.7)$$

$$\int_{\mathcal{I}} [y(t) - x_j(t)]^2 \, dt \;\lessgtr\; \int_{\mathcal{I}} [y(t) - x_i(t)]^2 \, dt, \qquad (3.8)$$

$$\int_{\mathcal{I}} y(t)x_i(t) \, dt - \frac{1}{2} \int_{\mathcal{I}} x_i^2(t) \, dt \;\lessgtr\; \int_{\mathcal{I}} y(t)x_j(t) \, dt - \frac{1}{2} \int_{\mathcal{I}} x_j^2(t) \, dt. \quad (3.9)$$

The last two relationships merit special attention:

Relationship (3.8) states that the decision regions that minimize error probability are *minimum distance* regions, that is, the decision rule selects the hypothesis corresponding to the waveform x_i that is closest to the observed waveform in mean-square distance.[2]

Relationship (3.9) states that the set of m observables,

$$\int_{\mathcal{I}} y(t)x_i(t) \, dt, \quad i = 1, \ldots, m, \qquad (3.10)$$

is a *sufficient statistic*, that is, it contains all the information in the original observations relevant to make an optimum decision.

More formally and generally, in a statistical inference problem where a parameter θ is to be inferred[3] on the basis of observations y, we say that the function of the data $g(y)$ is a sufficient statistic for θ if the conditional distribution of y given $g(y)$ does not depend on θ. In other words, the residual randomness in y to an observer that knows a sufficient statistic for θ contains

---

[2] The square root of the energy of the difference between two signals is commonly referred to as the *mean-square, L_2, L^2,* or *Euclidean* distance, with the latter term sometimes reserved for the special case when the signals belong to a finite-dimensional vector space.

[3] In hypothesis testing θ takes a finite (or countably infinite) number of values, whereas in *estimation* problems it takes an uncountable number of values.

no information about θ. Note that the (a priori) distribution of θ does not enter the definition of sufficient statistic; in fact, the definition we have given is general enough that we do not require that such a distribution be defined. A *minimal sufficient statistic* is a function of every other sufficient statistic. In less precise, but perhaps more intuitive, terms, a minimal sufficient statistic is one with the smallest number of components. In the hypothesis testing problem of Proposition 3.2, (3.10) is a minimal sufficient statistic if and only if $x_1, \ldots, x_m$ are linearly independent (Problem 3.19).

A rigorous proof of Proposition 3.2 requires concepts such as the Wiener process, stochastic integrals, and Radon–Nykodim derivatives, which are beyond the scope of our coverage (see, e.g., Poor [336]). A heuristic justification of Proposition 3.2 is that if hypothesis H_i is indeed the true hypothesis, then the noise realization is

$$n(t) = y(t) - x_i(t);$$

thus, the solution given by Proposition 3.2 corresponds to choosing the hypothesis that explains the received waveform with the lowest-energy noise realization (i.e., with the best least-squares fit).

Another heuristic argument is based on the resemblance of the solution given by Proposition 3.2 to the solution (3.4) we obtained in the vector Gaussian problem. Let us assume (without loss of generality) that the observation interval $\mathcal{I}$ is the whole real line. Furthermore, let us assume that the signals $x_1, \ldots, x_m$ are band-limited to B. Because of the duality of the time and frequency domains we could solve the decision problem in the frequency domain, expressing our decision rule as a function of the spectrum of $y(t)$. In fact, we should be able to focus attention on the observed spectrum up to frequency B, as the observations contain no information related to the true hypothesis beyond that frequency. In other words, the observed spectrum up to frequency B is a sufficient statistic. One could think of a hypothetical spectrum analyzer that would consist of L nonoverlapping ideal bandpass filters of bandwidth B/L. The hypothesis testing problem at the output of the analyzer now fits the setting of the Gaussian vector decision problem we considered above (with complex-valued components). Indeed, since the bandpass filters are nonoverlapping and linear and have the same transfer function (shifted in frequency) their outputs are independent Gaussian with identical variances. Furthermore, as L gets larger and larger we would expect that the energy of the L-vector approximation of the spectrum will converge to the true energy of the spectrum, which by Parseval's identity is equal to the energy in the time domain. Thus, the distances in (3.4) will approach the continuous-time distances in (3.6) (or (3.8)). Alternatively, we could start by low-pass filtering the observations (getting rid of the noise beyond frequency B) and then taking L samples of the observations at the Nyquist rate $2B$. This will fit the foregoing Gaussian vector setting with independent components of identical variance. If the signals $x_1, \ldots, x_m$ are approximately time-limited, then we would expect that as L increases, the sampled-time average (cf. (3.4)) will approach the corresponding time-domain integral (cf. (3.6)).

3.1.3 COMPOSITE HYPOTHESIS TESTING

Sometimes we will come across problems of *composite hypothesis testing*. In the special case of binary composite hypothesis testing, the set of m hypotheses is partitioned into two subsets:

$$\mathcal{H}_J = \{H_i, i \in J\}, \tag{3.11}$$

$$\bar{\mathcal{H}}_J = \{H_1, \ldots, H_m\} \setminus \mathcal{H}_J. \tag{3.12}$$

Rather than selecting one hypothesis among the m possibilities, in this case we are interested in deciding whether the true hypothesis belongs to $\mathcal{H}_J$. One way to solve this problem would be to first select the individual hypothesis that minimizes error probability (according to the result in Proposition 3.1) and then simply check whether that hypothesis is in $\mathcal{H}_J$. However, this may not be the best strategy because we are disregarding the probabilities of all but the winning hypothesis. The following result gives the minimum probability of error decision rule in the special case where $\mathcal{H}_J$ and $\bar{\mathcal{H}}_J$ contain the same number of equiprobable hypotheses.

Proposition 3.3 *Consider m equiprobable hypotheses under which an observed random vector Z has density functions:*

$$H_1 : Z \sim f_{Z|1},$$

$$\vdots$$

$$H_m : Z \sim f_{Z|m}.$$

Let $\mathcal{H}_J$ and $\bar{\mathcal{H}}_J$ be the classes of hypotheses defined in (3.11) and (3.12), respectively. Assume that $|J| = m/2$. Then the following decision regions minimize the error probability of the composite hypothesis testing problem:

$$\Omega_J = \left\{ z : \sum_{i \in J} f_{Z|i}(z) > \sum_{i \notin J} f_{Z|i}(z) \right\}, \tag{3.13}$$

$$\bar{\Omega}_J = \left\{ z : \sum_{i \in J} f_{Z|i}(z) \leq \sum_{i \notin J} f_{Z|i}(z) \right\}. \tag{3.14}$$

The justification of Proposition 3.3 is very simple. Since, by assumption, $\mathcal{H}_J$ and $\bar{\mathcal{H}}_J$ have the same number of elements, the problem is a special case of Proposition 3.1 with two equiprobable hypotheses under which the observations are distributed according to

$$\mathcal{H}_J : Z \sim \frac{2}{m} \sum_{i \in J} f_{Z|i},$$

$$\bar{\mathcal{H}}_J : Z \sim \frac{2}{m} \sum_{i \notin J} f_{Z|i}.$$

Analogously, the following counterpart to Proposition 3.2 can be obtained:

Proposition 3.4 *Let $x_1, \ldots, x_m$ be finite-energy deterministic functions de-fined on an interval of the real line $\mathcal{I}$. Let $n(t)$ be white Gaussian noise with unit power spectral density. Consider m equiprobable hypotheses:*

$$H_1 : y(t) = x_1(t) + \sigma n(t), \ t \in \mathcal{I},$$

$$\vdots$$

$$H_m : y(t) = x_m(t) + \sigma n(t), \ t \in \mathcal{I}.$$

Let $\mathcal{H}_J$ and $\bar{\mathcal{H}}_J$ be the classes of hypotheses defined in (3.11) and (3.12), respectively, with $|J| = m/2$. Then the following decision regions minimize the error probability of the composite hypothesis testing problem:

$$\Omega_J = \left\{ y : \sum_{i \in J} f[y|x_i] > \sum_{i \notin J} f[y|x_i] \right\}, \tag{3.15}$$

$$\bar{\Omega}_J = \left\{ y : \sum_{i \in J} f[y|x_i] \leq \sum_{i \notin J} f[y|x_i] \right\}, \tag{3.16}$$

where

$$f[y|x_i] = \exp\left(-\frac{1}{2\sigma^2} \int_{\mathcal{I}} [y(t) - x_i(t)]^2 \, dt\right). \tag{3.17}$$

Furthermore,

$$P[\mathcal{H}_J|y = \{y(t), t \in \mathcal{I}\}] = \frac{1}{1 + \frac{\sum_{i \notin J} f[y|x_i]}{\sum_{i \in J} f[y|x_i]}}. \tag{3.18}$$

If $\|x_1\| = \cdots = \|x_m\|$ in Proposition 3.4, then it can be easily checked that the optimal decision rule is

$$\sum_{i \in J} \exp\left(\frac{1}{\sigma^2} \int_{\mathcal{I}} y(t) x_i(t) \, dt\right) \lessgtr \sum_{i \notin J} \exp\left(\frac{1}{\sigma^2} \int_{\mathcal{I}} y(t) x_i(t) \, dt\right).$$

3.2 OPTIMAL RECEIVER FOR THE SINGLE-USER CHANNEL

In this section we study the single-user version of the basic CDMA channel of Chapter 2. With only one user, the issue of synchronism among transmitters does not arise, and the channel in (2.1) becomes

$$y(t) = Abs(t) + \sigma n(t), \quad t \in [0, T], \tag{3.19}$$

where the deterministic signal s has unit energy, the noise is white and Gaussian, and $b \in \{-1, +1\}$.

Before we derive the demodulator that minimizes the probability of error for (3.19), it is instructive to consider the restricted class of *linear detectors*, which will figure prominently in future chapters. Consider a demodulator for (3.19) that outputs the sign of the correlation of the observed waveform with a deterministic signal h of duration T:

$$\hat{b} = \text{sgn}(\langle y, h \rangle) = \text{sgn}\left(\int_0^T y(t)h(t)\, dt \right). \tag{3.20}$$

This detector summarizes the information contained in the observed waveform $y(t)$, $t \in [0, T]$ by the scalar decision statistic: $\langle y, h \rangle$. To optimize the choice of h, we notice that the linearity of the decision statistic makes it easy to discern the respective contributions of signal and noise:

$$Y = \langle y, h \rangle = Ab\langle s, h \rangle + \sigma \langle n, h \rangle. \tag{3.21}$$

The distribution of the second term in the right side of (3.21) is given by the following result, which states the main properties of white noise we use throughout this text.

Proposition 3.5 *If h and g are finite-energy deterministic signals and $n(t)$ is white noise with unit spectral density, then:*

1. $E[\langle n, h \rangle] = 0.$
2. $E[\langle n, h \rangle^2] = \|h\|^2.$
3. *If $n(t)$ is a Gaussian process, then $\langle n, h \rangle$ is a Gaussian random variable.*
4. $E[\langle n, h \rangle \langle n, g \rangle] = \langle g, h \rangle.$
5. $E[\langle n, h \rangle n] = h.$

The fact that linear transformations of Gaussian processes are Gaussian random variables yields Proposition 3.5.3. Proposition 3.5.2 is a special case of Proposition 3.5.4. To justify Propositions 3.5.1, 3.5.4, and 3.5.5, we

93

just need to interchange integration and expectation:

$$E\left[\langle n, h\rangle\right] = \int_0^T E[n(t)]h(t)\,dt = 0,$$

$$E\left[\langle n, h\rangle\langle n, g\rangle\right] = \int_0^T \int_0^T E[n(t)n(\lambda)]h(t)g(\lambda)\,dt\,d\lambda$$

$$= \int_0^T h(t)g(t)\,dt,$$

$$E\left[\langle n, h\rangle n(\lambda)\right] = \int_0^T E[n(t)n(\lambda)]h(t)\,dt$$

$$= h(\lambda),$$

where we used the fact that white noise with unit power spectral density has zero mean and autocorrelation function

$$E[n(t)n(\lambda)] = \delta(t - \lambda).$$

According to Proposition 3.5, $\sigma \|h\|$ is the standard deviation of the contribution of the noise to the decision statistic in (3.21). Thus, we would like to choose h to make $\langle s, h\rangle$ big and $\sigma \|h\|$ small. A sensible way to choose h according to this tradeoff is to maximize the signal-to-noise ratio of the decision statistic Y:

$$\max_h \frac{A^2 \left(\langle s, h\rangle\right)^2}{\sigma^2 \|h\|^2}. \tag{3.22}$$

The solution to (3.22) is readily obtained from the Cauchy–Schwarz inequality:

$$\left(\langle s, h\rangle\right)^2 \leq \|h\|^2 \|s\|^2,$$

with equality if and only if h is a multiple of s. Therefore,

$$\frac{A^2 \left(\langle s, h\rangle\right)^2}{\sigma^2 \|h\|^2} \leq \frac{A^2}{\sigma^2} \|s\|^2, \tag{3.23}$$

with equality if and only if h is a (nonzero) multiple of s. Thus, any nonzero multiple of the signal s will maximize the signal-to-noise ratio (3.22). Among those h, we can discard the negative multiples of s, because they yield erroneous decisions in the absence of noise ($\sigma = 0$). All positive multiples of s are equivalent choices for h since they result in the same decisions:

$$\hat{b} = \operatorname{sgn}(\langle y, \alpha s\rangle) = \operatorname{sgn}\left(\int_0^T y(t)s(t)\,dt\right), \tag{3.24}$$

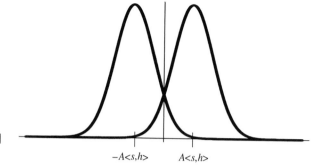

FIGURE 3.2.
Conditional
distributions of Y
given $b = -1$ and
$b = +1$.

$-A\langle s, h \rangle$ $A\langle s, h \rangle$

with $\alpha > 0$. The linear detector in (3.24) is known as the *matched filter*
detector for the signal s. It can be implemented either as a correlator (multiplication of the received waveform with s followed by integration) or as a
linear filter with impulse response $s(T - t)$ sampled at time T.[4]

We have seen that the matched filter is optimal in the sense that it maximizes the signal-to-noise ratio of the decision statistic Y. Notice that in this
derivation we did not invoke the fact that the noise is Gaussian. If we do use
that assumption, we can actually prove that among all linear detectors, the
matched filter minimizes the probability of error. To that end, Proposition
3.5 and (3.21) lead to the conclusion that the distribution of the decision
statistic Y conditioned on $b = -1$ is Gaussian with mean $-A\langle s, h \rangle$ and
variance $\sigma^2 \|h\|^2$, which is abbreviated as $\mathcal{N}(-A\langle s, h \rangle, \sigma^2 \|h\|^2)$. Analo-
gously, the distribution of the decision statistic conditioned on $b = +1$ is
$\mathcal{N}(A\langle s, h \rangle, \sigma^2 \|h\|^2)$. Figure 3.2 shows both conditional distributions $f_{Y|1}$
and $f_{Y|-1}$.

Therefore, we have the binary hypothesis testing problem:

$$H_1 : Y \sim f_{Y|1} = \mathcal{N}(A\langle s, h \rangle, \sigma^2 \|h\|^2),$$

$$H_{-1} : Y \sim f_{Y|-1} = \mathcal{N}(-A\langle s, h \rangle, \sigma^2 \|h\|^2),$$

which is a special case of the vector Gaussian problem considered in Section
3.1 with $L = 1$ and $m = 2$. Particularizing the solution obtained in (3.4),
the optimal decision regions are

$$\Omega_1 = \{x \in (-\infty, \infty) : f_{Y|1}(x) \geq f_{Y|-1}(x)\} = [0, +\infty), \quad (3.25)$$

$$\Omega_{-1} = (-\infty, 0), \quad (3.26)$$

---

[4] Originally the term *matched filter* was coined to refer to the latter notion, only. The
distinction between both implementations matters only when the output of the linear
filter with impulse response $s(T - t)$ is continuously observed (as in radar, synchronization, multipath delay estimation, etc.) or is sampled faster than the data rate (as in
some data communications settings).

where we assumed that $\langle s, h \rangle \geq 0$. Otherwise, the derivation is entirely analogous.[5]

We have shown that if the decision statistic is a linear transformation of the data, then it is optimal to compare it to a zero-threshold as in (3.20). Using (3.25), we can express the probability of error as

$$P = \frac{1}{2} \int_0^\infty f_{Y|-1}(v)\,dv + \frac{1}{2} \int_{-\infty}^0 f_{Y|1}(v)\,dv$$

$$= \frac{1}{2} \int_{A\langle s,h \rangle}^\infty \frac{1}{\sqrt{2\pi}\sigma \|h\|} \exp\left(-\frac{v^2}{2\sigma^2 \|h\|^2}\right) dv$$

$$+ \frac{1}{2} \int_{-\infty}^{-A\langle s,h \rangle} \frac{1}{\sqrt{2\pi}\sigma \|h\|} \exp\left(-\frac{v^2}{2\sigma^2 \|h\|^2}\right) dv$$

$$= \int_{\frac{A\langle s,h \rangle}{\sigma \|h\|}}^\infty \frac{1}{\sqrt{2\pi}} e^{-\frac{v^2}{2}}\,dv \tag{3.27}$$

$$= Q\left(\frac{A\langle s, h \rangle}{\sigma \|h\|}\right), \tag{3.28}$$

where (3.27) follows by symmetry and a change of integration variable, and (3.28) follows from the following notation for the complementary cumulative distribution function of the unit normal random variable:

$$Q(x) \overset{\text{def}}{=} \int_x^\infty \frac{1}{\sqrt{2\pi}} e^{-t^2/2}\,dt. \tag{3.29}$$

Section 3.3 gives a summary of the main properties of the Q-function. For now, the only property we need is that it is monotonically decreasing. Since we are assuming that the argument of (3.28) is nonnegative, the monotonicity of the Q-function implies that in order to minimize (3.28) we can maximize the square of the argument of (3.28). But that is exactly what we did in (3.22). Therefore, the sign of the matched filter output is shown to minimize error probability among the class of linear transformations. Using the result of (3.23) as the argument (squared) of (3.28) we see that the matched filter for s achieves error probability equal to

$$P^c = Q\left(\frac{A}{\sigma}\right). \tag{3.30}$$

Finally, let us proceed to drop the restriction that the decision statistic is a linear transformation of the received waveform and search for the detector

[5] Whether $0 \in \Omega_1$ or $0 \in \Omega_{-1}$ does not affect the error probability because $Y = 0$ with zero probability.

that achieves the minimum error probability among all detectors. This means that we can no longer assume that the observable is $\langle y, h \rangle$, and we have to work with the received process itself, $\{y(t), t \in [0, T]\}$. This problem is a special case of the problem solved in Proposition 3.2: $m = 2, \mathcal{I} = [0, T]$, and

$$x_1(t) = As(t),$$

$$x_2(t) = -As(t).$$

Because the energies of x_1 and x_2 are identical, particularizing (3.9) we see that the minimum error probability detector decides $\hat{b} = 1$ if

$$\int_{\mathcal{I}} y(t)x_1(t)\, dt \geq \int_{\mathcal{I}} y(t)x_2(t)\, dt.$$

Moreover,

$$\int_{\mathcal{I}} y(t)x_1(t)\, dt = -\int_{\mathcal{I}} y(t)x_2(t)\, dt = A \int_0^T y(t)s(t)\, dt,$$

which means that the matched-filter output $\langle s, y \rangle$ is a sufficient statistic and the detector in (3.24) achieves the lowest error probability among all detectors. Thus, the error probability achieved by the matched filter (3.30) is the minimum error probability. Note that the optimum demodulator for (3.19) does not need knowledge of the signal-to-noise ratio A/σ when, as we have assumed, both hypotheses are equiprobable (cf. Problem 3.14).

The minimum bit-error-rate (3.30) for the single-user channel (3.19) is determined by the signal-to-noise ratio. The shape of the transmitted signal does not influence the minimum bit-error-rate because of the inherent symmetry of the white Gaussian noise model: its projection along every direction has the same distribution (Proposition 3.5).

3.3 THE Q-FUNCTION

We now pause to review a number of properties of the complementary Gaussian cumulative distribution function:

$$Q(x) = \int_x^\infty \frac{1}{\sqrt{2\pi}} e^{-t^2/2}\, dt. \tag{3.31}$$

Even though the integral in (3.31) does not admit a closed-form solution, its analytical properties have been extensively studied.

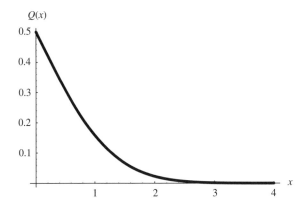

FIGURE 3.3.
Linear plot of
$Q(x)$ for $x > 0$.

1.

$$Q(x) = P[X > x], \qquad (3.32)$$

where X is a zero-mean, unit-variance, Gaussian random variable.

2. $Q(x)$ is monotonically decreasing.

3.

$$Q(x) + Q(-x) = 1.$$

4. $Q(x)$ is convex on the interval $(0, +\infty)$ (Figure 3.3).

5. $Q(\sqrt{x})$ is convex on the interval $(0, +\infty)$.

6.

$$\int_0^\infty Q(x)\,dx = \frac{1}{\sqrt{2\pi}}. \qquad (3.33)$$

7.

$$\frac{1}{\sqrt{2\pi}\,x}\left(1 - \frac{1}{x^2}\right)e^{-x^2/2} < Q(x), \quad x > 1, \qquad (3.34)$$

8.

$$Q(x) < \frac{1}{\sqrt{2\pi}\,x}e^{-x^2/2}, \quad x > 0, \qquad (3.35)$$

9.

$$Q(x) = \frac{1 - I(x)}{\sqrt{2\pi}\,x}e^{-x^2/2}, \quad x > 0, \qquad (3.36)$$

where $I(x)$ is the Laplace transform of the Rayleigh distribution (2.50):

$$I(x) = \int_0^\infty re^{-xr - r^2/2}\,dr.$$

10.

$$Q(x) \le \frac{1}{2} e^{-x^2/2}, \quad x \ge 0. \tag{3.37}$$

11.

$$Q(x) \le \frac{1}{2} e^{-\sqrt{2/\pi} x}, \quad x \ge 0. \tag{3.38}$$

12. The function

$$Q(x) e^{x^2/2}$$

is monotone decreasing on $[0, +\infty)$.

13.

$$\lim_{x \to \infty} Q(\alpha x) e^{x^2/2} = \begin{cases} +\infty, & \alpha < 1, \\ 0, & \alpha \ge 1. \end{cases} \tag{3.39}$$

14.[6]

$$\int_0^\infty Q(\alpha x) e^{x^2/2} \, dx = \frac{1}{\sqrt{8\pi}} \log\left(\frac{\alpha+1}{\alpha-1}\right), \quad \alpha > 1. \tag{3.40}$$

15.

$$\lim_{x \to \infty} \frac{Q(\alpha x)}{Q(\beta x)} = \begin{cases} +\infty, & [\alpha]^+ < \beta; \\ 2, & \alpha < \beta = 0; \\ 1, & \alpha = \beta \text{ or } \max\{\alpha, \beta\} < 0; \\ 1/2, & \beta < \alpha = 0; \\ 0, & [\beta]^+ < \alpha; \end{cases} \tag{3.41}$$

where

$$[z]^+ = \max\{0, z\}. \tag{3.42}$$

16.

$$2 \lim_{\sigma \to 0} \sigma^2 \log Q\left(\frac{x}{\sigma}\right) = -([x]^+)^2. \tag{3.43}$$

17.

$$2 \lim_{\sigma \to 0} \sigma^2 \log\left(1 - \prod_{k=1}^K Q\left(\frac{x_k}{\sigma}\right)\right) = -\min{}^2\{\max_k x_k, 0\}. \tag{3.44}$$

[6] Unless otherwise noted, logarithms are natural.

99

18. If $0 < a < b$, then

$$\lim_{\sigma \to 0} \frac{I_0 \left(\frac{ab}{\sigma^2}\right) e^{-(a^2+b^2)/(2\sigma^2)}}{Q \left(\frac{b-a}{\sigma}\right)} = \frac{b-a}{\sqrt{ab}}, \tag{3.45}$$

with I_0 equal to the Bessel function defined in Section 2.6.1.

19. If $0 < a < b$, then[7]

$$\lim_{\sigma \to 0} \frac{Q \left(\frac{a}{\sigma}, \frac{b}{\sigma}\right)}{Q \left(\frac{b-a}{\sigma}\right)} = \sqrt{\frac{b}{a}}. \tag{3.46}$$

20. If $0 < a < b$, then[8]

$$\lim_{\sigma \to 0} \frac{T \left(\frac{a}{\sigma}, \frac{b}{\sigma}\right)}{Q \left(\frac{b-a}{\sigma}\right)} = \frac{b+a}{2\sqrt{ab}}. \tag{3.47}$$

21.

$$\frac{Q(\sqrt{x+y})}{Q(\sqrt{x})} \le e^{-y/2}, \quad x \ge 0, y \ge 0. \tag{3.48}$$

22. The complementary error function, sometimes used in the statistics and communications literature, is equal to

$$\text{erfc}(x) \overset{\text{def}}{=} \frac{2}{\sqrt{\pi}} \int_x^\infty e^{-t^2} \, dt = 2Q(\sqrt{2}x). \tag{3.49}$$

23.

$$Q(x) = \frac{1}{\pi} \int_0^{\pi/2} e^{-x^2/(2 \sin^2(\theta))} d\theta, \quad x \ge 0. \tag{3.50}$$

24. If $-y \le x \le y$, then

$$Q(x-y) - Q(x+y) \ge \frac{1}{2} - Q(2y). \tag{3.51}$$

25.

$$(Q(x) - Q(y))^2 \le \frac{1}{2\pi} (x-y)^2. \tag{3.52}$$

26.

$$Q(x) = \frac{e^{-x^2/2}}{x\sqrt{2\pi}} \left(1 - \frac{1}{x^2} + \frac{1 \cdot 3}{x^4} - \frac{1 \cdot 3 \cdot 5}{x^6} + \cdots\right). \tag{3.53}$$

---

[7] $Q(x, y)$ is Marcum's Q-function defined in (3.155).
[8] $T(x, y)$ is defined in (3.153).

27.

$$Q(x) = \frac{1}{2} - \frac{1}{\sqrt{2\pi}} \left(x - \frac{x^3}{2} + \frac{x^5}{2 \cdot 4} - \frac{x^7}{2 \cdot 4 \cdot 8} + \cdots \right). \qquad (3.54)$$

28. If $N_1, \ldots, N_L$ are independent Gaussian random variables with zero-mean and standard deviation equal to σ, the complementary cumulative distribution function

$$P\left[\sqrt{N_1^2 + \cdots + N_L^2} \geq x \right], \quad x \geq 0 \qquad (3.55)$$

is equal to $2Q(x)$ if $L = 1$. Moreover, if L is even, then (3.55) admits a closed-form expression:

$$P\left[\sqrt{N_1^2 + \cdots + N_L^2} \geq x \right] = e^{-x^2/2\sigma^2} \sum_{n=0}^{L/2-1} \frac{1}{n!} \left(\frac{x^2}{2\sigma^2} \right)^n. \qquad (3.56)$$

29.

$$E[Q(X)] = \frac{1}{\sqrt{2\pi}} \int_{-\infty}^{\infty} P[X \leq x]\, e^{-x^2/2}\, dx. \qquad (3.57)$$

30.

$$\frac{1}{\alpha} \int_0^{\alpha} Q(x)\, dx = Q(\alpha) + \frac{1}{\alpha\sqrt{2\pi}}(1 - e^{-\alpha^2/2}). \qquad (3.58)$$

31.

$$\int_0^1 Q(\lambda\sqrt{x})\, dx = \left(1 - \frac{1}{\lambda^2} \right) Q(\lambda) + \frac{1}{2\lambda^2} - \frac{1}{\lambda\sqrt{2\pi}} e^{-\lambda^2/2}. \qquad (3.59)$$

32.

$$\int_0^{\infty} e^{-x}\, Q\left(\frac{x}{\sigma} \right) dx = \frac{1}{2} - e^{\sigma^2/2}\, Q(\sigma). \qquad (3.60)$$

33.

$$\int_0^{\infty} x\, e^{-x^2/2}\, Q\left(\frac{x}{\sigma} \right) dx = \frac{1}{2} \left(1 - \frac{1}{\sqrt{\sigma^2 + 1}} \right). \qquad (3.61)$$

34.

$$\int_0^{\infty} x\, e^{-x^2/2}\, Q\left(\frac{x + \mu}{\sigma} \right) dx$$

$$= Q\left(\frac{\mu}{\sigma} \right) - \frac{1}{\sqrt{1 + \sigma^2}}\, Q\left(\frac{\mu}{\sigma\sqrt{1 + \sigma^2}} \right) e^{-\mu^2/(2(1+\sigma^2))}. \qquad (3.62)$$

35.

$$\int_0^\infty x^{2n-1} e^{-x^2/2} Q\left(\frac{x}{\sigma}\right) dx$$

$$= \frac{(n-1)!}{2}(1-(\sigma^2+1)^{-1/2})^n$$

$$\times \sum_{k=0}^{n-1} 2^{-k}\binom{n-1+k}{k}\left(1+(\sigma^2+1)^{-1/2}\right)^k. \qquad (3.63)$$

36. If $\mathbf{X}$ is a complex-valued, zero-mean, Gaussian L-vector with independent real and imaginary parts and if the eigenvalues $\lambda_1, \ldots, \lambda_L$ of the covariance matrix $E[\mathbf{XX}^*]$ are distinct, then

$$E\left[Q\left(\|\mathbf{X}\|\right)\right] = \sum_{i=1}^L \frac{\alpha_i}{2}\left(1 - \frac{1}{\sqrt{1+\frac{2}{\lambda_i}}}\right), \qquad (3.64)$$

where

$$\alpha_j = \prod_{i=1,i\neq j}^L \frac{\lambda_j}{\lambda_j - \lambda_i}.$$

37. If $\mathbf{X}$ is a complex-valued, zero-mean, Gaussian L-vector with independent coefficients with variance $E[|X_i|^2] = 2\gamma^2$ and independent real and imaginary parts, then

$$E\left[Q\left(\|\mathbf{X}\|\right)\right] = \frac{1}{2} - \frac{1}{2}\frac{1}{\sqrt{1+\gamma^{-2}}}\left(1+\sum_{n=1}^{L-1}\frac{1\cdot 3\cdot 5\cdots(2n-1)}{n!\, 2^n\,(\gamma^2+1)^n}\right).$$
$$(3.65)$$

38. If X is a zero-mean, unit-variance, normal random variable, then

$$E\left[Q(\mu+\lambda X)\right] = Q\left(\frac{\mu}{\sqrt{1+\lambda^2}}\right). \qquad (3.66)$$

39. If X is a zero-mean, unit-variance, normal random variable, then

$$E\left[Q(\mu+\lambda X)X\right] = -\frac{1}{\sqrt{2\pi}}\sqrt{\frac{\lambda^2}{1+\lambda^2}}e^{-\mu^2/(2+2\lambda^2)}. \qquad (3.67)$$

40. For all $n > 0$,

$$\lim_{\sigma\to 0} E\left[Q^n\left(\frac{\mu+X}{\sigma}\right)\right] = \frac{1}{2^n}P[X=-\mu] + P[X<-\mu]. \qquad (3.68)$$

41. If X is a symmetric random variable taking values on the interval $[-\delta, \delta]$, and its characteristic function is denoted by

$$\Phi_X(\omega) \stackrel{\text{def}}{=} E[\cos(\omega X)],$$

then

$$E[Q(\mu + X)] = C_{0,L} + 2 \sum_{l=1}^{L-1} C_{l,L} \Phi_X \left(\frac{l\pi}{\delta} \right) + o(L), \qquad (3.69)$$

where the coefficients $[C_{0,L}, \ldots, C_{L-1,L}]$ depend only on μ, δ, and L and are defined as the discrete Fourier transform of $[F_{0,L}, \ldots, F_{L-1,L}]$,

$$C_{i,L} = \frac{1}{L} \sum_{k=0}^{L-1} F_{k,L} \cos\left(\frac{(2k+1)i\pi}{2L} \right),$$

where

$$F_{i,L} \stackrel{\text{def}}{=} \frac{1}{2} Q\left(\mu + \frac{2i+1}{2L} \delta \right) + \frac{1}{2} Q\left(\mu - \frac{2i+1}{2L} \delta \right).$$

Figure 3.3 shows a linear plot of the Q-function for positive arguments. Naturally, it is more useful to examine the behavior of the Q-function in the region where its values are more representative of bit-error-rates of interest in practice. Figure 3.4 shows a log–linear plot of the Q-function, along with the bounds in (3.34), (3.35), (3.37), and (3.38).

The upper bound in (3.35) is an excellent approximation to the Q-function unless the argument is small (or negative). But, in view of (3.39), it is sometimes preferable to view the asymptotic behavior of $Q(x)$ as that of the simpler upper bound (3.37).

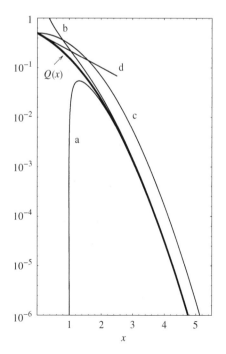

FIGURE 3.4.
Log–linear plot of $Q(x)$ and its bounds: (a) (3.34), (b) (3.35), (c) (3.37), and (d) (3.38).

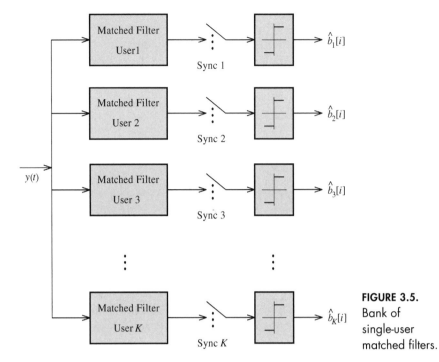

FIGURE 3.5.
Bank of single-user matched filters.

3.4   THE MATCHED FILTER IN THE CDMA CHANNEL

In this section we analyze the error probability of the single-user matched filter when used in the Gaussian CDMA channel. Figure 3.5 depicts a natural strategy to demodulate the transmitted streams in the basic CDMA models (2.1) and (2.6). Each filter is matched to one of the signature waveforms. As in any digital demodulator, it is necessary for the receiver to acquire (and track) synchronism with the bit epochs of the streams to be demodulated. If the CDMA channel is synchronous, then a single synchronizer provides the timing necessary to sample all the matched filters. In an asynchronous system, the individual timings of each of the transmitters must be acquired. In carrier-modulated systems, coherent detection requires synchronization with the frequency and phase of each of the modulating carriers. In an asynchronous direct-sequence spread-spectrum system where all users employ the same chip waveform, the continuous-time-to-discrete-time conversion of the bank of K matched filters can be carried out by a single chip-matched filter sampled at K times the chip rate, with sampling instants determined by the synchronizers. Needless to say, if the chip waveform is (almost) strictly band-limited, then sampling at twice its bandwidth is sufficient. It is not infrequent that only one or a subset of users is to be demodulated at a particular receiver. In that case, the bank of matched filters need not concern itself with users other than the ones of interest, because each of its branches operates autonomously. For the same reason, the analysis of the error probability need only focus on the output of one of the matched filters. We will examine the synchronous CDMA channel first.

3.4.1 PROBABILITY OF ERROR FOR SYNCHRONOUS USERS

As we saw in Section 2.9.1, the kth user matched-filter output is equal to

$$y_k = \int_0^T y(t)s_k(t)\,dt = A_k b_k + \sum_{j \neq k} A_j b_j \rho_{jk} + n_k, \qquad (3.70)$$

where

$$n_k = \sigma \int_0^T n(t)s_k(t)\,dt \qquad (3.71)$$

is a Gaussian random variable with zero mean and variance equal to σ^2.

If the signature waveform of the kth user is orthogonal to the other signature waveforms, then $\rho_{jk} = 0$, $j \neq k$ and the matched filter output (3.70) reduces to that obtained in the single-user problem:

$$y_k = A_k b_k + n_k.$$

The probability of error (cf. (3.30)) of a threshold comparison of y_k is

$$P_k^c(\sigma) = Q\left(\frac{A_k}{\sigma}\right), \qquad (3.72)$$

which is the same error probability we would obtain in the absence of other users. Since the presence of other users cannot decrease the error probability,[9] we conclude that the bank of single-user matched filters is optimal in the special case of synchronous orthogonal CDMA.

Let us now return to the nonorthogonal CDMA channel. It is best to consider the two-user case first (Figure 3.6).

The probability of error of user 1 is

$$P_1^c(\sigma) = P[b_1 \neq \hat{b}_1]$$

$$= P[b_1 = +1]P[y_1 < 0|b_1 = +1] \qquad (3.73)$$

$$+ P[b_1 = -1]P[y_1 > 0|b_1 = -1]. \qquad (3.74)$$

---

[9] cf. Problem 3.5.

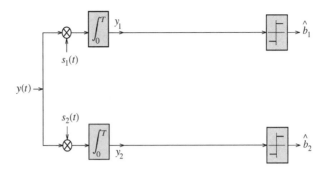

FIGURE 3.6.
Bank of
single-user
matched filters
$(K = 2)$.

Yet, y_1 is not Gaussian conditioned on b_1, so we will further condition on b_2:

$$P[y_1 > 0|b_1 = -1] = P[y_1 > 0|b_1 = -1, b_2 = +1]P[b_2 = +1]$$

$$+ P[y_1 > 0|b_1 = -1, b_2 = -1]P[b_2 = -1]$$

$$= P[n_1 > A_1 - A_2\rho]P[b_2 = +1]$$

$$+ P[n_1 > A_1 + A_2\rho]P[b_2 = -1]$$

$$= \frac{1}{2}Q\left(\frac{A_1 - A_2\rho}{\sigma}\right) + \frac{1}{2}Q\left(\frac{A_1 + A_2\rho}{\sigma}\right), \quad (3.75)$$

where we have used the independence of (b_1, b_2, n_1) and the simplified notation $\rho_{12} = \rho$, which we adopt throughout in the case of two synchronous users. By symmetry, we get the same expression for $P[y_1 < 0|b_1 = +1]$. Therefore, the bit-error-rate of the conventional receiver for user 1 in the presence of one interfering user is

$$P_1^c(\sigma) = \frac{1}{2}Q\left(\frac{A_1 - A_2\rho}{\sigma}\right) + \frac{1}{2}Q\left(\frac{A_1 + A_2\rho}{\sigma}\right) \quad (3.76)$$

$$= \frac{1}{2}Q\left(\frac{A_1 - A_2|\rho|}{\sigma}\right) + \frac{1}{2}Q\left(\frac{A_1 + A_2|\rho|}{\sigma}\right). \quad (3.77)$$

Interchanging the roles of users 1 and 2 in (3.77) we obtain

$$P_2^c(\sigma) = \frac{1}{2}Q\left(\frac{A_2 - A_1\rho}{\sigma}\right) + \frac{1}{2}Q\left(\frac{A_2 + A_1\rho}{\sigma}\right). \quad (3.78)$$

Let us further examine (3.77). Since the Q-function is monotonically decreasing, we readily obtain the upper bound

$$P_1^c(\sigma) \leq Q\left(\frac{A_1 - A_2|\rho|}{\sigma}\right). \quad (3.79)$$

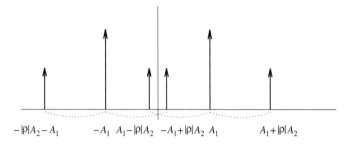

FIGURE 3.7.
Output of
matched filter for
user 1 with one
interfering user.

$-|\rho|A_2 - A_1$ $-A_1$ $A_1 - |\rho|A_2$ $-A_1 + |\rho|A_2$ A_1 $A_1 + |\rho|A_2$

This bound is smaller than $1/2$ provided that the interferer is not dominant:

$$\frac{A_2}{A_1} < \frac{1}{|\rho|}. \tag{3.80}$$

In that case, because of the asymptotic behavior ($\sigma \to 0$) of the Q-function (3.41), Equation (3.77) is dominated by the term with the smallest argument, and the upper bound (3.79) is an excellent approximation (modulo a factor of 2) to $P_1^c(\sigma)$ for all but low signal-to-noise ratios. This implies that the bit-error-rate of the conventional receiver behaves as that of a single-user system with reduced signal-to-noise ratio:

$$\left(\frac{A_1 - A_2|\rho|}{\sigma} \right)^2.$$

If the relative amplitude of the interferer is such that

$$\frac{A_2}{A_1} > \frac{1}{|\rho|}, \tag{3.81}$$

then the conventional receiver exhibits a highly anomalous behavior: the *near–far problem*. For example, the error probability is not monotonic with σ. To see this, first note from (3.77) that

$$\lim_{\sigma \to \infty} P_1^c(\sigma) = \frac{1}{2}, \tag{3.82}$$

a property we would expect from any detector. At the other extreme, we get

$$\lim_{\sigma \to 0} P_1^c(\sigma) = \frac{1}{2} \tag{3.83}$$

because, due to (3.81), as $\sigma \to 0$, the polarity of the output of the matched filter for user 1 is governed by b_2 rather than b_1.

Figure 3.7 shows the location of the noiseless output of the matched filter of user 1 under no interference (A_1 and $-A_1$) and with one interfering user satisfying (3.81). In this situation, some background noise is better than no noise: in the absence of background noise erroneous decisions are made with probability $1/2$, as the interference shifts the matched filter output to the wrong side of the threshold with probability $1/2$. The presence of a noise component either (a) has no effect on the decision, (b) prevents the detector from making an error, or (c) introduces an error. The noise excursion necessary for (c) is at least $|\rho|A_2 + A_1$, whereas the noise excursion required for (b) is only $|\rho|A_2 - A_1$. Thus, in the presence of noise the bit-error-rate is strictly smaller than $1/2$.

The noise level that minimizes the error probability under the condition (3.81) is (Problem 3.28)

$$\sigma^2 = \frac{A_1 A_2 \rho}{\operatorname{arctanh}\left(\frac{A_1}{A_2\rho}\right)}. \tag{3.84}$$

In the remaining case:

$$\frac{A_2}{A_1} = \frac{1}{|\rho|}, \tag{3.85}$$

the error probability of the single-user matched filter (3.77) reduces to

$$P_1^c(\sigma) = \frac{1}{4} + \frac{1}{2}Q\left(\frac{2A_1}{\sigma}\right),$$

which admits a simple interpretation: with probability $1/2$, the signal of user 2 exactly cancels the signal of user 1 at the matched filter output, which then becomes a zero-mean Gaussian random variable; with probability $1/2$, the signal of user 2 doubles the contribution of the desired signal to the matched filter output.

Figure 3.8 plots the bit-error-rate in (3.77) with $\rho = 0.2$ as a function of the signal-to-noise ratio A_1/σ for several values of the relative amplitude of the interferer. We can see how the bit-error-rate degrades fairly rapidly as the interfering user's amplitude grows. The upper curve is an example of the behavior under condition (3.81).

It is interesting to assess the mutual interfering effect of two users by examining the region of signal-to-noise ratios that result in a guaranteed level of bit-error-rate. Indeed, in practice it is common to fix a maximum tolerable bit-error-rate level and then find the necessary energies to satisfy that requirement. Using as low transmitted energy as possible is a primary design goal. (For example, in wireless cellular communication systems, this is important for increased battery life and reduction of intercell interference.) Figure 3.9 plots the power-tradeoff region so that both users achieve a bit-error-rate not higher than 3×10^{-5}, parametrized by the crosscorrelation between the signature waveforms. If the signature waveforms are mutually orthogonal, then we have two independent systems that achieve the desired objective as long as their signal-to-noise ratios are greater than or equal to

$$Q^{-1}(3 \times 10^{-5}) = 12 \text{ dB}.$$

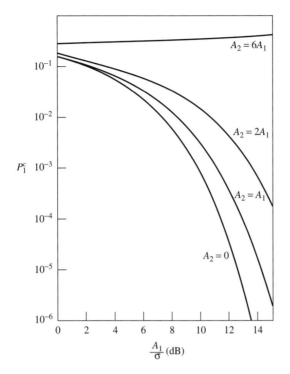

FIGURE 3.8.
Bit-error-rate of single-user matched filter with two synchronous users and $\rho = 0.2$.

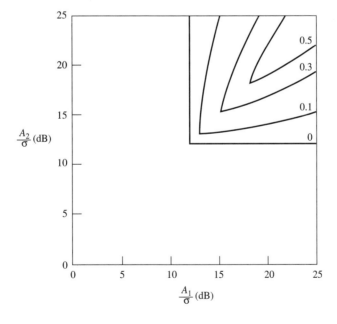

FIGURE 3.9.
Signal-to-noise ratios necessary to achieve bit-error-rate not higher than 3×10^{-5} for both users, parametrized by ρ.

As we increase the crosscorrelation between the signals we notice: (a) even if both amplitudes are identical the necessary energy increases rapidly and (b) the sensitivity to imbalances in the received energies grows. This is particularly important in mobile systems where the received amplitudes may vary over wide ranges. The sensitivity of the conventional detector illustrated in Figure 3.9 (in the two-user case) dictates the need for strict *power control* so that the received amplitudes are very similar.

It is useful to visualize the operation of the conventional detector (or, any detector, for that matter) in a signal space diagram. In Section 3.1 we specified a decision rule by the partition of the observation space into decision regions corresponding to each hypothesis. In the K-user synchronous channel, there are 2^K hypotheses and the original observation space is the set of real-valued functions on $[0, T]$, which is an infinite-dimensional space. However the K-user demodulator in the detector in Figure 3.5 bases its decisions on the K-dimensional vector (3.70)

$$(y_1, \ldots, y_K) = \left(\int_0^T y(t)s_1(t)\,dt, \ldots, \int_0^T y(t)s_K(t)\,dt \right)$$

computed from the original observations. Therefore, we can actually view the observations as the K-vector $(y_1, \ldots, y_K)$, instead of the original received waveform which belongs to an infinite-dimensional space. At this point, we opt for this simplified view not because the K-dimensional observations contain the same information as the original observations (which we have not yet proved) but simply because it is a convenient way to visualize the operation of the multiuser detector, in particular when $K = 2$. In this case, conditioned on (b_1, b_2), (y_1, y_2) is a Gaussian vector with mean (cf. (3.70) and (3.71))

$$(A_1b_1 + A_2b_2\rho, \ A_2b_2 + A_1b_1\rho) \tag{3.86}$$

and covariance matrix

$$\mathrm{cov}(y_1, y_2) = \sigma^2 \begin{bmatrix} 1 & \rho \\ \rho & 1 \end{bmatrix}.$$

In Figure 3.10 we have depicted the mean vectors for each of the four hypotheses in the (y_1, y_2) space, in the case where $A_1 = A_2 = 1$ and $\rho = 0.2$. We can view the "received" vector (y_1, y_2) as the sum of a "transmitted" vector (3.86) and a zero-mean Gaussian vector (n_1, n_2).

The decision regions in the (y_1, y_2) space of the single-user matched filter detector (Figure 3.6) are simply the four quadrants. In the case shown in Figure 3.10 ($A_1 = A_2 = 1$, $\rho = 0.2$), and in the absence of noise, the detector selects the correct hypothesis since it falls within its decision region. The probability of error found in (3.77) is the average of the probabilities that the received vector satisfies $y_1 < 0$ given that $(+, +)$ and $(+, -)$ are transmitted respectively. Accordingly, for the purposes of the decisions corresponding to user 1, the two-dimensional view of Figure 3.10 can be further simplified to the one-dimensional view of Figure 3.11. Nevertheless, for the sake of

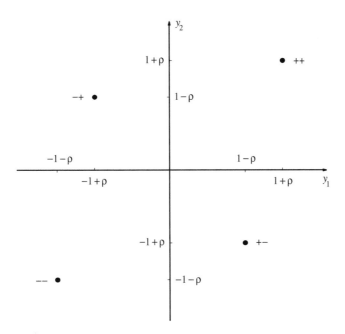

FIGURE 3.10.
Decision regions
in the
two-dimensional
space of matched
filter outputs.

comparison with other multiuser detectors considered in future chapters
(where decisions do involve interaction of y_1 and y_2), it is advantageous to
keep our focus on a two-dimensional observation space.

A shortcoming of the (y_1, y_2) diagram in Figure 3.10 is that the noise
components n_1 and n_2 added to the transmitted vector are correlated:

$$E[n_1 n_2] = \sigma^2 \rho;$$

consequently, the distribution of the noise vector is not circularly symmetric
and the norm of the noise vector does not determine the likelihood of that
realization. This impairs the intuition one would be expected to gain from
the comparison of the respective decision regions (in (y_1, y_2) space) of
different detectors. A better choice than the (y_1, y_2) diagram is a signal
space diagram whose components $(\tilde{y}_1, \tilde{y}_2)$ are equal to the correlations of
the received waveform with an (arbitrary) orthonormal basis (ϕ_1, ϕ_2) that
spans the linear space generated by the signals (s_1, s_2). For example, a choice
for that orthonormal basis is

$$\phi_1 = s_1, \tag{3.87a}$$

$$\phi_2 = \frac{1}{\sqrt{1 - \rho^2}} s_2 - \frac{\rho}{\sqrt{1 - \rho^2}} s_1. \tag{3.87b}$$

FIGURE 3.11.
Decision regions
in the
one-dimensional
space of matched
filter output.

$$\begin{array}{ccccc} -1-\rho & -1+\rho & & 1-\rho & 1+\rho \\ \bullet & \bullet & + & \bullet & \bullet \\ -- & -+ & & +- & ++ \end{array} \quad y_1$$

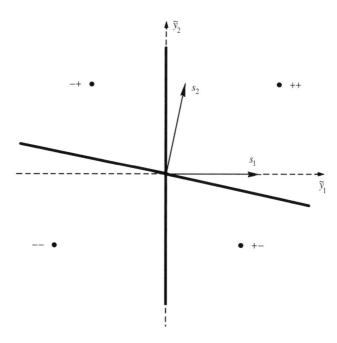

FIGURE 3.12.
Decision regions
of matched filter
detector
(orthogonal
space); $A_1 = A_2$.

Conditioned on (b_1, b_2), $(\tilde{y}_1, \tilde{y}_2)$ is Gaussian with mean

$$(A_1 b_1 \langle s_1, \phi_1 \rangle + A_2 b_2 \langle s_2, \phi_1 \rangle, A_1 b_1 \langle s_1, \phi_2 \rangle + A_2 b_2 \langle s_2, \phi_2 \rangle)$$

$$= (A_1 b_1 + A_2 b_2 \rho, A_2 b_2 \sqrt{1 - \rho^2}) \qquad (3.88)$$

and covariance matrix equal to

$$\text{cov}(\tilde{y}_1, \tilde{y}_2) = \begin{bmatrix} \sigma^2 & 0 \\ 0 & \sigma^2 \end{bmatrix}.$$

The counterpart to Figure 3.10 in the alternative orthogonal representation $(\tilde{y}_1, \tilde{y}_2)$ using the basis in (3.87) is shown in Figure 3.12 where the decision regions are defined by the lines (hyperplanes in K-dimensional space) orthogonal to s_1 and s_2 respectively. In contrast to Figure 3.10 the inner product between the vectors representing the signature waveforms s_1, s_2 in Figure 3.12 is indeed equal to their crosscorrelation.

Even though $(\tilde{y}_1, \tilde{y}_2)$ are not computed by the demodulator, it is useful to visualize the received vector as belonging to the two-dimensional space depicted in Figure 3.12. Indeed, the decisions of the detector in Figure 3.6 are transparent to all those (infinite) components in $y(t)$ orthogonal to ϕ_1 and ϕ_2.

The anomalous behavior of the single-user matched filter detector in the near–far situation (3.81) is illustrated in Figure 3.13, which corresponds to the same signature waveforms as those used in Figure 3.12, but with $A_1 =$

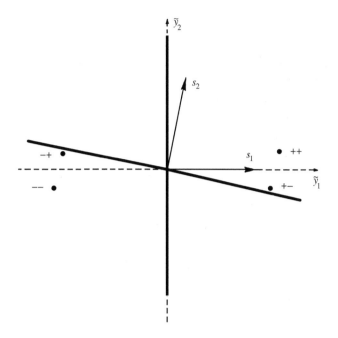

FIGURE 3.13.
Decision regions
of matched filter
detector
(orthogonal
space);
$A_1 = 6A_2$.

$6A_2$. The decision regions remain exactly as in Figure 3.12; however, the transmitted vectors corresponding to $(+, -)$ and $(-, +)$ have now migrated outside their original decision regions. Therefore, when those bit pairs are transmitted, an error will occur in the decision of user 2 unless the noise realization prevents it. In fact, in the absence of background noise, the decisions for users 1 and 2 are identical and equal to the data transmitted by user 1.

The generalization of the bit-error-rate of the single-user matched filter from two users to an arbitrary number of users is straightforward. Following the same reasoning as before, we can write the bit-error-rate of the kth user as

$$P_k^c(\sigma) = P[b_k = +1]P[y_k < 0|b_k = +1]$$

$$+ P[b_k = -1]P[y_k > 0|b_k = -1]$$

$$= \frac{1}{2}P\left[n_k > A_k - \sum_{j \neq k}A_j b_j \rho_{jk}\right] + \frac{1}{2}P\left[n_k < -A_k - \sum_{j \neq k}A_j b_j \rho_{jk}\right]$$

$$= P\left[n_k > A_k - \sum_{j \neq k}A_j b_j \rho_{jk}\right] \qquad (3.89)$$

$$= \frac{1}{2^{K-1}}\sum_{e_1 \in \{-1,1\}}\cdots\sum_{\substack{e_j \in \{-1,1\} \\ j \neq k}}\cdots\sum_{e_K \in \{-1,1\}}Q\left(\frac{A_k}{\sigma} + \sum_{j \neq k}e_j\frac{A_j}{\sigma}\rho_{jk}\right),$$

$$(3.90)$$

113

where (3.89) follows by symmetry and (3.90) is obtained by conditioning on all the interfering bits. We see that the error probability of the single-user matched filter in the CDMA Gaussian channel depends on the shape of the signature waveforms only through their crosscorrelations. This property is due to both the nature of the receiver and the fact that the background noise is white and Gaussian (cf. Problem 3.29). Moreover, the error probability depends on the received amplitudes and noise level σ only through the ratios A_k/σ, as decisions are invariant to scaling of the received waveform.

As in (3.77), the average of Q-functions in (3.90) is upper bounded by

$$P_k^c \leq Q\left(\frac{A_k}{\sigma} - \sum_{j \neq k} \frac{A_j}{\sigma} |\rho_{jk}|\right). \tag{3.91}$$

In order to generalize condition (3.81) to the K-user case we note that (3.90) goes to 0 as $\sigma \to 0$ if and only if the argument of each of the Q-functions therein is positive, that is, if

$$A_k > \sum_{j \neq k} A_j |\rho_{jk}|. \tag{3.92}$$

Condition (3.92) for error-free decisions in the absence of background noise is commonly referred to as the *open-eye* condition. Under this condition, the bound (3.91) becomes tight (modulo a factor independent of σ) as $\sigma \to 0$.

The number of operations required for the computation of (3.90) grows exponentially in the number of users. For this reason, a number of authors have approximated (3.90) by replacing the binomial random variable

$$\sum_{j \neq k} A_j b_j \rho_{jk}$$

by a Gaussian random variable with identical variance. The approximated bit-error-rate becomes

$$\tilde{P}_k^c(\sigma) = Q\left(\frac{A_k}{\sqrt{\sigma^2 + \sum_{j \neq k} A_j^2 \rho_{jk}^2}}\right). \tag{3.93}$$

Whereas at low signal-to-noise ratios the approximation of (3.90) by (3.93) is generally good, for high signal-to-noise ratios it may be unreliable. The behavior of the Gaussian approximation is illustrated in Figures 3.14 and 3.15. In those figures, the crosscorrelations between every pair of signature waveforms are equal to 0.08 and all users have the same power (cf. Problem 3.10). The cases of 10 users (Figure 3.14) and 14 users (Figure 3.15) are representative of the open-eye and closed-eye situations, respectively. In the latter case, we observe the nonmonotonic behavior of the bit-error-rate of the

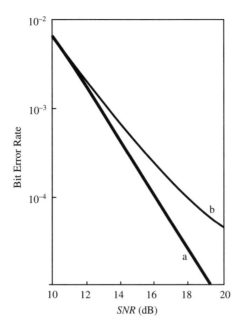

FIGURE 3.14.
Bit-error-rate of the single-user matched filter with 10 equal-energy users and identical crosscorrelations $\rho_{kl} = 0.08$; (a) exact, (b) Gaussian approximation.

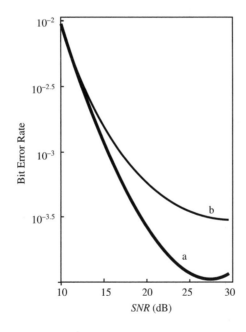

FIGURE 3.15.
Bit-error-rate of the single-user matched filter with 14 equal-energy users and identical crosscorrelations $\rho_{kl} = 0.08$; (a) exact, (b) Gaussian approximation.

single-user matched filter with signal-to-noise ratio, already observed in the two-user case. In the limit as $\sigma \to 0$, Equations (3.90) and (3.93) behave differently. For example, (3.93) has a nonzero limit even if the open-eye condition is satisfied. The reason is that the error in replacing the binomial distribution by the Gaussian distribution is greatest in the tails, which determine the bit-error-rate (unless the background noise is the dominant factor). When performance is averaged with respect to random carrier phases (cf. Section 2.5), the multiuser interference is no longer binomially distributed but remains amplitude limited – in contrast to a Gaussian random variable with the same variance.

Suppose that the random direct-sequence model is used and bit-error-rate is averaged with respect to the choice of binary sequences with spreading gain N. If $K \to \infty$ and $N \to \infty$ but their ratio is kept constant,

$$\frac{K}{N} = \beta,$$

then, the averaged bit-error-rate converges to

$$\lim_{K \to \infty} E\big[P_1^c(\sigma)\big] = Q\left(\frac{A_1}{\sqrt{\sigma^2 + \beta \bar{A}^2}}\right), \qquad (3.94)$$

where

$$\bar{A}^2 \stackrel{\text{def}}{=} \lim_{K \to \infty} \frac{1}{K} \sum_{j=2}^{K} A_j^2.$$

A sufficient condition for the validity of (3.94) is that the amplitudes A_j be bounded (Problem 3.39). Here we will justify (3.94) under the condition that the received energies are identical for all users $A_k = A$. According to (3.89) we need to compute the limit of

$$P\left[n_1 + A \sum_{j=2}^{K} b_j \rho_{1j} > A\right]$$

$$= P\left[n_1 + A \sum_{j=2}^{K} b_j \frac{1}{N} \sum_{n=1}^{N} d_{jn} > A\right] \qquad (3.95)$$

$$= P\left[n_1 + A \sqrt{\frac{K-1}{N}} \frac{1}{\sqrt{(K-1)N}} \sum_{j=2}^{K} \sum_{n=1}^{N} d_{jn} > A\right], \qquad (3.96)$$

where the random variables d_{jn} in (3.95) are independent equally likely to be $+1$ or -1. The (De Moivre–Laplace) Central Limit Theorem dictates convergence in distribution (as $K \to \infty$) of the random variable

$$\frac{1}{\sqrt{(K-1)N}} \sum_{j=2}^{K} \sum_{n=1}^{N} d_{jn}$$

to a zero-mean, unit-variance, Gaussian random variable. Thus, the right side of (3.96) converges to

$$Q\left(\frac{A}{\sqrt{\sigma^2 + \beta A^2}}\right),$$

as we wanted to verify.

The limiting result in (3.94) can be strengthened to show that even if the bit-error-rate is not averaged with respect to the random sequences, it converges as $K = \beta N \to \infty$ to the right side of (3.94) with probability one for any signal-to-noise ratio (Verdú and Shamai [504]). It should be noted that for high signal-to-noise ratios, convergence may be very slow with K (cf. the case of $K = 100$ depicted in Figure 6.7).

An easy-to-compute upper bound to P_k^c can be found by partitioning the set of users into

$$\{1, \ldots, K\} = \{k\} \cup G \cup \bar{G},$$

where G is a subset of interferers that satisfies the "partial" open-eye condition

$$A_k > \sum_{j \in G} A_j |\rho_{jk}|. \tag{3.97}$$

Then, as shown below, the error probability of the single-user matched filter is bounded by

$$P_k^c \leq \exp\left(-\frac{(A_k - \sum_{j \in G} A_j |\rho_{jk}|)^2}{2(\sigma^2 + \sum_{j \in \bar{G}} A_j^2 \rho_{jk}^2)}\right). \tag{3.98}$$

The freedom to choose G (subject to (3.97)) can be exploited to minimize the upper bound (3.98). Two special cases of (3.98) deserve special mention. First, if the (full) open-eye condition (3.92) condition is satisfied, then

$$P_k^c \leq \exp\left(-\frac{(A_k - \sum_{j \neq k} A_j |\rho_{jk}|)^2}{2\sigma^2}\right),$$

because we can take $\bar{G} = \emptyset$. Second, we can always take $G = \emptyset$, in which case (3.98) becomes (cf. (3.93))

$$P_k^c \leq \exp\left(-\frac{A_k^2}{2(\sigma^2 + \sum_{j \neq k} A_j^2 \rho_{jk}^2)}\right).$$

To justify (3.98), denote for brevity

$$\gamma = A_k - \sum_{j \in G} A_j |\rho_{jk}| > 0.$$

Then, if we choose an arbitrary $\lambda > 0$ and we denote the unit step by $u(t)$, we can write

$$P_k^c \leq E\left[u\left(n_k + \sum_{j \in \bar{G}} A_j \rho_{jk} b_j - \gamma\right)\right] \tag{3.99}$$

$$\leq E\left[\exp\left(\lambda n_k + \sum_{j \in \bar{G}} \lambda A_j \rho_{jk} b_j - \lambda \gamma\right)\right] \tag{3.100}$$

$$= \exp(-\lambda \gamma) E[\exp(\lambda n_k)] \prod_{j \in \bar{G}} E[\exp(\lambda A_j \rho_{jk} b_j)] \tag{3.101}$$

$$= \exp\left(-\lambda \gamma + \frac{\lambda^2 \sigma^2}{2}\right) \prod_{j \in \bar{G}} \cosh(A_j \lambda |\rho_{jk}|) \tag{3.102}$$

$$\leq \exp\left(\frac{\lambda^2}{2}\left(\sigma^2 + \sum_{j \in \bar{G}} A_j^2 \rho_{jk}^2\right) - \lambda \gamma\right), \tag{3.103}$$

where (3.100)[10] and (3.103) follow from

$$u(t) \leq \exp(\lambda t), \quad \lambda > 0,$$

$$\cosh(t) \leq \exp\left(\frac{t^2}{2}\right),$$

respectively. The desired bound (3.98) is obtained by choosing $\lambda > 0$ to minimize (3.103).

3.4.2 PROBABILITY OF ERROR FOR ASYNCHRONOUS USERS

The analysis in the asynchronous case is entirely similar. The main difference is that now each bit is affected by $2K - 2$ interfering bits. This doubles the number of terms in (3.90):

$$P_k^c(\sigma) = \frac{1}{4^{K-1}} \sum_{(e_1, d_1) \in \{-1,1\}^2} \cdots \sum_{\substack{(e_j, d_j) \in \{-1,1\}^2 \\ j \neq k}} \cdots \sum_{(e_K, d_K) \in \{-1,1\}^2}$$

$$Q\left(\frac{A_k}{\sigma} + \sum_{j \neq k} \frac{A_j}{\sigma}(e_j \rho_{jk} + d_j \rho_{kj})\right). \tag{3.104}$$

The generalization of condition (3.92) to the asynchronous case is

$$A_k > \sum_{j \neq k} A_j(|\rho_{jk}| + |\rho_{kj}|). \tag{3.105}$$

---

[10] $E[\exp(\lambda X)]$, $\lambda > 0$, is known as the *Chernoff upper bound* to $P[X > 0]$.

Recall that the asynchronous crosscorrelations in (3.104) depend on the offsets between the users' symbol periods (as well as on their carrier phases for example). Consequently, those parameters are random variables that may actually be time varying. Given a set of signature waveforms it is possible to compute the distribution (or, simply, the expectation) of (3.104). However, this is a computationally intensive task.

We see that for any number of users and no matter which signature waveforms are used, there is always a set of offsets and received energies for which errors occur even in the absence of background noise because the decisions of the single-user matched filter are dominated by the multiaccess interference. Moreover, this may happen even if perfect power control ensures equal received energies. This phenomenon becomes more critical when the crosscorrelation between the signature waveforms increases, as, for example, when the spectral efficiency increases.

The infinite-user random sequence analysis in the previous section can be carried over to the asynchronous case incorporating two fictitious interferers per actual interferer. Averaging over the received delays and using the result in (2.27) for rectangular chip waveforms (see also Problem 2.15), we see that the bit-error-rate is equivalent to that of a synchronous system with $(2/3)(K - 1)$ interferers.

3.5 ASYMPTOTIC MULTIUSER EFFICIENCY AND RELATED MEASURES

The main performance measure of interest in digital communications in general, and in multiuser detection in particular, is the bit-error-rate $P_k(\sigma)$. In addition, there are several performance measures derived from the bit-error-rate that turn out to be useful in the analysis, design, and understanding of the various detectors. We have already seen an example of such a measure: the power-tradeoff region of signal-to-noise ratios that results in a guaranteed bit-error-rate level. The analysis of the conventional detector motivates the introduction of several other performance measures.

It is common to average error probability with respect to random quantities such as delays, phases, and received signal-to-noise ratios. When the unknown quantities are slowly time-varying (relative to the data rate), averaging bit-error-rate may be misleading because it may be dominated by particularly unfavorable, but rare, channel conditions. In practice it is common to design the communication system to guarantee that *outage* (performance falling below a certain threshold) occurs no more than a certain percentage of time. In that case, the cumulative distribution function (computed for a value or range of values of interest) of the error probability is more insightful than its average.

Although not a foolproof predictor of error probability, the *signal-to-interference ratio* (SIR) is useful in the assessment of the quality of multiuser detectors, particularly when used in conjunction with error-control codes (Section 2.4). SIR gives the ratio of powers due to the desired user and due to all other components at a soft-decision variable such as the output of the single-user matched filter. In the absence of interfering users, the output signal-to-noise ratio of the kth matched filter is A_k^2/σ^2, whereas in the presence of synchronous interferers it equals

$$\frac{A_{ck}^2}{\sigma^2} \stackrel{\text{def}}{=} \frac{A_k^2}{\sigma^2 + \sum_{j\neq k} A_j^2 \rho_{jk}^2}. \tag{3.106}$$

The presence of other users in the channel can only increase the bit-error-rate, and it is of interest to quantify the multiuser error probability relative to the optimum single-user error probability (3.30). To that end, we define the *effective energy* of user k, $e_k(\sigma)$, as the energy that user k would require to achieve bit-error-rate equal to $P_k(\sigma)$ in a single-user Gaussian channel with the same background noise level, that is (cf. (3.30)),

$$P_k(\sigma) = Q\left(\frac{\sqrt{e_k(\sigma)}}{\sigma}\right). \tag{3.107}$$

Since the error probability of any multiuser detector is lower bounded by the single-user error probability

$$P_k(\sigma) \geq Q\left(\frac{A_k}{\sigma}\right),$$

the effective energy is always upper bounded by the actual energy:

$$e_k(\sigma) \leq A_k^2. \tag{3.108}$$

The effective energy normalized by σ^2 can be seen as an alternative way to give the information contained in the bit-error-rate; note the one-to-one correspondence between the functions $e_k(\sigma)$ and $P_k(\sigma)$:

$$e_k(\sigma) = \sigma^2 (Q^{-1}(P_k(\sigma)))^2.$$

The *power-tradeoff* region for a given permissible bit-error-rate P (same for all users), such as that depicted in Figure 3.9, is the set of signal-to-noise ratios $\sigma^{-2}(A_1^2, \ldots, A_K^2)$ such that

$$\max_k P_k(\sigma) \leq P, \tag{3.109}$$

or, equivalently,

$$\min_{k} \frac{e_k(\sigma)}{\sigma^2} \geq (Q^{-1}(\mathsf{P}))^2. \tag{3.110}$$

The *multiuser efficiency* or ratio between the effective and actual energies, $e_k(\sigma)/A_k^2$, is an alternative way to characterize the multiuser bit-error-rate. From (3.108) it follows that the efficiency belongs to the interval $[0, 1]$ (or $[-\infty, 0]$ in dB) and quantifies the performance loss due to the existence of other users in the channel. Multiuser efficiency depends on the signature waveforms, received signal-to-noise ratios, and the detector employed.

The *asymptotic multiuser efficiency* is defined as

$$\eta_k = \lim_{\sigma \to 0} \frac{e_k(\sigma)}{A_k^2} \tag{3.111}$$

and measures the slope with which $\mathsf{P}_k(\sigma)$ goes to 0 (in logarithmic scale) in the high signal-to-noise ratio region, that is (Problem 3.10),

$$\eta_k = \sup \left\{ 0 \leq r \leq 1 : \lim_{\sigma \to 0} \mathsf{P}_k(\sigma)/Q\left(\frac{\sqrt{r}A_k}{\sigma}\right) = 0 \right\} \tag{3.112}$$

$$= \frac{2}{A_k^2} \lim_{\sigma \to 0} \sigma^2 \log 1/\mathsf{P}_k(\sigma). \tag{3.113}$$

We can immediately conclude that in those situations (such as the single-user matched filter detector under the closed-eye condition) where the bit-error-rate does not vanish as the background noise level goes to 0, the asymptotic multiuser efficiency is equal to 0. Conversely, if the asymptotic multiuser efficiency is positive, then the bit-error-rate not only vanishes but it does so exponentially in the signal-to-noise ratio.

Except for low signal-to-noise ratios, usually the multiuser efficiency $e_k(\sigma)/A_k^2$ is very close to η_k. Moreover, in some cases, the asymptotic multiuser efficiency is easier to compute than the bit-error-rate and it allows intuitive and easy-to-grasp depictions of the effects of unequal received energies.

The *worst asymptotic effective energy* is defined as

$$\omega(A_1, \ldots, A_K) \stackrel{\text{def}}{=} \min_{k=1,\ldots,K} \lim_{\sigma \to 0} e_k(\sigma) \tag{3.114}$$

$$= \min_{k=1,\ldots,K} A_k^2 \eta_k \tag{3.115}$$

$$= 2 \lim_{\sigma \to 0} \sigma^2 \log 1/P\left[\bigcup_{k=1}^{K} \{b_k \neq \hat{b}_k\} \right], \tag{3.116}$$

where (3.115) follows from (3.111), and (3.116) can be verified upon taking

$\lim_{\sigma \to 0} \sigma^2 \log(\cdot)$ of both sides of

$$\max_{k=1,\ldots,K} P_k(\sigma) \leq P\left[\bigcup_{k=1}^{K} \{b_k \neq \hat{b}_k\}\right] \leq \sum_{k=1}^{K} P_k(\sigma).$$

According to (3.110), the power-tradeoff region for bit-error-rate P is (asymptotically) the set of $\sigma^{-2}(A_1^2, \ldots, A_K^2)$ such that

$$\sigma^{-2}\omega(A_1, \ldots, A_K) \geq (Q^{-1}(P))^2.$$

Notice that

$$\frac{1}{A^2}\omega(A, \ldots, A) = \min_{k=1,\ldots,K} \eta_k$$

quantifies the power penalty experienced by equal-power users in the high signal-to-noise ratio region.

It is important to quantify the degree of robustness against the near–far problem achieved by multiuser detectors. To that end, we define the *near–far resistance* as the multiuser asymptotic efficiency minimized over the received energies of all the other users:

$$\bar{\eta}_k = \inf_{\substack{A_j > 0 \\ j \neq k}} \eta_k. \tag{3.117}$$

The near–far resistance depends on the signature waveforms and on the demodulator.

The near–far problem is most common in systems with mobile transmitters, since their received energies do not remain constant. To take this fact into account we introduce a slightly different version of the *near–far resistance* measure for the asynchronous CDMA channel by dropping the assumption that the received energies remain constant throughout the transmission period (cf. Section 2.6). This results in the more restrictive definition of near–far resistance as

$$\bar{\eta}_k = \inf_{\substack{A_j[i] > 0 \\ (i,j) \neq (0,k)}} \eta_k, \tag{3.118}$$

which will turn out to be more analytically tractable than (3.117).

Whenever the bit-error-rate is given as the weighted sum of Q-functions (e.g., (3.90)) it is very easy to compute the asymptotic multiuser efficiency. The single-user matched-filter error probability $P_k^c(\sigma)$ serves as a simple illustration of the procedure. In the two-user synchronous case, if

$$A_1 \leq A_2|\rho| \tag{3.119}$$

then the bit-error-rate in (3.77) does not vanish as $\sigma \to 0$, and thus, $\eta_1^c = 0$. If

$$A_1 > A_2|\rho|, \tag{3.120}$$

then (3.41) yields

$$\lim_{\sigma \to 0} \frac{P_1^c(\sigma)}{Q\left(\frac{\sqrt{r}A_1}{\sigma}\right)} = \lim_{\sigma \to 0} \frac{\frac{1}{2}Q\left(\frac{A_1-A_2\rho}{\sigma}\right) + \frac{1}{2}Q\left(\frac{A_1+A_2\rho}{\sigma}\right)}{Q\left(\frac{\sqrt{r}A_1}{\sigma}\right)}$$

$$= \begin{cases} 0, & \sqrt{r}A_1 < A_1 - A_2|\rho|, \\ +\infty, & \sqrt{r}A_1 > A_1 - A_2|\rho|, \end{cases} \tag{3.121}$$

which implies that under (3.120),

$$\eta_1^c = \left(1 - \frac{A_2}{A_1}|\rho|\right)^2.$$

Putting together the asymptotic multiuser efficiency in both regions, (3.119) and (3.120), we obtain

$$\eta_1^c = \max{}^2\left\{0, 1 - \frac{A_2}{A_1}|\rho|\right\}, \tag{3.122}$$

which is plotted in Figures 3.16 and 3.17 as a function of the relative amplitude of the interferer.

Proceeding analogously in the K-user synchronous and asynchronous channels we obtain, respectively,

$$\eta_k^c = \max{}^2\left\{0, 1 - \sum_{j\neq k}\frac{A_j}{A_k}|\rho_{jk}|\right\} \tag{3.123}$$

FIGURE 3.16.
Asymptotic
multiuser
efficiency of
conventional
detector as a
function of the
amplitude of the
interferer; $\rho = 0.2$
(linear plot).

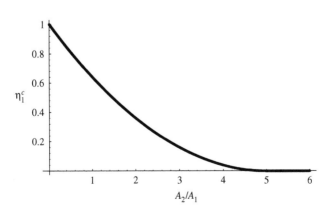

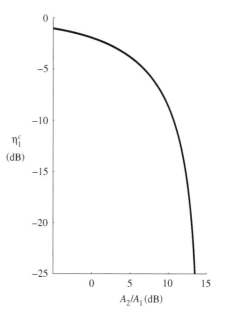

FIGURE 3.17.
Asymptotic
multiuser
efficiency of
conventional
detector as a
function of the
amplitude of the
interferer; $\rho = 0.2$
(log–log plot).

and

$$\eta_k^c = \max{}^2\left\{0, 1 - \sum_{j\neq k}\frac{A_j}{A_k}(|\rho_{jk}| + |\rho_{kj}|)\right\}. \qquad (3.124)$$

Note that the asymptotic efficiency of the conventional detector can be seen as a normalized measure of the eye-opening (cf. conditions (3.92) and (3.105)).

Minimizing (3.123) or (3.124) over $\{A_j, \, j \neq k\}$ we see that the near–far resistance of the kth user is equal to 0 unless $\rho_{jk} = \rho_{kj} = 0$ for all $j \neq k$. In other words, the kth user signature waveform must be orthogonal to each of the partially overlapping waveforms of every interferer. Because this condition cannot be satisfied for all offsets in an asynchronous channel, we conclude that the conventional receiver is not near–far resistant, except in the trivial case of synchronous orthogonal signature waveforms (in which case it is optimal).

3.6   COHERENT SINGLE-USER MATCHED FILTER IN RAYLEIGH FADING

This section is devoted to the analysis of the single-user matched filter when the received signals are subject to independent frequency-flat Rayleigh fading. The receiver is assumed to know the fading coefficients of the user of interest. We consider the cases of scalar observations and diversity reception.

3.6.1 SCALAR RECEPTION

It is convenient to adopt the complex-valued model in (2.48):

$$y(t) = \sum_{i=-M}^{M} \sum_{k=1}^{K} A_k[i] b_k[i] s_k(t - iT) + \sigma n(t), \tag{3.125}$$

where $A_k[i]$ are complex with independent Gaussian real and imaginary components having zero mean and standard deviation equal to A_k. A basic assumption we make in this section is that the receiver knows the fading coefficients $A_k[i]$ (both amplitude and phase) for the desired user.

Let us first examine the single-user case:

$$y(t) = Abs(t) + \sigma n(t), \quad t \in [0, T], \tag{3.126}$$

where we have considered the one-shot model without loss of optimality because of the assumption that the fading coefficients are perfectly known. In parallel to the real-valued setting, the optimum decision rule selects $b \in \{-1, 1\}$ that minimizes

$$\int_0^T |y(t) - Abs(t)|^2 \, dt$$

$$= \int_0^T |y(t)|^2 \, dt + \int_0^T |As(t)|^2 \, dt - 2\Re \left\{ \int_0^T y^*(t) Abs(t) \, dt \right\}.$$

This means that the optimum decision rule in the single-user case is

$$\hat{b} = \mathrm{sgn}\left(\Re \left\{ A \int_0^T y^*(t) s(t) \, dt \right\} \right). \tag{3.127}$$

The inner product of $y(t)$ and $s(t)$ is a sufficient statistic denoted by

$$y = \int_0^T y(t) s^*(t) \, dt.$$

Note that

$$\mathrm{sgn}(\Re\{Ay^*\})$$

is equal to $+1$ if and only if the angle between the complex numbers A and y is acute (i.e., if their absolute phase difference is less than $\pi/2$).

To find the error probability of (3.127), we condition on the transmitted bit and the received faded coefficient:

$$P[\hat{b} = 1|b = -1, A] = P[-|A|^2 + \sigma \Re \left\{ A \int_0^T n^*(t) s(t) \, dt \right\} > 0|A]$$

$$= P[-|A|^2 + \sigma \Re\{A\} N_\Re - \sigma \Im\{A\} N_\Im > 0|A]$$

$$= Q\left(\frac{|A|}{\sigma} \right), \tag{3.128}$$

where

$$N_{\Re} \stackrel{\text{def}}{=} \Re\left\{ \int_0^T \mathsf{n}^*(t)\mathsf{s}(t)\, dt \right\},$$

$$N_{\Im} \stackrel{\text{def}}{=} \Im\left\{ \int_0^T \mathsf{n}^*(t)\mathsf{s}(t)\, dt \right\}$$

are independent Gaussian random variables with zero mean and unit variance.

By symmetry, the conditional probability of making an error when $+1$ is sent is the same. We need to average (3.128) with respect to the Rayleigh distributed $|\mathsf{A}|$. If the real and imaginary components of A have zero mean and standard deviation each equal to A then $\mathsf{A} = AR$, where

$$f_R(r) = r^2 e^{-r^2/2}, \quad r \geq 0.$$

Accordingly,

$$\mathsf{P}^{Fc}(\sigma) = E\left[Q\left(\frac{|\mathsf{A}|}{\sigma} \right) \right]$$

$$= \int_0^\infty r e^{-r^2/2} Q\left(\frac{Ar}{\sigma} \right) dr$$

$$= \frac{1}{2}\left(1 - \frac{1}{\sqrt{1 + \sigma^2/A^2}} \right) \tag{3.129}$$

$$\leq \frac{\sigma^2}{4A^2}, \tag{3.130}$$

where (3.129) is obtained from (3.61).

It is interesting to contrast the single-user bit-error-rate with and without Rayleigh fading. Whereas the error probability with deterministic amplitude (3.30) decreases exponentially in $-A^2/\sigma^2$, (3.129) has a much slower hyperbolic decay with A^2/σ^2:

$$\mathsf{P}^{Fc}(\sigma) = \frac{\sigma^2}{4A^2} + O\left(\frac{\sigma^4}{A^4} \right). \tag{3.131}$$

This is due to the fact that when averaging error probability with respect to the received amplitude, performance is dominated by very small amplitude values, which a Rayleigh-distributed random variable takes with nonnegligible probability. For high enough σ, $\mathsf{P}^{Fc}(\sigma) < \mathsf{P}^c(\sigma)$ (Problem 3.43). This is just an artifact of the different roles of A/σ in the notations we have employed with and without fading. For simplicity, take $\mathsf{s}(t)$ to be real in (3.126). Since A is the standard deviation of both the real and imaginary components of A, and the real and imaginary components of $\mathsf{n}(t)$ have unit spectral density, the real and imaginary channels in (3.126) each have the same signal-to-noise ratio as the real-valued channel (3.19).

In view of (3.131) the multiuser asymptotic efficiency in Rayleigh fading of a detector with bit-error-rate $P_k^F(\sigma)$ should be defined as

$$\eta_k^F = \lim_{\sigma \to 0} \frac{\sigma^2}{4A_k^2 P_k^F(\sigma)}. \tag{3.132}$$

The performance analysis of the decision rule (3.127) in the presence of multiaccess interference can be carried out by conditioning with respect to the transmitted bits and fading coefficients. Analogously to (3.129) we obtain

$$P_k^{Fc}(\sigma) = \frac{1}{2^{K-1}} \sum_{\substack{e_1 \in \{-1,1\}}} \cdots \sum_{\substack{e_j \in \{-1,1\} \\ j \neq k}} \cdots \sum_{\substack{e_K \in \{-1,1\}}}$$

$$E\left[Q\left(\frac{|A_k|}{\sigma} + \sum_{j \neq k} e_j \frac{\Re\{A_j^* A_k\}}{\sigma |A_k|} \rho_{jk}\right)\right]$$

$$= E\left[Q\left(\frac{|A_k|}{\sigma} + \sum_{j \neq k} \frac{\Re\{A_j\}}{\sigma} \rho_{jk}\right)\right] \tag{3.133}$$

$$= E\left[Q\left(\frac{|A_k|}{\sqrt{\sigma^2 + \sum_{j \neq k} A_j^2 \rho_{jk}^2}}\right)\right] \tag{3.134}$$

$$= \frac{1}{2}\left(1 - \frac{A_k}{\sqrt{\sigma^2 + \sum_j A_j^2 \rho_{jk}^2}}\right), \tag{3.135}$$

where the sum in the right side of (3.135) is over all users and $\rho_{kk} = 1$. In (3.133) the phase term $A_k/|A_k|$ and binary coefficients have been dropped since they do not affect the distribution of the random variable inside the Q-function. Equation (3.134) follows from (3.66) because $\Re\{A_j\}$ are independent Gaussian random variables. To obtain (3.135), we simply solve the same integral that led to the single-user result (3.129). The conclusion is that for the purposes of the bit-error-rate of the single-user matched-filter receiver, the Rayleigh-distributed interferers act like an additional source of Gaussian background noise.

Unless the signature waveforms of the interferers are orthogonal to that of the desired user, $P_k^{Fc}(\sigma)$ does not go to 0 as $\sigma \to 0$. According to (3.132), this implies that the single-user matched filter suffers from zero asymptotic multiuser efficiency in the presence of Rayleigh fading, in contrast to the asymptotic multiuser efficiency of the single-user matched filter found in (3.123) in the absence of fading.

3.6.2 DIVERSITY RECEPTION

Let us consider first the particular case of (2.74) in which $K = 1$:

$$y_1(t) = A_{11}b\,s_{11}(t) + \sigma n_1(t),$$

$$\vdots \qquad\qquad (3.136)$$

$$y_D(t) = A_{D1}b\,s_{D1}(t) + \sigma n_D(t),$$

where $\{A_{11}, \dots, A_{D1}\}$ are independent complex Gaussian random variables with variance $E[|A_{d1}|^2] = 2A_1^2$, and the D noise processes are independent. It is easy to generalize previous results on binary hypothesis testing in Gaussian noise to show that the optimum decision rule selects $b \in \{-1, 1\}$ that minimizes

$$\int_0^T \sum_{d=1}^D |y_d(t) - A_{d1}b\,s(t)_{d1}|^2\,dt$$

$$= \sum_{d=1}^D \int_0^T |y_d(t)|^2\,dt + \sum_{d=1}^D \int_0^T |A_{d1}s_{d1}(t)|^2\,dt$$

$$- 2b \sum_{d=1}^D \Re\left\{ \int_0^T y_d^*(t)A_{d1}s_{d1}(t)\,dt \right\}. \qquad (3.137)$$

Accordingly, the optimum decision rule is (cf. (3.127))

$$\hat{b} = \operatorname{sgn}\left(\Re\left\{ \sum_{d=1}^D A_{d1}y_{d1}^* \right\} \right) \qquad (3.138)$$

with

$$y_{d1} = \int_0^T y_d(t)s_{d1}^*(t)\,dt.$$

In the context of diversity, the decision rule in (3.138) is referred to as *maximal-ratio combining* (Problem 2.34). Conditioning on the fading coefficients, the bit-error-rate is

$$P[\hat{b} = 1 | b = -1, A_{11}, \dots, A_{D1}]$$

$$= P\left[-\sum_{d=1}^D |A_{d1}|^2 + \sigma \sum_{d=1}^D \Re\{A_{d1}n_d^*\} > 0 | A_{11}, \dots A_{D1} \right]$$

$$= Q\left(\frac{1}{\sigma} \sqrt{ \sum_{d=1}^D |A_{d1}|^2 } \right), \qquad (3.139)$$

where we have used the fact that, conditioned on the fading coefficients, $\sum_{d=1}^{D} A_{d1} n_d^*$ is a complex Gaussian random variable with independent real and imaginary parts each with variance equal to $\sigma^2 \sum_{d=1}^{D} |A_{d1}|^2$.

In order to average (3.139) with respect to independent complex Gaussian A_{d1} with

$$E[|A_{d1}|^2] = 2A^2$$

we just need to invoke (3.65) with $L = D$ and $\gamma = A/\sigma$ to yield

$$P_1^{Dc}(\sigma) = \frac{1}{2} - \frac{1}{2} \frac{1}{\sqrt{1 + \sigma^2/A^2}} \left(1 + \sum_{n=1}^{D-1} \frac{1 \cdot 3 \cdot 5 \cdots (2n - 1)}{n! \, 2^n \, (A^2/\sigma^2 + 1)^n} \right). \quad (3.140)$$

Now, let us investigate the more general bit-error-rate analysis of

$$\hat{b}_k = \text{sgn} \left(\Re \left\{ \sum_{d=1}^{D} A_{dk} y_{dk}^* \right\} \right) \quad (3.141)$$

with

$$y_{dk} = \int_0^T y_d(t) s_{dk}^*(t) \, dt, \quad (3.142)$$

where the received processes are now subject to multiaccess interference (2.74):

$$y_1(t) = \sum_{k=1}^{K} A_{1k} b_k s_{1k}(t) + \sigma n_1(t),$$

$$\vdots \quad (3.143)$$

$$y_D(t) = \sum_{k=1}^{K} A_{Dk} b_k s_{Dk}(t) + \sigma n_D(t),$$

and where the fading coefficients are complex, Gaussian, and independent with

$$E[|A_{dk}|^2] = 2A_k^2.$$

The inner product in (3.142) is affected by multiaccess interference:

$$y_{dk} = A_{dk} b_k + \sum_{l \neq k} A_{dl} b_l \rho_{lk}(d) + \sigma n_{dk}, \quad (3.144)$$

where the crosscorrelations of the signals in the dth branch are denoted by

$$\rho_{kj}(d) = \int_0^T s_{dk}^*(t) s_{dj}(t) \, dt. \quad (3.145)$$

Conditioning on b_k and on $(A_{1k}, \ldots, A_{Dk})$, the decision statistic in (3.141) is equal to

$$\Re\left\{\sum_{d=1}^{D} A_{dk} y_{dk}^*\right\} = b_k \sum_{d=1}^{D} |A_{dk}|^2 + \Re\{n\}, \qquad (3.146)$$

where n is a complex Gaussian random variable with independent real and imaginary components and variance given by

$$E[|n|^2] = 2\sigma^2 \sum_{d=1}^{D} |A_{dk}|^2 + 2 \sum_{d=1}^{D} |A_{dk}|^2 \sum_{l \neq k} A_l^2 |\rho_{lk}^2(d)|^2. \qquad (3.147)$$

Therefore, the probability of error is

$$P_k^{Dc}(\sigma) = E\left[Q\left(\frac{\sum_{d=1}^{D} |A_{dk}|^2}{\sqrt{\sum_{d=1}^{D} |A_{dk}|^2 \left(\sigma^2 + \sum_{l \neq k} A_l^2 |\rho_{lk}^2(d)|^2\right)}}\right)\right], \qquad (3.148)$$

where the expectation is with respect to the independent Rayleigh random variables $|A_{dk}|$. In the special case in which

$$\sum_{l \neq k} A_l^2 |\rho_{lk}^2(d)|^2 = \vartheta_k$$

is independent of d, the result in (3.140) applies verbatim replacing σ^2 by $\sigma^2 + \vartheta_k$.

3.7 DIFFERENTIALLY-COHERENT DEMODULATION

If the fading coefficients of the desired user can be considered to be constant in consecutive intervals, that is,

$$A_k[i] = A_k[i+1] = A_k e^{j\theta_k},$$

and the data streams are differentially encoded, then a simple detector can be used that (unlike that in Section 3.6.1) does not require knowledge of the received phase of the desired user.

As we saw in Section 2.4, when the data streams are differentially encoded, the transmitted bits are obtained from the original data stream via

$$e_k[i] = e_k[i-1]b_k[i].$$

Then, the received waveform on two consecutive bit intervals $t \in [iT - T, iT + T]$ is

$$y(t) = \sum_{k=1}^{K} A_k \left(e_k[i-1]s_k(t - iT + T) + e_k[i]s_k(t - iT)\right) + \sigma n(t).$$

$$(3.149)$$

130

A sensible way to decide on $b_k[i] = e_k[i-1]e_k[i]$ is to compare the phases of consecutive matched filter outputs:

$$\hat{b}_k[i] = \text{sgn}(\Re\{y_k^*[i-1]y_k[i]\}), \tag{3.150}$$

where

$$y_k[i] = \int_{iT}^{iT+T} y(t)s_k^*(t-iT)\,dt. \tag{3.151}$$

The detector in (3.150) does not require knowledge of the received amplitude/phase of the desired user.

We proceed with the analysis of the bit-error-rate in the single-user case ($K = 1$) as a function of A_1 and θ_1. Recalling that $n(t)$ is white Gaussian noise with independent real and imaginary parts, each with unit spectral level, and using (3.149) it is straightforward to show that

$$\Re\{y_1^*[i-1]y_1[i]\} = \Re\{X_1^*[i-1]X_1[i]\}, \tag{3.152}$$

where $X_1[i]$ is a complex Gaussian random variable with mean $A_1 e^{j\theta_1}e_1[i]$ and independent real/imaginary parts, each with variance σ^2. The bit-error-rate analysis will follow from the following general result:

Proposition 3.6 *(Stein [420]) Let X and Y be independent complex-valued Gaussian random variables with means* m_X, m_Y, *and independent real/imaginary parts each with variance* $\gamma^2/2$. *Then*

$$P[\Re\{XY^*\} < 0] = T(a, b)$$

$$\stackrel{\text{def}}{=} Q(a, b) - \frac{1}{2}I_0(ab)e^{-(a^2+b^2)/2}, \tag{3.153}$$

where I_0 *is the zero-order Bessel function*

$$I_0(x) = \frac{1}{2\pi}\int_0^{2\pi} e^{x\cos(\alpha+\beta)}\,d\alpha, \tag{3.154}$$

$$a = \frac{1}{\sqrt{2}\gamma}|m_X - m_Y|,$$

$$b = \frac{1}{\sqrt{2}\gamma}|m_X + m_Y|,$$

and the following function is known as Marcum's Q-function (Marcum [266]):

$$Q(a, b) \stackrel{\text{def}}{=} \int_b^{\infty} \exp\left(-\frac{a^2+x^2}{2}\right) I_0(ax)x\,dx. \tag{3.155}$$

Applying Proposition 3.6 to the case of interest, we equate $\gamma^2 = 2\sigma^2$,

$$m_X = A_1 e^{j\theta_1} e_1[i],$$

and

$$m_Y = A_1 e^{j\theta_1} e_1[i-1].$$

Furthermore, by symmetry we may condition on $b_1[i] = 1$, which means that $e_1[i] = e_1[i-1]$. Thus,

$$a = 0,$$

$$b = \frac{A_1}{\sigma}.$$

Since $I_0(0) = 1$ and

$$Q(0, b) = e^{-b^2/2},$$

we obtain that the error probability of the differentially-coherent detector in the single-user case admits the simple expression

$$P_1^{dc}(\sigma) = \frac{1}{2} e^{-A_1^2/(2\sigma^2)}. \tag{3.156}$$

As was to be expected, P_1^{dc} is not a function of θ_1. If the amplitude is not the deterministic constant A_1 but rather is Rayleigh distributed: $A_1 R$ with

$$f_R(r) = r e^{-r^2/2},$$

then the above analysis holds verbatim conditioning on R. The final averaging of (3.156) yields

$$\frac{1}{2} E\left[e^{-A_1^2 R^2/(2\sigma^2)}\right] = \frac{1}{2} \frac{1}{1 + \frac{A_1^2}{\sigma^2}}. \tag{3.157}$$

The single-user bit-error-rate analysis is fairly straightforward to generalize to K users by conditioning on all amplitudes/phases, on $b_1[i] = 1$, and on $e[i-1]$ and $e[i]$ where

$$e[i] = [1, e_2[i], \ldots, e_K[i]].$$

If we let

$$a \overset{\text{def}}{=} \frac{1}{2\sigma} [A_1 e^{j\theta_1}, A_2 e^{j\theta_2} \rho_{12}, \ldots, A_K e^{j\theta_K} \rho_{1K}],$$

then

$$P_1^{dc}(\sigma) = 4^{1-K} \sum_{e[-1]} \sum_{e[0]} T(|a^T(e[-1] - e[0])|, |a^T(e[-1] + e[0])|),$$

$$\tag{3.158}$$

where the function T is defined in (3.153). At this point, rather than attempting an average of (3.158) with respect to phases, we show that the near–far resistance of the differentially-coherent single-user detector is zero. First we lower bound $\mathsf{P}_1^{dc}(\sigma)$ by

$$\mathsf{P}_1^{dc}(\sigma) \geq 4^{1-K} T\left(0, \frac{1}{\sigma}\left|\sum_{k=1}^{K} A_k e^{j\theta_k} \rho_{1k}\right|\right) \tag{3.159}$$

$$= 2^{1-2K} e^{-\left|\sum_{k=1}^{K} A_k e^{j\theta_k} \rho_{1k}\right|^2/(2\sigma^2)}, \tag{3.160}$$

where (3.159) follows by retaining only one term in (3.158): that corresponding to

$$\mathbf{e}[-1] = \mathbf{e}[0] = [1 \cdots 1].$$

According to the result in Problem 3.44, the near–far resistance corresponding to the average of P_1^{dc} over independent uniformly distributed $\theta_1, \ldots, \theta_K$ is dominated by the worst-case realization. We conclude that

$$\bar{\eta}_1^{dc} \leq \min_{A_2,\ldots,A_K} \min_{\theta_2,\ldots,\theta_K} \left|1 + \sum_{k=2}^{K} A_k e^{j\theta_k} \rho_{1k}\right|^2 = 0.$$

3.8   NONCOHERENT DEMODULATION

The differentially-coherent modulation format allows partial knowledge of the incoming phase assuming it stays fairly constant over consecutive symbol intervals. If the phase changes rapidly or the receiver is constrained not to extend its observation window beyond the symbol of interest, we are led to consider a completely noncoherent one-shot channel with no information about the received phase, θ:

$$y(t) = |A|e^{j\theta} s(t; b) + \sigma n(t), \tag{3.161}$$

where for all $b \in \{1, \ldots, m\}$, $\|s(\cdot; b)\| = 1$, and the real and imaginary components of $n(t)$ are white Gaussian with unit spectral density.

In (3.161) we have formulated a nonlinearly modulated (cf. Section 2.5.2) m-ary single-user channel. For every $b \in \{1, \ldots, m\}$ we define the complex matched-filter output:

$$y_b = \int_0^T y(t) s^*(t; b)\, dt. \tag{3.162}$$

We now show that $|y_1|^2, \ldots, |y_m|^2$ are sufficient statistics, and in fact, the optimum demodulator for (3.161) selects $\hat{b}$ corresponding to the maximum

$|y_b|^2$. Conditioned on the data and received amplitude/phase, the likelihood function of the observations is

$$f[y \mid b, |A|, \theta] = \exp\left(\frac{-1}{2\sigma^2} \int_0^T |y(t) - |A|s(t; b)e^{j\theta}|^2 \, dt \right) \quad (3.163)$$

$$= C \exp\left(\frac{-1}{\sigma^2} \Re\{|A|e^{-j\theta} y_b\} \right), \quad (3.164)$$

where C does not depend on either b or θ. Averaging (3.164) with respect to uniformly distributed θ we get

$$\frac{1}{2\pi} \int_0^{2\pi} \exp\left(\frac{-1}{\sigma^2} \Re\{|A|e^{-j\theta} y_b\} \right) d\theta$$

$$= \frac{1}{2\pi} \int_0^{2\pi} \exp\left(\frac{-1}{\sigma^2} |A| \, |y_b| \cos \alpha \right) d\alpha \quad (3.165)$$

$$= I_0\left(\frac{|A| \, |y_b|}{\sigma^2} \right), \quad (3.166)$$

where I_0 is the zero-order Bessel function (3.154). Since I_0 is a monotone increasing function, the average of (3.166) with respect to $|A|$ will be a monotone increasing function of $|y_b|$ regardless of the distribution of $|A|$. Therefore, the optimum demodulator for (3.161) computes the complex matched-filter outputs (3.162) for each of the hypotheses and selects the one with maximum magnitude.

We have shown that the optimum noncoherent demodulator for the single-user channel (3.161) does not depend on the crosscorrelations $\langle s(\cdot; b), s(\cdot; d) \rangle$. However, the probability of error obviously depends on those parameters. Instead of providing a full analysis, we simply give the probability of (symbol) error conditioned on $|A|$ (Proakis [345]):

$$P^{\text{noncoh}} = \sum_{n=1}^{m-1} (-1)^{n+1} \binom{m-1}{n} \frac{1}{n+1} \exp\left(-\frac{n|A|^2}{2(n+1)\sigma^2} \right), \quad (3.167)$$

which holds in the special case of deterministic amplitude and orthogonal signaling:

$$\langle s(\cdot; b), s(\cdot; d) \rangle = \delta_{bd}.$$

If $m = 2$, then (3.167) reduces to

$$P^{\text{noncoh}} = \frac{1}{2} \exp\left(-\frac{|A|^2}{4\sigma^2} \right). \quad (3.168)$$

In the case of binary nonorthogonal signaling, we denote

$$\sin \beta \stackrel{\text{def}}{=} \langle s(\cdot; 1), s(\cdot; 2) \rangle;$$

then the error probability averaged with respect to Rayleigh $|A|$ is equal to (cf. (3.153) and Proakis [345])

$$P^{\text{noncoh}} = T\left(\frac{A}{\sqrt{2\sigma}}\sin\frac{\beta}{2}, \frac{A}{\sqrt{2\sigma}}\cos\frac{\beta}{2}\right). \qquad (3.169)$$

An exact analysis of the noncoherent single-user demodulator in the presence of multiuser interference generalizing (3.167) or (3.169) has not been reported in the literature. In order to grasp the essential features of the multiuser performance of the noncoherent single-user demodulator when the received process is

$$y(t) = \sum_{k=1}^{K} |A_k| e^{j\theta_k} s_k(t; b_k) + \sigma n(t), \qquad (3.170)$$

where θ_k are independent and uniformly distributed, we will restrict attention to binary orthogonal modulation: $b_k \in \{+, -\}$,

$$\langle s_k(\cdot, +), s_k(\cdot, -)\rangle = 0,$$

and we will discuss two different scenarios: the two-user noiseless channel and K-user Rayleigh fading.

Two-user noiseless channel. In this case (3.170) becomes

$$y(t) = |A_1|e^{j\theta_1} s_1(t; b_1) + |A_2|e^{j\theta_2} s_2(t; b_2),$$

where we denote the crosscorrelations

$$\rho[b_1, b_2] \stackrel{\text{def}}{=} \int_0^T s_1(t, b_1) s_2^*(t, b_2) \, dt.$$

· Conditioned on $|A_1|$ and $|A_2|$, the probability of error is given by (Problem 3.47)

$$P_1^{\text{noncoh}} = \frac{1}{2} - \frac{1}{4}[g(+, +) + g(+, -) + g(-, +) + g(-, -)] \quad (3.171)$$

with

$$g(b_1, b_2) = \frac{1}{\pi}\arcsin(h(\delta(b_1, b_2))),$$

$$h(x) \stackrel{\text{def}}{=} \begin{cases} 1, & x > 1, \\ x, & -1 \le x \ge 1, \\ -1 & x < -1, \end{cases}$$

$$\delta(b_1, b_2) \stackrel{\text{def}}{=} \frac{|A_1|^2 + |A_2|^2(|\rho[b_1, b_2]|^2 - |\rho[-b_1, b_2]|^2)}{2|A_1||A_2||\rho[b_1, b_2]|}.$$

135

The interesting aspect of the error probability (3.171) is that as long as there is (b_1, b_2) such that

$$\delta(b_1, b_2) < 1, \tag{3.172}$$

then the error probability is nonzero in the absence of noise. Condition (3.172) depends not only on the crosscorrelations but on the received amplitudes. Examining the definition of the function $\delta(b_1, b_2)$ we see that, unless the pairs of signals of both users are mutually orthogonal (in which case the single-user error probability in (3.168) holds), then we can find a pair (b_1, b_2) such that (3.172) is satisfied for sufficiently small ratio $|A_1|/|A_2|$. This means that whenever the noncoherent error probability is averaged with respect to an amplitude distribution that places nonzero mass in every interval $(0, \epsilon)$ (such as all those mentioned in Section 2.6.1), then it will show an error floor as the background noise level decreases. Comparing this behavior to the case of no multiuser interference (e.g., Problem 3.46) we conclude that asymptotic multiuser efficiency (and near–far resistance) are zero for the single-user noncoherent demodulator in frequency-flat fading.

K-*user Rayleigh fading.* If $|A_1|, \ldots, |A_K|$ are independent Rayleigh-distributed, with second moments denoted by

$$E[|A_k|^2] = A_k^2,$$

then given $(b_1, \ldots, b_K)$ the decision variables

$$y_{1+} = \langle y, s_k(\cdot, +) \rangle, \tag{3.173}$$

$$y_{1-} = \langle y, s_k(\cdot, -) \rangle \tag{3.174}$$

are correlated, zero-mean, complex Gaussian random variables with independent real and imaginary components. Assuming $b_1 = -1$, the second moments of the observables are

$$E[|y_{1+}|^2] = \sigma^2 + \sum_{k=2}^{K} A_k^2 |\rho_{1k}[+, b_k]|^2, \tag{3.175}$$

$$E[|y_{1-}|^2] = \sigma^2 + A_1^2 + \sum_{k=2}^{K} A_k^2 |\rho_{1k}[-, b_k]|^2, \tag{3.176}$$

$$E[y_{1+}y_{1-}^*] = \sum_{k=2}^{K} A_k^2 \rho_{1k}[+, b_k] \rho_{1k}^*[-, b_k]. \tag{3.177}$$

Using these expressions, we can readily compute the conditional probability of error,

$$P[|y_{1+}|^2 > |y_{1-}|^2 | b_1 = -1, b_2, \ldots, b_K],$$

by particularizing the following fact:

136

Proposition 3.7 *Let X and Y be zero-mean complex Gaussian random variables with independent real and imaginary components. Then*[11]

$$P[|X|^2 < |Y|^2] = \frac{1}{2} - \frac{E[|X|^2] - E[|Y|^2]}{2\sqrt{(E[|X|^2 + |Y|^2])^2 - 4|E[XY^*]|^2}}. \quad (3.178)$$

3.9 BIBLIOGRAPHICAL NOTES

General references on detection and estimation dealing with the material in Sections 3.1 and 3.2 are Van Trees [462] and Poor [336]. The matched filter was originally found in North [303] as the maximal signal-to-noise ratio solution in a radar problem. The signal space approach to digital communications popularized in the textbook Wozencraft and Jacobs [543], dates back to Shannon [407].

The single-user matched filter receiver has been used for CDMA demodulation since the inception of direct-sequence spread-spectrum for multiuser applications (Simon et al. [410]). There have been a number of works devoted to the computation or approximation of the expected value of the bit-error-rate of the conventional single-user matched filter receiver. All early works, for example Kaiser et al. [192], Pursley [347], and Yao [552] approximate multiaccess interference as a white Gaussian process. More refined analyses have been reported in Geraniotis and Pursley [108], Pursley et al. [348], Laforgia et al. [226], McDowell and Lehnert [272], and Letaief [238] for given sets of pseudonoise signature waveforms, and in Lehnert and Pursley [236], Morrow and Lehnert [287], Sadowsky and Bahr [372], Holtzman [147, 148], van Rooyen and Solms [461], Sun and Polydoros [436], and Verdú and Shamai [504] for random signature sequences. An analysis of the capabilities of the single-user matched filter with random direct-sequence signature waveforms dates back to Savage [383]. The Chernoff bound technique in (3.98) is originally due to Saltzberg [375]. Discussion and analysis of the effects of the choice of specific families of pseudonoise sequences on the bit-error-rate of the conventional receiver can be found in the spread-spectrum literature (Simon et al. [410]). The sensitivity of the single-user matched filter bit-error-rate to unequal received signal-to-noise ratios is studied in Kudoh and Matsumoto [224], and for fading channels in Vojcic et al. [515], Vojcic et al. [516], Vojcic et al. [514], and Cameron and Woerner

---

[11] A more general problem is solved in Appendix B of Proakis [345].

[42]. Very accurate and fast power control (with a maximum tolerable error of less than 1 dB) is required in the IS-95 system with single-user matched filtering (e.g., Vojcic et al. [515] and Kohno et al. [220]).

The bit-error-rate analysis of the single-user matched filter in the presence of asynchronous interferers and Rician fading has been investigated for various spread-spectrum systems in Geraniotis [107] and Vandendorpe [463]. A mixture of Rayleigh and Rician fading is considered in Vojcic et al. [516] and Vojcic et al. [514]. Analyses in the presence of frequency-selective fading appear in Geraniotis and Pursley [109], Geraniotis [107], Letaief [237], and Goeckel and Stark [120]. The effects of unequalized multipath with Rayleigh-faded coefficients and asynchronous interferers are studied in Kavehrad [202]. The single-user receiver is analyzed in Caire et al. [41] with trellis-coded modulation and the homogeneous fading model of (2.66). Homogeneous fading in multicarrier CDMA is considered in Schnell [394]. The effect of unequal carrier frequencies is analyzed in Sousa [418].

The numerically efficient approximation to bit-error-rates in (3.69) is due to Levy [239]. A table of integrals of the Q-function is given in Ng and Geller [301]. Its average with respect to the Rice distribution (2.51) is not known to admit a closed-form expression but can be represented in terms of Marcum's Q-function (Lindsey [244]).

The performance of spread-spectrum correlation receivers in impulsive non-Gaussian channels is studied in Aazhang and Poor [2] and Aazhang and Poor [3].

The differentially-coherent single-user matched filter in the presence of multiaccess interference is analyzed in Varanasi and Aazhang [475]. The formulas given for the bit-error-rate of the noncoherent single-user matched filter are justified in Proakis [345]. The case of noncoherent multipath reception with orthogonal m-ary modulation is analyzed in Viterbi [510] and Jalloul and Holtzman [175] approximating the multiaccess interference as white Gaussian noise. An approximation of the error probability of binary frequency-shift-keying for direct-sequence spread-spectrum multiaccess is given in Geraniotis [106] substituting the multiple-access interference at the output of the matched filter by Gaussian random variables with exact second-order moments.

Asymptotic multiuser efficiency was first proposed as a performance measure of multiuser demodulators in Verdú [486, 487, 490]. The asymptotic multiuser efficiency of the single-user matched filter was given in Verdú [491]. For Rayleigh fading channels, Equation (3.132) was defined in Zvonar and Brady [570]. The asymptotic efficiency in the presence of Rician fading is found in Vasudevan and Varanasi [479]. The zero asymptotic efficiency of the noncoherent single-user receiver was noted in Russ and Varanasi [370] for Rayleigh fading channels. The relevance of the worst asymptotic effective energy for the minimization of the sum of transmitted powers is recognized in Varanasi [469]. Power-tradeoff regions are introduced in this text as an analysis tool for multiuser detectors.

The near–far resistance measure in (3.117) was proposed in Verdú [489], Lupas-Golaszewski and Verdú [256], and Lupas and Verdú [254] in the context of synchronous channels. The definition in (3.118) was proposed in Lupas and Verdú [253, 255] in order to study the near–far problem in asynchronous channels.

3.10 PROBLEMS

PROBLEM 3.1. Show that the minimum error probability for equiprobable binary hypothesis

$$H_0 : Z \sim f_{Z|0},$$
$$H_1 : Z \sim f_{Z|1}$$

is equal to

$$P = \frac{1}{2} \int \min\{f_{Z|0}(z), f_{Z|1}(z)\} \, dz \qquad (3.179)$$

$$= \frac{1}{2} - \frac{1}{4} \int |f_{Z|0}(z) - f_{Z|1}(z)| \, dz. \qquad (3.180)$$

Note that (3.179) implies the *Bhattacharyya bound*:

$$P \leq \frac{1}{2} \int \sqrt{f_{Z|0}(z) f_{Z|1}(z)} \, dz. \qquad (3.181)$$

PROBLEM 3.2. Consider the binary hypothesis testing problem between a standard Cauchy distribution and a zero-mean Gaussian with variance equal to 2 depicted in Figure 3.1.

(a) Find the minimum probability of error P if both hypothesis are equiprobable.
(b) Find the test that minimizes the maximum of the probability of deciding H_1 when H_2 is true and the probability of deciding H_2 when H_1 is true.

139

PROBLEM 3.3. Consider an m-ary hypothesis testing problem:

$$H_1 : Z \sim f_{Z|1},$$

$$\vdots$$

$$H_m : Z \sim f_{Z|m}.$$

Show that the vector of $m - 1$ a posteriori probabilities,

$$P[H_1|Z], \cdots P[H_{m-1}|Z],$$

is a sufficient statistic. (You may assume for convenience that Z is countably valued.)

PROBLEM 3.4. Let $\mathbf{n}$ be a Gaussian vector with zero mean and covariance matrix equal to a positive-definite matrix $\mathbf{H}$. Let $\mathbf{x}_i$ be a deterministic vector with

$$\|\mathbf{x}_i\|^2 = \mathbf{x}_i^T \mathbf{x}_i < \infty,$$

and consider the hypothesis testing problem:

$$H_1 : \mathbf{y} = \mathbf{x}_1 + \sigma \mathbf{n},$$

$$\vdots$$

$$H_m : \mathbf{y} = \mathbf{x}_m + \sigma \mathbf{n}.$$

(a) Find the maximum-likelihood decision rule.
(b) Simplify the decision rule in (a) in the case: $m = 2$, $\|\mathbf{x}_1\| = \|\mathbf{x}_2\|$.
(c) Generalize the decision rule in (b) to make it the MAP rule when $P[H_1] = \alpha$.

PROBLEM 3.5. Consider the following m-ary hypothesis testing problem with L-dimensional observations:

$$H_1 : \mathbf{y} = \mathbf{C}\mathbf{x}_1 + \mathbf{d},$$

$$\vdots$$

$$H_m : \mathbf{y} = \mathbf{C}\mathbf{x}_m + \mathbf{d},$$

where $\mathbf{C}$ is an invertible $L \times L$ matrix. Furthermore, we assume that $\mathbf{C}$ and $\mathbf{d}$ are independent of the random vector $\mathbf{x}$, and their joint distribution is independent of the true hypothesis. Show that if $\mathbf{C}$ and $\mathbf{d}$ are known to the detector, then the maximum-likelihood decision rule is the maximum-likelihood decision rule for

$$H_1 : \mathbf{z} = \mathbf{x}_1,$$

$$\vdots$$

$$H_m : \mathbf{z} = \mathbf{x}_m$$

applied to

$$\mathbf{z} = \mathbf{C}^{-1}(\mathbf{y} - \mathbf{d}).$$

PROBLEM 3.6. Independent observations $(Y_1, Y_2, \ldots)$ with identical distribution are obtained. The common distribution is determined by

$$\mathrm{H}_0 : Y_i \sim f_{Y|0},$$
$$\mathrm{H}_1 : Y_i \sim f_{Y|1}.$$

Denote the loglikelihood ratio by

$$\lambda(y) = \log \frac{f_{Y|0}(y)}{f_{Y|1}(y)},$$

and denote the a posteriori probability after j observations by

$$\pi_1[j] = P[\mathrm{H}_1|y_1, \ldots, y_j].$$

In particular, $\pi_1[0]$ denotes the a priori probability of hypothesis H_1.

(a) Show that

$$\pi_1[j] = \frac{1}{1 + \exp(l[j])}$$

with

$$l[j] = l[j-1] + \lambda[y_j],$$
$$l[0] = \log\left(\pi_1^{-1}[0] - 1\right). \qquad (3.182)$$

Note that the sign of $l[j]$ determines the maximum-a-posteriori decision, whereas its absolute value quantifies the reliability of the decision.

(b) Assume that the probability density functions $f_{Y|0}$ and $f_{Y|1}$ are such that

$$D(f_{Y|0}\|f_{Y|1}) < \infty,$$
$$D(f_{Y|1}\|f_{Y|0}) < \infty,$$

where the divergence functional is defined as

$$D(f\|g) \overset{\text{def}}{=} \int f(x) \log \frac{f(x)}{g(x)} dx.$$

Show that if H_1 is true, then $l[j]/j$ converges almost surely to $-D(f_{Y|1}\|f_{Y|0})$, and if H_0 is true, then $l[j]/j$ converges almost surely to $D(f_{Y|0}\|f_{Y|1})$.

PROBLEM 3.7. Consider the m-ary hypothesis testing problem:

$$H_1 : y(t) = As_1(t) + \sigma n(t),$$

$$\vdots$$

$$H_m : y(t) = As_m(t) + \sigma n(t),$$

where $\langle s_i, s_j \rangle = 0$ if $i \neq j$, and $\|s_i\| = 1$ for all i.

(a) Find the minimum probability of error.

(b) Define the energy per bit as

$$E_b = \frac{A^2}{\log_2 M}.$$

Find the minimum value of E_b/σ^2 such that the probability of error goes to 0 as $M \to \infty$.

PROBLEM 3.8. Consider a single-user discrete-time channel:

$$\mathbf{y} = Ab\mathbf{s} + \mathbf{n},$$

where $\mathbf{s}$ is a deterministic L-vector and $\mathbf{n}$ is a Gaussian L-vector with zero mean and covariance matrix given by the positive-definite matrix Σ.

(a) Show that the optimum detector for equiprobable $b \in \{-1, 1\}$ is

$$\hat{b} = \text{sgn}(\mathbf{y}^T \mathbf{h}),$$

where $\mathbf{h}$ is the *discrete-time matched filter*:

$$\mathbf{h} = \Sigma^{-1} \mathbf{s}.$$

(b) Show that among all transmitted vectors $\mathbf{s}$, such that $\|\mathbf{s}\| = 1$, the one that minimizes the error probability of the matched filter detector is a minimum-eigenvalue eigenvector of Σ.

PROBLEM 3.9. Consider a single-user complex-valued discrete-time channel:

$$\mathbf{y} = Ab\mathbf{s} + \mathbf{n},$$

where $\mathbf{s}$ is a deterministic complex-valued L-vector and $\mathbf{n}$ is a Gaussian L-vector with zero mean, independent real and imaginary parts, and covariance matrix given by the positive-definite matrix Σ. Show that the optimum detector for equiprobable $b \in \{-1, 1\}$ is

$$\hat{b} = \text{sgn}(\Re\{\mathbf{y}^* \Sigma^{-1} \mathbf{s}\}),$$

where $\mathbf{y}^*$ denotes the conjugate transpose of $\mathbf{y}$.

PROBLEM 3.10. A single-user white Gaussian channel

$$y(t) = Abs(t) + \sigma n(t), \quad t \in [0, T],$$

is demodulated with a correlator that instead of being sampled at time T is sampled at time τ uniformly distributed between 0 and T, that is,

$$\hat{b} = \text{sgn}\left(\int_0^\tau y(t)s(t)\, dt \right).$$

If $|s(t)| = 1/\sqrt{T}$, show that the increase in average error probability due to the timing jitter is equal to

$$\frac{\sigma^2}{A^2}\left[\frac{1}{2} - Q\left(\frac{A}{\sigma}\right) \right] - \frac{\sigma}{A\sqrt{2\pi}} e^{-A^2/2\sigma^2}.$$

PROBLEM 3.11. Consider a collection of L-dimensional positive-definite matrices $(\Xi_1, \ldots, \Xi_L)$ and an m-ary hypothesis testing problem where the observations are an L-dimensional zero-mean Gaussian random vector:

$$H_1 : \mathbf{Z} \sim \mathcal{N}(\mathbf{0}, \Xi_1),$$
$$\vdots$$
$$H_m : \mathbf{Z} \sim \mathcal{N}(\mathbf{0}, \Xi_m).$$

Show that the maximum-likelihood decision rule is

$$\text{argmin}_i \{ \mathbf{Z}^T \Xi_i^{-1} \mathbf{Z} + \log \det \Xi_i \}.$$

PROBLEM 3.12. Two demodulators for a communication system that transmits independent binary equiprobable data have the specifications shown in the table

	p_{00}	p_{11}
NOKIA	0.95	0.67
MOTOROLA	0.80	0.80

where p_{00} [resp. p_{11}] stands for the probability that if 0 [resp. 1] is transmitted, the demodulator makes the correct decision.

(a) Which demodulator achieves better bit-error-rate?
(b) The demodulator is now used in conjunction with an error-correction code: 0 is encoded as 000 and 1 is encoded as 111; the decoder selects 0 if the output of the demodulator belongs to {000, 001, 010, 100} and 1 otherwise. Which demodulator achieves better bit-error-rate?

143

(c) Can you get the same answers to both (a) and (b) if both de-modulators satisfy $p_{00} = p_{11}$?

PROBLEM 3.13. For the composite hypothesis testing problem of Proposition 3.4, find

$$P[\mathcal{H}_J | y(t), t \in \mathcal{I}].$$

PROBLEM 3.14. Consider the binary, nonequiprobable, antipodal, single-user, white Gaussian noise channel

$$y(t) = Abs(t) + \sigma n(t), \quad t \in [0, T],$$

where the a priori probabilities satisfy

$$\pi_0 = P[b = 0] \neq P[b = 1] = \pi_1.$$

Find the detector that minimizes the probability of error.

PROBLEM 3.15. Consider the binary, equiprobable, antipodal, single-user, white Gaussian noise channel

$$y(t) = Abs(t) + \sigma n(t), \quad t \in [0, T],$$

where $s(t) = 1/\sqrt{T}, t \in [0, T]$. Find the probability of error achieved by the following two-stage detector: in the first stage, the detector uses $\{y(t), t \in [0, T/2]\}$ to compute

$$\tilde{\pi}_0 = P[b = -1 | \{y(t), t \in [0, T/2]\}],$$

$$\tilde{\pi}_1 = P[b = 1 | \{y(t), t \in [0, T/2]\}].$$

The second stage is an optimum detector (Problem 3.14) for b using the observations $\{y(t), t \in [T/2, T]\}$ and the "a priori" probabilities $\tilde{\pi}_0$ and $\tilde{\pi}_1$ computed by the first stage.

PROBLEM 3.16. Consider the binary, equiprobable, antipodal, single-user, white-noise channel

$$y(t) = Abs(t) + \sigma n(t),$$

where s is unknown to the receiver other than the fact that

$$s \in \mathcal{S} \stackrel{\text{def}}{=} \left\{ s : \int_0^T (s(t) - s_0(t))^2 \, dt \leq \delta \right\},$$

for some known signal s_0 and δ.

(a) Show that $h = s_0$ achieves

$$\max_h \min_{s \in \mathcal{S}} \text{SNR}(s, h) \tag{3.183}$$

with the signal-to-noise ratio defined as

$$\text{SNR}(s, h) \overset{\text{def}}{=} \frac{(\langle s, h \rangle)^2}{\sigma^2 \|h\|^2}.$$

(b) Find the value of (3.183) as a function of s_0, σ, and δ.

PROBLEM 3.17. Consider the binary, equiprobable, antipodal, one-shot single-user channel

$$y(t) = Abs(t) + \sigma n(t),$$

where n is white Gaussian noise with unit spectral density and s is the raised-cosine pulse in Problem 1.9:

$$s(t) = \frac{g_\alpha(t)}{\|g_\alpha(t)\|}.$$

Suppose that y is processed by an ideal low-pass filter with bandwidth

$$\frac{1+\alpha}{2},$$

followed by sampling at time 0 and by a comparison to a zero-threshold. Compare the error probability of this detector with that of the optimum detector, in terms of the signal-to-noise ratio loss as a function of α. Why is there no loss of optimality when $\alpha = 0$?

PROBLEM 3.18. Consider the binary, equiprobable, antipodal, single-user channel

$$y(t) = Abs(t) + m(t),$$

where the unit-energy signal s is time-limited to $[0, T]$ and $b \in \{-1, 1\}$. Assume that the demodulator is

$$\hat{b} = \text{sgn}\,(\langle y, s \rangle),$$

when $\langle y, s \rangle \neq 0$. If $\langle y, s \rangle = 0$, then $\hat{b}$ takes either value with equal probability. Furthermore, suppose that $m(t)$ is a random process chosen by a jammer under the constraint:

$$E\left[\int_0^T m^2(t)\,dt \right] \leq \sigma^2 \leq A^2.$$

The highest error probability that such a jammer can cause to the above demodulator is known (Shamai and Verdú [406]) to be equal to

$$P_{\max} = \frac{\sigma^2}{4A^2}.$$

(a) Find the jamming process, $m(t)$, that achieves P_{max}.

(b) Assuming the jamming process you found in (a), is there a demodulator that achieves error probability smaller than P_{max}?

PROBLEM 3.19. Find a minimal sufficient statistic for the hypothesis testing problem in Proposition 3.2.

PROBLEM 3.20. A signal s is present on the interval $[0, T]$ with probability $1/2$. When the signal is present, it is equally likely that the detector receives s or $-s$. Find the rule to decide whether the signal is present or absent with minimum probability of error assuming that the observations are contaminated by additive white Gaussian noise.

PROBLEM 3.21. Generalize Proposition 3.3 to the case when $\mathcal{H}_J$ and $\bar{\mathcal{H}}_J$ do not have the same number of elements.

PROBLEM 3.22. Consider the composite hypothesis testing problem in Proposition 3.4 with $\|x_1\| = \cdots = \|x_m\|$. Consider the test that selects the most likely hypothesis among the m hypotheses and then checks whether that hypothesis belongs to J.

(a) Show that this test is not optimal by means of an example.

(b) Show that the decision regions of this test converge to the optimal ones as $\sigma \to 0$.

PROBLEM 3.23. (*Bounds of the Q-function*)

(a) Prove bounds (3.34) and (3.35). [*Hint*: Integrate the definition of Q by parts, and use the definite integral

$$\int_x^\infty v e^{-v^2/2} dv = e^{-x^2/2}.]$$

(b) Prove bound (3.38). [*Hint*: Work with the logarithm of the Q-function.]

(c) Prove bound (3.37). [*Hint*: Compare to the upper bounds (3.35) and (3.38).]

PROBLEM 3.24. Show (3.64). [*Hint*: Use a partial-fraction expansion of the characteristic function of $\|X\|^2$ to write its probability density function as a weighted sum of exponential probability density functions.]

PROBLEM 3.25. Find $E[Q(\|X\|)]$ if X is a complex-valued, zero-mean, Gaussian three-dimensional vector with independent coefficients with independent real and imaginary parts and variances

$$E[|X_1|^2] = 2\gamma^2,$$

$$E[|X_2|^2] = 2\gamma^2,$$

$$E[|X_3|^2] = 4\gamma^2.$$

(See e.g., Schlegel [386].)

PROBLEM 3.26. Use (3.56) and (3.57) to show (3.65).

PROBLEM 3.27.

(a) Conditioned on amplitudes and crosscorrelations, find the characteristic function of the random variable

$$X = \sum_{k=2}^{K} A_k b_k \rho_{1k}.$$

(b) Find the characteristic function of the random variable

$$X = \sum_{k=2}^{K} A_k b_k \rho_{1k} \cos(\theta_k),$$

where the phases θ_k are independent and uniformly distributed on $[0, 2\pi)$.

PROBLEM 3.28. Consider the two-user synchronous CDMA channel (2.5).

(a) Show that the signal-to-noise ratio of user 1 that minimizes the error probability of the matched filter for user 1 is

$$\frac{A_1^2}{\sigma^2} = \begin{cases} \gamma \operatorname{arctanh}(\gamma), & \text{if } |\gamma| < 1; \\ +\infty, & \text{if } |\gamma| \geq 1; \end{cases}$$

where

$$\gamma = \frac{A_1}{A_2 \rho}.$$

(b) Show that the unique signal-to-noise ratio of user 1 that maximizes the error probability of the matched filter for user 1 is

$$\frac{A_1^2}{\sigma^2} = 0,$$

regardless of the value of γ.

PROBLEM 3.29. Find the error probability of the single-user matched filter receiver for user 1 in a two-user synchronous CDMA channel (2.5) if the background noise $n(t)$ is

(a) Gaussian with autocorrelation function $R(\tau)$.

(b)

$$n(t) = \tan(\theta), \quad t \in [0, T],$$

where θ is uniformly distributed on $[-\pi, \pi]$.

PROBLEM 3.30. Derive expressions (3.112) and (3.113) for the asymptotic multiuser efficiency.

PROBLEM 3.31. Find the signal-to-interference ratio at the output of the single-user matched filter for the basic asynchronous CDMA model.

PROBLEM 3.32. Consider a synchronous CDMA system using a bank of matched filters for which the open-eye condition is satisfied for all users. Initially, all users have the same received signal-to-noise ratio γ. Then user 1 increases its power by α_1 dB. To compensate, users $2, \ldots, K$ increase their power by α dB. Find the minimum α necessary so that none of the error probabilities increases. Assume $\gamma \to \infty$.

PROBLEM 3.33. Show that if the random variable X is such that for all $x > 0$,

$$P[-x < X \leq 0] \geq P[0 < X \leq x],$$

then the Chernoff bound can be strengthened as:

$$P[X > 0] \leq \frac{1}{2} E[e^{\lambda X}]$$

for all $\lambda \geq 0$.

PROBLEM 3.34. Find the probability of error of the single-user matched filter for user 1 in the absence of background noise $P_1^c(0)$ in a 5-user synchronous CDMA channel with crosscorrelations

$$(\rho_{12}, \rho_{13}, \rho_{14}, \rho_{15}) = (-0.2, 0.1, 0.3, 0.2)$$

and received amplitudes

$$(A_2, A_3, A_4, A_5) = A_1(2, 1, 2, 1).$$

PROBLEM 3.35. Show that determining $P_1^c(0)$ is NP-hard in K. [*Hint*: By reduction to PARTITION (Garey and Johnson [102]).]

PROBLEM 3.36. Consider a two-user synchronous channel with crosscorrelation ρ where the received amplitudes are random, independent, identically distributed:

$$P[A_k = \alpha A] = \frac{1}{4},$$

$$P[A_k = A] = \frac{1}{2},$$

$$P[A_k = \alpha^{-1}A] = \frac{1}{4}.$$

Find the average asymptotic multiuser efficiency of the single-user matched filter (3.122).

PROBLEM 3.37. Consider a synchronous direct-sequence spread-spectrum system where binary signatures are chosen independently and equally likely. Suppose that K and N grow without bound while keeping their ratio constant and equal to $\beta = K/N$.

(a) Assume that all users have the same received power $A_k^2 = A^2$. Show that the signal-to-interference ratio at the output of the single-user matched filter converges in mean-square sense to

$$\lim_{K\to\infty} \frac{A_c^2}{\sigma^2} = \frac{A^2}{\sigma^2 + \beta A^2}. \tag{3.184}$$

(b) Generalize the result in (a) to the case where $A_{2k-1} = \alpha A_{2k} = \alpha A$.

PROBLEM 3.38. Consider a K-user synchronous CDMA channel where users are assigned direct-sequence spread-spectrum waveforms with N chips per symbol. Assume that all K users are received with equal amplitudes and there is no background noise ($\sigma = 0$). The demodulator is a bank of single-user matched filters and the binary spreading codes are chosen randomly (equiprobably over all choices) and independently from user to user.

(a) Find the following function of K:

$$\lim_{N\to\infty} \frac{1}{N} \log 1/P_1^c(0). \tag{3.185}$$

(b) Show that if K grows with N as $K = o(N)$, then

$$\lim_{N\to\infty} P_1^c(0) = 0.$$

[*Hint*: It follows from standard large-deviations results (Grimmett and Stirzaker [128]) that if a fair coin shows H_n heads in n tosses, then

$$P[2H_n - n > an]^{1/n} \to \frac{1}{\sqrt{(1+a)^{(1+a)}(1-a)^{(1-a)}}}$$

for $0 < a < 1$, and

$$P\left[\frac{1}{n}H_n \geq \frac{1}{2} + \delta\right] \leq e^{-n\delta^2/4},$$

for $0 < \delta$.]

PROBLEM 3.39. Consider a K-user synchronous CDMA channel where users are assigned direct-sequence spread-spectrum waveforms with N chips per symbol. The binary spreading codes are chosen randomly (equiprobably over all choices) and independently from user to user. Let $\beta = K/N$. Verify that if $A_j = o(K)$ for all $j = 2, \ldots, K$, then the bit-error-rate of the single-user matched filter averaged over the choice of sequences converges to

$$\lim_{K \to \infty} E[P_1^c(\sigma)] = Q\left(\frac{A_1}{\sqrt{\sigma^2 + \beta \bar{A}^2}}\right), \tag{3.186}$$

where

$$\bar{A}^2 \overset{\text{def}}{=} \lim_{K \to \infty} \frac{1}{K} \sum_{j=2}^{K} A_j^2.$$

[*Hint*: Use the Lindeberg–Feller Central Limit Theorem (Feller [91])].

PROBLEM 3.40. Using the same random sequence model as in Problem 3.39, show that the variance of the random variable

$$1\left\{\sum_{j=2}^{K} \rho_{1j} > 1\right\}$$

converges as $K \to \infty$ to

$$Q\left(\frac{1}{\sqrt{\beta}}\right)\left(1 - Q\left(\frac{1}{\sqrt{\beta}}\right)\right).$$

PROBLEM 3.41. Consider a K-user synchronous channel where the signature waveforms are *equicorrelated* $\rho_{jk} = \rho$ for all j and k and

$$A_1 = \cdots = A_K = A.$$

(a) Find the range of ρ for which $\mathbf{R}$ is positive-definite.
(b) Show that the bit-error-rate of the single-user matched filter is given by

$$P_k^c(\sigma) = 2^{1-K} \sum_{n=0}^{K-1} \binom{K-1}{n} Q\left(\frac{A}{\sigma}(1 + \rho(K - 1 - 2n))\right).$$

(c) Particularize the Gaussian approximation (3.93) to the present case.

150

(d) For every K, find a value of ρ such that

$$\lim_{\sigma \to 0} \tilde{P}_k^c(\sigma) < \lim_{\sigma \to 0} P_k^c(\sigma).$$

(e) For every K, find a value of ρ so that $P_k^c(\sigma)$ is nonmonotonic. Give an explanation for this behavior.

PROBLEM 3.42. Extend the bound (3.98) to the carrier-modulated case (Section 2.5). [*Hint* (Viterbi [510]): Note that

$$E[\exp(\lambda \bar{\rho}_{kj})] = I_0(\lambda \rho_{kj})$$

where the zero-order Bessel function satisfies

$$I_0(x) < \exp\left(\frac{x^2}{4}\right).]$$

PROBLEM 3.43. Show that for sufficiently small A/σ,

$$P^{Fc}(\sigma) = \frac{1}{2}\left(1 - \frac{1}{\sqrt{1+\sigma^2/A^2}}\right) < Q\left(\frac{A}{\sigma}\right) = P^c.$$

PROBLEM 3.44. A certain bit-error-rate function parametrized by α admits the representation

$$P(\sigma, \alpha) = g(\sigma, \alpha)e^{-f(\alpha)/2\sigma^2}$$

with $f(\alpha) \geq 0$ and there exist $\sigma_0 > 0$, $\epsilon_1 > 0$, $\epsilon_2 > 0$ such that if $0 < \sigma < \sigma_0$, then

$$0 < \epsilon_1 \leq g(\sigma, \alpha) \leq \epsilon_2 < +\infty.$$

Show that the asymptotic efficiency corresponding to $E[P(\sigma, A)]$, where the average is with respect to a given distribution of the random parameter A, is given by

$$2 \lim_{\sigma \to 0} \sigma^2 \log 1/E[P(\sigma, A)] = f^*,$$

where f^* is the essential infimum of $f(A)$. [*Note*: The essential infimum of a random variable X is defined as

$$\sup\{x : P[X \leq x] = 0\}.]$$

PROBLEM 3.45.

(a) Find the worst asymptotic effective energy $\omega^c(A_1, \ldots, A_K)$ (cf. (3.115)) for the synchronous single-user matched filter.

(b) Verify that the equal-power power penalty is

$$\frac{1}{A^2}\omega^c(A, \ldots, A) = \max{}^2\left\{0, 1 - \max_k \sum_{j\neq k} |\rho_{jk}|\right\}.$$

PROBLEM 3.46. Find the average error probability of the single-user channel with binary orthogonal modulation and noncoherent demodulation (3.168) when $|A| = AR$ and R is Rice-distributed with parameter d.

PROBLEM 3.47. Verify the error probability (3.171) of the noncoherent single-user demodulator in the absence of noise and in the presence of one interferer.

PROBLEM 3.48. Show that the error probability of the coherent single-user matched filter with synchronous or asynchronous interferers satisfies

$$\lim_{\sigma\to\infty}\left[\frac{1}{2} - \mathsf{P}_k^c(\sigma)\right]\sigma = \frac{A_k}{\sqrt{2\pi}}.$$

PROBLEM 3.49. For the K-user error probability of the noncoherent synchronous single-user detector in the presence of Rayleigh-distributed interferers, find

$$\lim_{\sigma\to\infty}\left[\frac{1}{2} - \mathsf{P}_k^{\mathrm{noncoh}}(\sigma)\right]\sigma^2.$$

PROBLEM 3.50. (Varanasi [466]) A certain multiuser communication system has error probability equal to

$$\mathsf{P}_k(\sigma) = T\left(\frac{\alpha A_k}{\sigma}, \frac{\beta A_k}{\sigma}\right),$$

where the function T is defined in (3.153). Find the asymptotic multiuser efficiency η_k.

PROBLEM 3.51. A certain m-ary single-user communication channel is

$$\begin{bmatrix} z_1 \\ \vdots \\ z_m \end{bmatrix} = \begin{bmatrix} A_1\, 1\{b=1\} \\ \vdots \\ A_m\, 1\{b=m\} \end{bmatrix} + \sigma \begin{bmatrix} n_1 \\ \vdots \\ n_m \end{bmatrix}, \qquad (3.187)$$

where the vector $[n_1 \cdots n_m]$ is complex Gaussian with independent real and imaginary parts. We consider the problem of designing a noncoherent receiver that knows $[|A_1| \cdots |A_m|]$ but has no information on the phases of $[A_1 \cdots A_m]$.

(a) (Russ and Varanasi [370]) Show that the maximum-likelihood decision rule is

$$\hat{b} = \arg\max_i \left\{ \exp\left(-\frac{|A_i|^2 P_{ii}}{\sigma^2} \right) I_0 \left(2\frac{|A_i|}{\sigma^2} \left| \sum_{j=1}^m P_{ij} z_j \right| \right) \right\},$$

(3.188)

where the matrix $\mathbf{P}$ is the inverse of the $m \times m$ covariance matrix of the noise vector $[n_1 \cdots n_m]$.

(b) Find the probability of error achieved by (3.188) in the special case of equiprobable b, $\mathbf{P} = \mathbf{I}$ and $|A_i| = |A|$.

CHAPTER FOUR

OPTIMUM MULTIUSER DETECTION

A very simple demodulator for the CDMA channel was analyzed in Chapter 3. We turn our attention now to the derivation and analysis of optimum strategies. The analysis of optimum multiuser detectors yields the minimum achievable probability of error (and optimum asymptotic multiuser efficiency, as well as optimum near–far resistance) in CDMA channels. This serves as a baseline of comparison for suboptimum multiuser detectors.

4.1 OPTIMUM DETECTOR FOR SYNCHRONOUS CHANNELS

The conventional single-user matched filter receiver requires no knowledge beyond the signature waveforms and timing of the users it wants to demodulate. In the derivation of an optimum receiver, we will assume that the receiver not only knows the signature waveform and timing of every active user, but it also knows (or can estimate) the received amplitudes of all users and the noise level.

For some time, it was widely believed that the decisions of the conventional single-user matched filter were, if not optimal, almost optimal for channels with a large number of equal-power users. The reasoning that led to this belief used the central limit theorem to conclude that the second (binomially distributed) term in (cf. (2.76))

$$y_1 = A_1 b_1 + \sum_{k=2}^{K} A_k b_k \rho_{1k} + \sigma n_1 \tag{4.1}$$

is accurately approximated by a Gaussian random variable, and therefore that the matched filter is quasi optimal. In fact, it is not necessary to invoke the central limit theorem, but the open-eye condition (3.92) is sufficient (Problem 4.6) to conclude that zero-thresholding y_1 is optimal *if the receiver is only allowed to observe* y_1. The wrong conclusion of the near optimality of the single-user matched filter originates from the implicit assumption that the observable used to demodulate user 1 must be restricted to its matched filter output. Although $(y_1, \ldots, y_K)$ is a sufficient statistic for the data $(b_1, \ldots, b_K)$, it is not true that y_k is a sufficient statistic for b_k.

4.1.1 TWO-USER SYNCHRONOUS CHANNEL

The two-user synchronous channel (2.5) is

$$y(t) = A_1 b_1 s_1(t) + A_2 b_2 s_2(t) + \sigma n(t), \quad t \in [0, T].$$

According to Section 3.1, the minimum probability of error decision for user 1 is obtained by selecting the value of $b_1 \in \{-1, +1\}$ that maximizes the a posteriori probability

$$P[b_1 | \{y(t), 0 \le t \le T\}], \tag{4.2}$$

and analogously for user 2. We could pose a different optimum detection problem by requiring that the receiver selects the pair (b_1, b_2) that maximizes the joint a posteriori probability

$$P[(b_1, b_2) | \{y(t), 0 \le t \le T\}]. \tag{4.3}$$

We can write (4.2) in terms of (4.3):

$$P[b_1 | \{y(t), 0 \le t \le T\}] = P[(b_1, +1) | \{y(t), 0 \le t \le T\}]$$
$$+ P[(b_1, -1) | \{y(t), 0 \le t \le T\}].$$

However, those optimum detection strategies, which we will refer to as *individually optimum* and *jointly optimum*, respectively, need not result in the same decisions. To see why the two criteria are indeed different, consider an example where the noise realization is such that the a posteriori probabilities take the following values:

$$P[(+1, +1) | \{y(t), 0 \le t \le T\}] = 0.26,$$

$$P[(-1, +1) | \{y(t), 0 \le t \le T\}] = 0.26,$$

$$P[(+1, -1) | \{y(t), 0 \le t \le T\}] = 0.27,$$

$$P[(-1, -1) | \{y(t), 0 \le t \le T\}] = 0.21.$$

Then, the jointly optimum decisions are $(b_1, b_2) = (+1, -1)$, whereas the individually optimum decisions are $(b_1, b_2) = (+1, +1)$. However, this distinction is a relatively minor technical point because we can expect that unless the signal-to-noise ratio is such that the error probability is extremely poor, both types of decisions will agree with very high probability: typically there is one value of (b_1, b_2) for which the a posteriori probability is very close to 1. The underlying reason why the jointly optimum and individually optimum decisions need not coincide is that b_1 and b_2 are not independent when conditioned on the observed waveform.

Let us solve first the case of jointly optimum decisions. Since the four possible values of (b_1, b_2) are equiprobable, we can use Proposition 3.2 in the special case of $m = 4$ hypotheses:

$$x_1 = A_1 s_1 + A_2 s_2,$$

$$x_2 = A_1 s_1 - A_2 s_2,$$

$$x_3 = -A_1 s_1 + A_2 s_2,$$

$$x_4 = -A_1 s_1 - A_2 s_2.$$

Particularizing the likelihood functions (3.6), we see that the optimum decision rule in Proposition 3.2 selects the pair (b_1, b_2) that maximizes

$$f[\{y(t), 0 \le t \le T\}|(b_1, b_2)]$$

$$= \exp\left(-\frac{1}{2\sigma^2} \int_0^T [y(t) - A_1 b_1 s_1(t) - A_2 b_2 s_2(t)]^2 dt\right). \quad (4.4)$$

Since the data are equiprobable and independent, the jointly optimum decisions are the *maximum-likelihood* decisions $(\hat{b}_1, \hat{b}_2)$ chosen such that $A_1 \hat{b}_1 s_1(t) + A_2 \hat{b}_2 s_2(t)$ is closest to the received signal in the mean-square sense. In other words, we are choosing the hypothesis that corresponds to the noise realization with minimum energy. The minimum-distance decision regions are shown in Figure 4.1 in the same two-dimensional space that was used in Figure 3.12, namely, $(\tilde{y}_1, \tilde{y}_2)$ are the projections of the received waveform along orthonormal signals that form a basis for the space spanned by the signature waveforms s_1 and s_2. As far as deciding which of the four signals $A_1 b_1 s_1 + A_2 b_2 s_2$ is closest to the received signal, we may disregard the (infinite number of) dimensions we are not representing in Figure 4.1 because the received waveform contains no information about b_1 and b_2 along those dimensions. Figure 4.1 should be compared to the decision regions, depicted in Figure 3.12, of the single-user matched filter detector for the same set of signature waveforms. Because it always selects the closest hypothesis to the received observations (Figure 4.2), the jointly optimum detector does

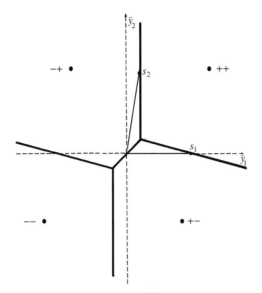

FIGURE 4.1.
Decision regions
of jointly optimum
detector for
$A_1 = A_2$,
$\rho = 0.2$.

not suffer from the anomalous behavior of the single-user matched filter (Figure 3.13) when there are imbalances in the received amplitudes.

Since the individually optimum multiuser detector achieves the minimum probability of error for each user, why would one ever consider the jointly optimum (maximum-likelihood) multiuser detector? As we will see, the complexity of the maximum-likelihood solution is lower, and, as we mentioned, unless the signal-to-noise ratio is very low, the probability of error achieved by the maximum-likelihood solution is extremely close to the

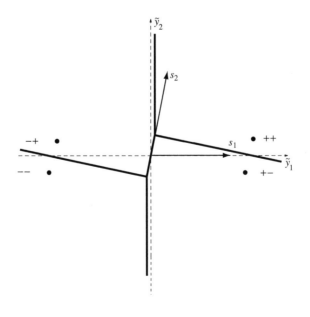

FIGURE 4.2.
Decision regions
of jointly optimum
detector for
$A_1 = 6A_2$,
$\rho = 0.2$.

157

minimum. If the transmitted bits are not equiprobable, then the maximum-likelihood decisions chosen to maximize (4.4) no longer select the jointly optimum decisions (that maximize (4.4)). In some situations (e.g., when certain types of error-correcting codes are used), it is required that the detector give more information than just a (*hard*) decision on each trans-mitted bit. *Soft* decisions inherently carry information that enable the as-sessment of their reliability. A rough way to do that is to further partition the decision regions, with the subdecision regions bordering other decision regions corresponding to areas of lower reliability. A finer way to provide soft decisions is to give the a posteriori probability of each bit (up to a scale factor); computing that information for the CDMA channel turns out to be (as we will see) no more complex than obtaining the minimum probability of error (individually optimum) decisions.

Let us see now how to implement the optimum decision rule that selects the closest hypothesis to the observations. The likelihood function in (4.4) can be expressed as

$$f[\{y(t), 0 \le t \le T\}|(b_1, b_2)]$$

$$= \exp\left(\frac{1}{\sigma^2}\Omega_2(b_1, b_2)\right) \exp\left(-\frac{A_1^2 + A_2^2}{2\sigma^2}\right) \exp\left(-\frac{1}{2\sigma^2}\int_0^T y^2(t)\, dt\right),$$

$$(4.5)$$

where

$$\Omega_2(b_1, b_2) = b_1 A_1 y_1 + b_2 A_2 y_2 - b_1 b_2 A_1 A_2 \rho, \qquad (4.6)$$

and as in (2.76) we denote the matched filter outputs by

$$y_k = \int_0^T y(t) s_k(t)\, dt.$$

To maximize the right-hand side of (4.5), we can disregard the factors that do not depend on (b_1, b_2). Thus, the maximum-likelihood (or jointly opti-mum) decisions are those that maximize the function Ω_2. It can be checked (Problem 4.3) that the optimum decisions are such that if $\min\{A_1|y_1|, A_2|y_2|\} \ge A_1 A_2 |\rho|$, then

$$\hat{b}_1 = \text{sgn}(y_1),$$

$$\hat{b}_2 = \text{sgn}(y_2).$$

Otherwise,

$$\hat{b}_1 = \text{sgn}(A_1 y_1 - \text{sgn}(\rho)A_2 y_2),$$

$$\hat{b}_2 = \text{sgn}(A_2 y_2 - \text{sgn}(\rho)A_1 y_1).$$

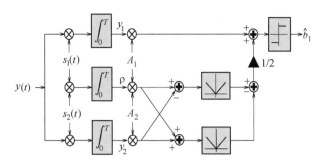

FIGURE 4.3.
Maximum-
likelihood
detection for user
1 with one
synchronous
interferer.

Referring to the case depicted in Figure 4.1 ($\rho > 0$), the condition min $\{A_1|y_1|, A_2|y_2|\} \geq A_1 A_2 \rho$ checks whether or not the observations are closest to either $(+1, +1)$ or $(-1, -1)$, in which case the decisions of the conventional detector are optimal. Note that the decision regions for $(+1, +1)$ and $(-1, -1)$ in Figure 4.1 are a subset of the corresponding regions in Figure 3.12, and vice versa for the regions for $(+1, -1)$ and $(-1, +1)$. If the observations do not fall in the regions of either $(+1, +1)$ or $(-1, -1)$, then it is just a matter of checking whether the observations are closest to $(+1, -1)$ or to $(-1, +1)$, and this is equivalent (because $\rho > 0$) to testing whether $A_1 y_1 \geq A_2 y_2$.

Another way to express the jointly optimal decisions is (cf. Figure 4.3 and Problem 4.3)

$$\hat{b}_1 = \text{sgn}\left(A_1 y_1 + \frac{1}{2}|A_2 y_2 - A_1 A_2 \rho| - \frac{1}{2}|A_2 y_2 + A_1 A_2 \rho| \right), \quad (4.7)$$

$$\hat{b}_2 = \text{sgn}\left(A_2 y_2 + \frac{1}{2}|A_1 y_1 - A_1 A_2 \rho| - \frac{1}{2}|A_1 y_1 + A_1 A_2 \rho| \right). \quad (4.8)$$

It is important to note that the received signal affects the optimum decisions only through the observables y_1 and y_2, which means that (y_1, y_2) is a sufficient statistic. Although this is the observable produced by the bank of single-user matched filters analyzed in Chapter 3, optimum decisions require a more elaborate processing than simply taking the signs of y_1 and y_2. As we can see from the foregoing solution, (4.7) and (4.8), the jointly optimum decisions depend on the values of A_1, A_2, and ρ but not σ. Note that even though knowledge of the background noise level is not required, the maximum-likelihood multiuser detector is not scale invariant (i.e., it is not enough to know the ratio A_2/A_1).

Let us now turn to the case of individually optimum decisions. Since b_1 is a priori equally likely to be $+1$ or -1, minimizing the probability of error $P[b_1 \neq \hat{b}_1]$ is a composite hypothesis-testing problem that fits the framework of Proposition 3.4 with

$$m = 4,$$

$$x_1 = A_1 s_1 + A_2 s_2,$$

$$x_2 = A_1 s_1 - A_2 s_2,$$

$$x_3 = -A_1 s_1 + A_2 s_2,$$

$$x_4 = -A_1 s_1 - A_2 s_2,$$

$$\mathcal{H}_J = \{H_1, H_2\}.$$

It follows that the individually optimal decision $\hat{b}_1$ should be the argument $b_1 \in \{-1, +1\}$ that maximizes

$$\exp\left(-\frac{1}{2\sigma^2} \int_0^T [y(t) - A_1 b_1 s_1(t) - A_2 s_2(t)]^2 dt\right)$$

$$+ \exp\left(-\frac{1}{2\sigma^2} \int_0^T [y(t) - A_1 b_1 s_1(t) + A_2 s_2(t)]^2 dt\right).$$

Straightforward algebra verifies that the individually optimum decision for the bit of user 1 is (Figure 4.4)

$$\hat{b}_1 = \text{sgn}\left(y_1 - \frac{\sigma^2}{2A_1} \log \frac{\cosh\left[\frac{A_2 y_2 + A_1 A_2 \rho}{\sigma^2}\right]}{\cosh\left[\frac{A_2 y_2 - A_1 A_2 \rho}{\sigma^2}\right]}\right). \qquad (4.9)$$

The analogous result for $\hat{b}_2$ is obtained simply by interchanging the roles of (y_1, A_1) and (y_2, A_2) in (4.9).

Note that in contrast to the jointly optimum detector, the optimum decisions depend now on the noise level σ^2. The nonlinearity in Figure 4.4 is the function

$$f_\sigma(x) = \sigma^2 \log(\cosh(x/\sigma^2)).$$

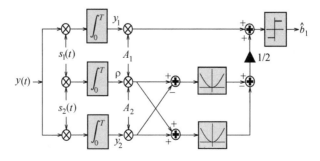

FIGURE 4.4.
Minimum bit-error-rate detector for user 1 with one synchronous interferer.

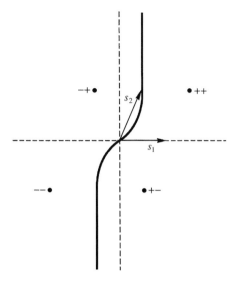

FIGURE 4.5.
Decision regions
of minimum
bit-error-rate
detector for user
1. $A_1 = A_2$,
$\rho = 0.2$.

Since

$$\lim_{\sigma \to 0} f_\sigma(x) = |x|,$$

the individually optimum decision regions converge to the jointly optimal decision regions as the signal-to-noise ratio increases (cf. (4.7) and (4.8)), in accordance with the intuition we put forth at the beginning of the section. The decision regions of the individually optimum detector (or minimum bit-error-rate detector) are shown in Figure 4.5. Comparing both diagrams, we see that the regions in Figure 4.5 are a smooth version of the minimum distance regions in Figure 4.1.

4.1.2 K-USER CHANNEL

Let us study now the K-user basic synchronous CDMA channel:

$$y(t) = \sum_{k=1}^{K} A_k b_k s_k(t) + \sigma n(t), \quad t \in [0, T]. \tag{4.10}$$

The problem of jointly optimum demodulation of

$$\mathbf{b} = [b_1, \ldots, b_K]^T$$

is a special case of Proposition 3.2 with $m = 2^K$ and every signal x_j given by $\sum_{k=1}^{K} A_k b_k s_k$ for a different value of $\mathbf{b}$. Therefore, the solution of Proposition 3.2 dictates that the most likely $\mathbf{b}$ maximizes

$$\exp\left(-\frac{1}{2\sigma^2} \int_0^T [y(t) - \sum_{k=1}^{K} b_k A_k s_k(t)]^2 dt\right),$$

or, equivalently, maximizes

$$\Omega(\mathbf{b}) = 2 \int_0^T \left[\sum_{k=1}^K A_k b_k s_k(t) \right] y(t) \, dt - \int_0^T \left[\sum_{k=1}^K A_k b_k s_k(t) \right]^2 dt$$

$$= 2\mathbf{b}^T \mathbf{A} \mathbf{y} - \mathbf{b}^T \mathbf{H} \mathbf{b}, \tag{4.11}$$

where we have denoted the column vector of matched filter outputs (cf. (2.78)) by

$$\mathbf{y} = [y_1, \ldots, y_K]^T; \tag{4.12}$$

the $K \times K$ diagonal matrix of received amplitudes by

$$\mathbf{A} = \text{diag}\{A_1, \ldots, A_K\}; \tag{4.13}$$

and the unnormalized crosscorrelation matrix by

$$\mathbf{H} = \mathbf{A} \mathbf{R} \mathbf{A}, \tag{4.14}$$

where $\mathbf{R}$ is the normalized crosscorrelation matrix (2.4) whose diagonal elements are equal to 1 and whose (i, j) element is equal to the crosscorrelation ρ_{ij}. The expression (4.11) reveals that the dependence of the likelihood function on the received signals is through the vector of matched filter outputs $\mathbf{y}$, which is therefore a sufficient statistic for demodulating the transmitted data.

The maximization of (4.11) is a combinatorial optimization problem. This means that the set of possible arguments comprises a finite set. Unlike optimization problems over the real line or Euclidean spaces where analytical techniques based on differentiation can often be used, combinatorial optimization problems can always be solved by exhaustive search, namely, compute the function for every possible argument and select the one that maximizes the function. But there are many combinatorial optimization problems for which there exist algorithms whose complexity grows polynomially in the size of the problem, and, thus, are much more efficient than exhaustive search. The computational complexity of any detector can be quantified by its time complexity per bit, that is, the number of operations required by the detector to demodulate the transmitted information divided by the total number of demodulated bits. A time complexity per bit of $f(K)$ is written as $O(g(K))$ if there exists a constant $c > 0$ such that for large enough K, $f(K) \le cg(K)$.

For the maximization of (4.11) it is possible to improve upon exhaustive search because the quantities $\mathbf{b}^T \mathbf{H} \mathbf{b}$ can be precomputed, and the generation of all the values of $\Omega(\mathbf{b})$ can be done in a tree structure that takes $O(2^K)$ operations (Problem 4.8). The selection of the optimum $\mathbf{b}$ can then be done in $O(2^K)$ operations, and so the time complexity per bit is $O(2^K/K)$. No algorithm is known for optimum multiuser detection whose computational complexity is polynomial in the number of users, regardless of the crosscorrelation matrix. Moreover, it is known (Verdú [496]) that if such algorithm existed for arbitrary crosscorrelation matrices, then a polynomial algorithm would exist for longstanding combinatorial problems such as the *traveling salesman* and *integer linear programming* problems, for which no polynomial solution has ever been found. It should be emphasized that this computational complexity result is worst case in nature. In synchronous CDMA (not subject to unknown channel distortion), the signature waveforms can be designed to facilitate the task of a multiuser detector. Suppose that the dimensionality N of the signal space and the number of users K are fixed. If $K \leq N$, orthogonal signature waveforms render single-user detection optimal; if $K > N$, then signature waveforms can be designed in a tree structure where the orthogonality of subsets of signature waveforms can be exploited to find jointly optimum detectors whose complexity grows polynomially with K (Learned et al. [230]).

If all the crosscorrelations are nonpositive, then maximum-likelihood decisions can be obtained with time complexity per bit of $O(K^2)$. The reason is that the maximization of $2\mathbf{b}^T \mathbf{Ay} - \mathbf{b}^T \mathbf{Hb}$ is equivalent to a well-known problem in combinatorial optimization. Let us construct a directed graph with nodes $\{0, 1, \ldots, K, K+1\}$ and edges whose *capacities* are defined by (Figure 4.6)

$$c_{ij} = \begin{cases} -A_i A_j \rho_{ij}, & \text{if } (i, j) \in \{1, \ldots, K\}^2, \\ \gamma, & \text{if } j = K+1 \text{ and } i \in \{1, \ldots, K\}, \\ A_i y_i + \gamma, & \text{if } i = 0 \text{ and } j \in \{1, \ldots, K\}, \\ 0, & \text{if } i = 0 \text{ and } j = K+1, \end{cases} \tag{4.15}$$

where γ is, for now, an arbitrary scalar. A *cut* that separates nodes 0 and $K+1$ is a partition of the nodes of the graph:

$$\{0, 1, \ldots, K, K+1\} = \{I \cup \{0\}\} \cup \{J \cup \{K+1\}\}$$

and the capacity of the cut is defined as

$$C(I, J) \stackrel{\text{def}}{=} \sum_{i \in I \cup \{0\}} \sum_{j \in J \cup \{K+1\}} c_{ij}.$$

Cuts and data vectors $\mathbf{b} \in \{-1, 1\}^K$ can be put in one-to-one correspondence, by letting $i \in I \subset \{1, \ldots, K\}$ if and only if $b_i = 1$. With this correspondence

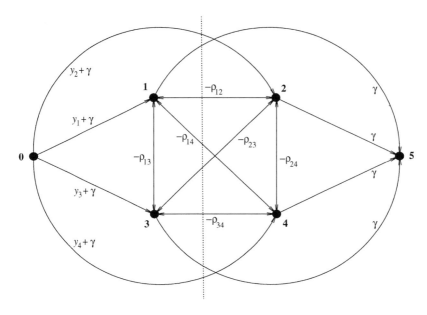

FIGURE 4.6.
Directed graph for maximum-likelihood detection in the special case of four users with unit amplitudes.

and (4.15) we obtain

$$4C(I, J) + \Omega(\mathbf{b}) = -4\sum_{i \in I}\sum_{j \in J} A_i A_j \rho_{ij} + 4\sum_{j \in J}(A_j y_j + \gamma)$$

$$+ 4\sum_{i \in I}\gamma + 2\sum_{i \in I} A_i y_i - 2\sum_{j \in J} A_j y_j$$

$$- \sum_{i \in I}\sum_{j \in I} A_i A_j \rho_{ij} - \sum_{i \in J}\sum_{j \in J} A_i A_j \rho_{ij}$$

$$+ 2\sum_{i \in I}\sum_{j \in J} A_i A_j \rho_{ij}$$

$$= 4K\gamma + 2\sum_{i=1}^{K} A_i y_i - \sum_{i=1}^{K}\sum_{j=1}^{K} A_i A_j \rho_{ij}, \quad (4.16)$$

where the right-hand side does not depend on the data-vector/cut. Therefore, maximizing the likelihood function $\Omega(\mathbf{b})$ is equivalent to finding a cut with minimum capacity: if $\{I^* \cup \{0\}\} \cup \{J^* \cup \{K+1\}\}$ is the cut with minimum capacity, then the decisions

$$\hat{b}_k = \begin{cases} +1, & \text{if } k \in I^*, \\ -1, & \text{if } k \in J^*, \end{cases}$$

are maximum-likelihood decisions.

The min-capacity-cut problem is very well known in combinatorial optimization because it is equivalent to the maximization of the flow from the origin node 0 to the destination node $K+1$. Although polynomial algorithms are not known for graphs with general capacities, if all the edge capacities are nonnegative, then algorithms with $O(K^3)$ computational complexity are known (Papadimitrou and Steiglitz [314]) All we need to ensure the nonnegativity of all the edge capacities is that the crosscorrelations satisfy $\rho_{ij} \leq 0$ because we can always select

$$\gamma = \max\left\{0, -\min_j A_j y_j\right\}.$$

According to Proposition 3.4, the individually optimum decision that minimizes the probability of error of the kth user is the element $\hat{b}_k \in \{-1, 1\}$ that maximizes

$$L_k(b) = \sum_{\substack{\mathbf{b} \\ b_k = b}} \exp(\Omega(\mathbf{b})/2\sigma^2). \tag{4.17}$$

Note that the a posteriori probabilities can be readily computed from (4.17) and via (3.18):

$$P[b_k = 1 | \{y(t), t \in [0, T]\}] = \frac{1}{1 + \frac{L_k(-1)}{L_k(1)}}. \tag{4.18}$$

Minimum probability of error detection leads to smooth decision regions in the K-dimensional decision space of the observables in (4.12), whereas the decision regions corresponding to the jointly optimum criterion are polytopes (cf. Figures 4.1 and 4.5). As we mentioned in the two-user case, in the majority of applications, jointly optimum decisions would be preferable to minimum bit-error-rate decisions due to their lower complexity and the fact that as $\sigma \to 0$ the minimum bit-error-rate decisions converge to the jointly optimum decisions that maximize (4.11). Using the linear term in the Taylor series expansion of the exponential function, we see that as $\sigma \to \infty$ maximizing (4.17) is equivalent to maximizing

$$L_k^*(b) = \sum_{\substack{\mathbf{b} \\ b_k = b}} \Omega(\mathbf{b})$$

$$= 2^K b A_k y_k - \frac{1}{2} \sum_{\mathbf{d} \in \{-1, +1\}^K} \mathbf{d}^T \mathbf{H} \mathbf{d}. \tag{4.19}$$

Since the last term in the right-hand side of (4.19) does not depend on b, the minimum bit-error-rate decisions converge as $\sigma \to \infty$ to those of the single-user matched filter detector: $\text{sgn}(y_k)$. This is to be expected because as $\sigma \to \infty$ the multiaccess interference becomes negligible with respect to the background white Gaussian noise, against which the matched filter is optimum.

4.2   OPTIMUM DETECTOR FOR ASYNCHRONOUS CHANNELS

The first crucial observation in the derivation of the optimum detector for the asynchronous channel is that optimum decisions require the observation of the whole frame of transmitted bits. Consider the timing diagram, depicted in Figure 4.7, of a two-user asynchronous channel with five bits in each frame (i.e., $M = 2$ in (2.6)). Let us focus on the demodulation of $b_1[0]$. Recall that the conventional detector takes a one-shot approach where only the received signal in the interval of $b_1[0]$ affects the decision. Let us see why this approach is suboptimal. The signal of $b_1[0]$ overlaps those of bits $b_2[-1]$ and $b_2[0]$. So it would be to the detector's advantage to learn as much as possible about those two interfering bits. This can be accomplished by extending the observation interval from $[0, T]$ to $[\tau - T, \tau + T]$. Once we do that, the new observation interval contains signals modulated by bits $b_1[-1]$ and $b_1[1]$; so, again, it would be to the detector's advantage to learn as much as possible about those two bits by further extending the observation interval to $[-T, 2T]$. It is now clear that this reasoning can be repeated as many times as necessary to conclude that unless the whole frame of data is observed, $[-2T, \tau + 3T]$, the decision on bit $b_1[0]$ (or any other bit) will be suboptimal.

As in the synchronous case, we have to deal with the issue of whether we require individually optimum or jointly optimum decisions. The former criterion results in the minimum bit-error-rate. Jointly optimum decisions are obtained by a *maximum-likelihood sequence detector* that selects the most likely sequence of transmitted bits given the observations. In fact, there are other conceivable intermediate optimality criteria. For example, we could think of a receiver interested in demodulating only one user, which selects the most likely sequence of bits for that user, rather than making individually optimal decisions.

As we discussed in Chapter 2, we could view the K-user, M-frame, asynchronous channel as a $K(2M + 1)$-user synchronous channel. Let us define a $K(2M + 1)$-vector **b** with components

$$b_{k+iK} = b_k[i], \quad k = 1, \ldots, K, \quad i = -M \ldots, M, \qquad (4.20)$$

FIGURE 4.7.
Suboptimality of one-shot approach in asynchronous channels.

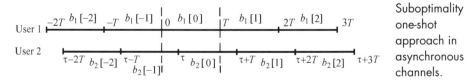

along with the signals

$$v_{k+iK}(t) = s_k(t - iT - \tau_k), \tag{4.21}$$

and the $K(2M + 1)$-vector of matched-filter outputs, $\mathbf{y}$, with components

$$y_j = \int_{-\infty}^{\infty} v_j(t)y(t)\,dt. \tag{4.22}$$

This means that y_{k+iK} is the output of the kth matched filter applied to the signal in the interval $[\tau_k + iT, \tau_k + iT + T]$, that is, the interval correspon-ding to $b_k[i]$.

As in the synchronous channel, the objective is to compute the $\mathbf{b}$ that maximizes

$$f[\{y(t), t \in [-MT, MT + 2T]\}|\mathbf{b}]$$

$$= \exp\left(-\frac{1}{2\sigma^2} \int_{-MT}^{MT+2T} (y(t) - S_t(\mathbf{b}))^2 dt\right), \tag{4.23}$$

where

$$S_t(\mathbf{b}) = \sum_{k=1}^{K}\sum_{i=-M}^{M} A_k b_k[i] s_k(t - iT - \tau_k). \tag{4.24}$$

Let $\mathbf{A}_M$ be the $K(2M + 1) \times K(2M + 1)$ diagonal matrix whose $k + iK$ diagonal element is equal to A_k. Let $\mathbf{R}$ be the $K(2M + 1) \times K(2M + 1)$ matrix with coefficients

$$r_{jl} = \int_{-\infty}^{\infty} v_j(t)v_l(t)\,dt. \tag{4.25}$$

The matrix $\mathbf{R}$ can be written as a function of the $K \times K$ crosscorrelation matrices in (2.106):

$$\mathbf{R} = \begin{bmatrix} \mathbf{R}[0] & \mathbf{R}^T[1] & \mathbf{0} & \cdots & \mathbf{0} & \mathbf{0} \\ \mathbf{R}[1] & \mathbf{R}[0] & \mathbf{R}^T[1] & \cdots & \mathbf{0} & \mathbf{0} \\ \mathbf{0} & \mathbf{R}[1] & \mathbf{R}[0] & \cdots & \mathbf{0} & \mathbf{0} \\ \cdots & \cdots & \cdots & \cdots & \cdots & \cdots \\ \mathbf{0} & \mathbf{0} & \mathbf{0} & \cdots & \mathbf{R}[1] & \mathbf{R}[0] \end{bmatrix}.$$

Furthermore, define

$$\mathbf{H} = \mathbf{A}_M \mathbf{R} \mathbf{A}_M. \tag{4.26}$$

Then the maximization of (4.23) is equivalent to selecting the $\mathbf{b}$ that maxi-mizes

$$\Omega(\mathbf{b}) = 2\int S_t(\mathbf{b})y(t)\,dt - \int S_t^2(\mathbf{b})\,dt$$

$$= 2\mathbf{b}^T \mathbf{A}_M \mathbf{y} - \mathbf{b}^T \mathbf{H} \mathbf{b}. \tag{4.27}$$

Once more, notice that the observations enter in the function to be maximized by the jointly optimum decisions (4.27) only through the matched-filter outputs. Therefore, **y** *is a sufficient statistic for* **b**.

If we were to maximize $\Omega(\mathbf{b})$ in the same fashion as in the synchronous case, we would find the complexity to be exponential in the product $K(2M+1)$, which is totally out of the question in practice due to the typically large values of M. At this point, it is necessary to exploit the structure of the matrix **H** in the asynchronous case in order to get a sensible solution.

To simplify ideas, let us first focus attention on the two-user case with $A_1 = A_2 = 1$. In this case,

$$
\mathbf{H} = \mathbf{R} = \begin{bmatrix}
1 & \rho_{12} & 0 & & \cdots & & & 0 \\
\rho_{12} & 1 & \rho_{21} & 0 & \cdots & & & 0 \\
0 & \rho_{21} & 1 & \rho_{12} & \ddots & & & \\
0 & 0 & \rho_{12} & 1 & \ddots & 0 & & \\
& & & & \ddots & \rho_{21} & 0 & 0 \\
& & & \ddots & & 1 & \rho_{12} & 0 \\
& & & & 0 & \rho_{12} & 1 & \rho_{21} \\
0 & & \cdots & & 0 & 0 & \rho_{21} & 1
\end{bmatrix}.
\tag{4.28}
$$

The fact that the symmetric matrix **H** is zero except along the main diagonal and the first subdiagonal is crucial for decreasing the complexity of the receiver. It means that we can write the payoff function as

$$
\Omega(\mathbf{b}) = \sum_{j=1-2M}^{2M+2} \lambda_j(b_{j-1}, b_j),
\tag{4.29}
$$

where $b_{-2M} = 0$ and

$$
\lambda_j(b_{j-1}, b_j) = \begin{cases} 2b_j y_j - 2\rho_{12} b_j b_{j+1} - 1, & \text{if } j \text{ is even;} \\ 2b_j y_j - 2\rho_{21} b_j b_{j+1} - 1, & \text{if } j \text{ is odd.} \end{cases}
$$

We see in (4.29) that the payoff function depends on each of its arguments sequentially, and only consecutive arguments are coupled. Consequently, to maximize (4.29), we can use the *dynamic programming* algorithm.

To explain the principle of operation of dynamic programming, we represent each possible sequence **b** as a path from the origin to the destination in the diagram in Figure 4.8. In this diagram we have labeled each arc with a "length." The payoff achieved by a given **b** is equal to the total length of its corresponding path. The dynamic programming algorithm is an efficient algorithm to find the longest (or the shortest) path from the origin to the destination in a layered graph. In a layered graph, nodes can

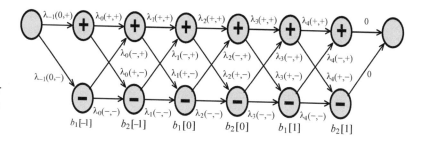

FIGURE 4.8.
Trellis diagram for two asynchronous users; $M = 1$.

be partitioned in layers or stages such that each node is connected only to nodes in adjacent layers (cf. Figure 4.8). This type of layered directed graph is commonly known in communication theory as a *trellis diagram* (Forney [97]). Dynamic programming is based on the following simple observation: if the longest path from the origin to the destination goes through a certain node N, then the subpath (within the longest path) that connects the origin with N must be the longest among those paths connecting the origin to N. So suppose we knew the longest path from the origin to each of the nodes at the ith stage. Denote the corresponding longest distances by $J_i(+)$ and $J_i(-)$. Then it is evident that the longest distances at the $(i + 1)$th stage satisfy:

$$J_{i+1}(+) = \max\{\lambda_i(+, +) + J_i(+), \lambda_i(-, +) + J_i(-)\}, \quad (4.30a)$$

$$J_{i+1}(-) = \max\{\lambda_i(+, -) + J_i(+), \lambda_i(-, -) + J_i(-)\}, \quad (4.30b)$$

where we have assumed that J is initialized to 0 at the first stage of the algorithm. Associated with the function J_i at each node is a longest path from the origin to that node whose length equals the value of the function J_i. Once the destination node is reached, we can read the longest path, which is the most likely sequence of data **b** given the observations.

As we anticipated at the beginning of the section, to make an optimal decision about the first received bit $b_1[-M]$ we must wait until we have observed the received waveform due to the last transmitted bit. In most practical applications, it is necessary to demodulate the transmitted symbols in real time, and since M is usually a very large integer, it is not feasible to wait until all the observables $\{y_i\}$ have been obtained before starting to make decisions. Therefore, a suboptimum version of the forward dynamic programming algorithm is adopted in practice whereby the decision on b_i is based on the path corresponding to the function J_{i+L} computed a fixed number L of steps ahead. The value of b_i is given by the node traversed by the longest path that leads to the maximum of $J_{i+L}(+)$ and $J_{i+L}(-)$. If L is sufficiently large (depending on signal-to-noise ratios, etc.), then with very high probability the paths associated with $J_{i+L}(+)$ and $J_{i+L}(-)$ traverse the same node at the ith stage. In that event, the maximum-likelihood decision on b_i can be made without awaiting the end of the transmission. Otherwise, the decisions are not necessarily optimal. This real-time version of forward dynamic programming (asymptotically optimal as $L \to \infty$) is known as the *Viterbi algorithm*.

Although there is no apparent reason to do so, we could have given a dual dynamic programming algorithm to optimize (4.29), where the recursion proceeds backwards (Problem 4.12).

In the general K-user case, we need to find a similar additive decomposition of the payoff function

$$\Omega(\mathbf{b}) = 2\mathbf{b}^T \mathbf{A}_M \mathbf{y} - \mathbf{b}^T \mathbf{H} \mathbf{b}. \tag{4.31}$$

Let us introduce the notation $\kappa(j) \in \{1, \ldots, K\}$ to represent the modulo-K remainder of j (i.e., for some i, $j = \kappa(j) + iK$). Then, we can write the first term in the right-hand side of (4.31) as

$$\mathbf{b}^T \mathbf{A}_M \mathbf{y} = \sum_{j=1-MK}^{MK+K} A_{\kappa(j)} b_j y_j. \tag{4.32}$$

In order to decompose the quadratic form $\mathbf{b}^T \mathbf{H} \mathbf{b}$ in (4.31), we will use the fact that the matrix $\mathbf{H}$ has bandwidth K and its elements are periodic along the diagonals (cf. (4.28)). Using (a) the definition in (4.25), (b) $\tau_1 \leq \cdots \leq \tau_K$, and (c) the fact that the signature waveforms are time-limited to the interval $[0, T]$, we can summarize the properties of $\mathbf{H}$ as follows:

(1) $h_{j,j} = A_{\kappa(j)}^2$.
(2) $h_{k+iK, n+iK} = h_{k,n}$.
(3) $h_{j,l} = 0$ unless $|j - l| < K$.
(4) $h_{i,j} = h_{j,i}$.
(5) $h_{j-n,j} = A_{\kappa(j-n)} A_{\kappa(j)} \rho_{\kappa(j-n), \kappa(j)}$, $\quad n = 1, \ldots, K - 1$.

Using these properties and letting $b_j = 0$ if $j < 1 - MK$ or $j > MK + K$ we can write

$$\mathbf{b}^T \mathbf{H} \mathbf{b} = \sum_{j=1-MK}^{MK+K} \sum_{l=1-MK}^{MK+K} b_j b_l h_{j,l}$$

$$= \sum_{j=1-MK}^{MK+K} b_j \left[A_{\kappa(j)}^2 b_j + 2 \sum_{l=j-K+1}^{j-1} b_l h_{l,j} \right]$$

$$= \sum_{j=1-MK}^{MK+K} b_j \left[A_{\kappa(j)}^2 b_j + 2 \sum_{n=1}^{K-1} b_{j-n} h_{j-n,j} \right]$$

$$= \sum_{j=1-MK}^{MK+K} A_{\kappa(j)} b_j \left[A_{\kappa(j)} b_j + 2 \sum_{n=1}^{K-1} b_{j-n} A_{\kappa(j-n)} \rho_{\kappa(j-n),\kappa(j)} \right].$$

(4.33)

We can now express $\Omega(\mathbf{b})$ as a sum of $(2M+1)K$ terms such that (a) each term depends on K components of $\mathbf{b}$ and (b) any consecutive terms share $K-1$ arguments. Specifically, we can write

$$\Omega(\mathbf{b}) = \sum_{j=1-MK}^{MK+K} \lambda_j(\mathbf{x}_j, b_j),$$

(4.34)

where

$$\lambda_j(\mathbf{x}, u) = A_{\kappa(j)} u \left[2y_j - u A_{\kappa(j)} - 2 \sum_{n=1}^{K-1} x(n) A_{\kappa(j-n)} \rho_{\kappa(j-n),\kappa(j)} \right]$$

(4.35)

and $\mathbf{x}_j$ is the state of a shift-register with $K-1$ registers:

$$\mathbf{x}_{j+1}^T = [x_{j+1}(1), \ldots, x_{j+1}(K-1)]$$

$$= [x_j(2), \ldots, x_j(K-1), b_j],$$

$$= f[\mathbf{x}_j, b_j],$$

(4.36)

which is started at $\mathbf{x}_{1-MK} = [0, \ldots, 0]$. Note that since $u \in \{-1, +1\}$ in (4.35), we can simplify the payoff function by dropping those terms that are independent of the matched-filter outputs or of $\mathbf{b}$, yielding the simplified metric:

$$\tilde{\lambda}_j(\mathbf{x}, u) = A_{\kappa(j)} u \left[y_j - \sum_{n=1}^{K-1} x(n) A_{\kappa(j-n)} \rho_{\kappa(j-n),\kappa(j)} \right].$$

(4.37)

As in the two-user case, we can find the sequence of bits that maximizes (4.34) by finding the longest path in a layered directed graph, which has as many layers (in addition to the origin and destination nodes) as the total number of bits (i.e., $K(2M+1)$). The number of nodes (or states) at layer j is the number of different values that the state $\mathbf{x}_j$ can take. Therefore, except for the first $K-1$ layers (which are affected by the zero-initial condition), the number of states at each layer is 2^{K-1} (Figure 4.9).

Note that each state is connected with two states in the previous layer, regardless of the number of users. (Generalizing this algorithm to nonbinary

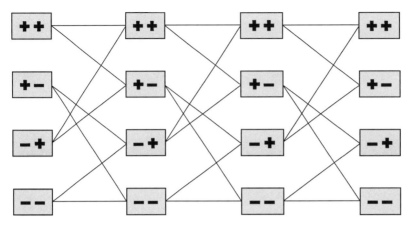

FIGURE 4.9.
Trellis diagram for three-user asynchronous channel.

modulation we would find that the number of connections with the previous layer is equal to the cardinality of the modulation alphabet.) Analogously to (4.30), the longest distance from origin to state $\mathbf{x}$ at stage $i + 1$ is given by

$$J_{j+1}(\mathbf{x}_{j+1}) = \max\{\tilde{\lambda}_j((-, x_{j+1}(1), \ldots, x_{j+1}(K-2)), x_{j+1}(K-1))$$

$$+ J_j(-, x_{j+1}(1), \ldots, x_{j+1}(K-2)),$$

$$\tilde{\lambda}_j((+, x_{j+1}(1), \ldots, x_{j+1}(K-2)), x_{j+1}(K-1))$$

$$+ J_j(+, x_{j+1}(1), \ldots, x_{j+1}(K-2))\}. \tag{4.38}$$

At each state, the algorithm performs the following computations:

- It computes the metrics $\tilde{\lambda}_j(\mathbf{x}, +1)$ and $\tilde{\lambda}_j(\mathbf{x}, -1)$.
- It calculates two sums.
- It compares two numbers.

The computation of $\tilde{\lambda}_j(\mathbf{x}, u)$ requires the computation of the term

$$\sum_{n=1}^{K-1} x(n) A_{\kappa(j-n)} \rho_{\kappa(j-n),\kappa(j)},$$

which can be computed once and stored for each state $\mathbf{x}$ because it is independent of the matched-filter outputs. Thus, the computational effort per state is independent of both K and M. Since there are as many iterations as transmitted bits and each iteration involves 2^{K-1} states, the time complexity per bit is $O(2^K)$. We have succeeded in obtaining an algorithm whose complexity is independent of M and is, in fact, similar to the complexity of the optimum detector for the synchronous channel. Figure 4.10 shows the general structure of the optimum receiver in the asynchronous problem. The

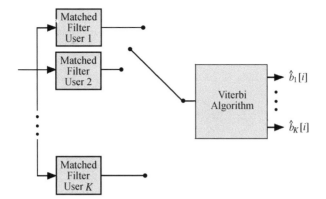

FIGURE 4.10.
Optimum
multiuser detector
for asynchronous
CDMA.

matched filters are sampled in round-robin fashion, at the corresponding bit-epochs dictated by the synchronization systems.

Alternatively, if the objective is to minimize the probability of error for each user (individually optimum decisions), then there is an optimum detection algorithm based on *backward–forward* dynamic programming, which is somewhat more computationally intensive than the foregoing algorithm but retains the desirable property of having a time complexity per bit that does not grow with the length of the frame. The objective is to evaluate $L_j(-1)$ and $L_j(1)$ for every $j = 1 - MK, \ldots, MK + K$ where (cf. (4.17))

$$L_j(b) = \sum_{\substack{\mathbf{b} \\ b_j = b}} \exp(\Omega(\mathbf{b})/2\sigma^2). \tag{4.39}$$

In contrast to maximum-likelihood detection, the problem is not one of maximization but one of summation. To do that efficiently, we will use the additive decomposition of $\Omega(\mathbf{b})$ in (4.34). In that way, the overall "length" of a given sequence $\mathbf{b}$ is now defined as the product of the individual lengths of the arcs in the trellis diagram:[1]

$$\exp(\Omega(\mathbf{b})/2\sigma^2) = \prod_{j=1-MK}^{MK+K} \exp(\lambda_j(\mathbf{x}_j, b_j)/2\sigma^2). \tag{4.40}$$

If the summation in (4.39) were over all sequences $\mathbf{b}$,

$$L = \sum_{\mathbf{b}} \exp(\Omega(\mathbf{b})/2\sigma^2), \tag{4.41}$$

---

[1] As in the maximum-likelihood detector we can replace λ_j by the simplified metrics $\tilde{\lambda}_j$.

we could use forward dynamic programming to solve for its value. Indeed, all we would need to do is to redefine (4.38) substituting *sum* instead of *maximum* and *product* instead of *sum*:

$$F_{j+1}(\mathbf{x}_{j+1})$$

$$= \exp(\lambda_j((-, x_{j+1}(1), \ldots, x_{j+1}(K-2)), x_{j+1}(K-1))/2\sigma^2)$$

$$\times F_j(-, x_{j+1}(1), \ldots, x_{j+1}(K-2))$$

$$+ \exp(\lambda_j((+, x_{j+1}(1), \ldots, x_{j+1}(K-2)), x_{j+1}(K-1))/2\sigma^2)$$

$$\times F_j(+, x_{j+1}(1), \ldots, x_{j+1}(K-2)), \tag{4.42}$$

with the function F initialized to 1 at the beginning of the first stage. The value $F_j(\mathbf{x}_j)$ is the sum of the lengths of the subpaths leading from the origin node to state $\mathbf{x}_j$ at the jth stage. As we had hinted, alternatively, we could run a backward iteration:

$$B_j(\mathbf{x}_j) = \exp(\lambda_j(\mathbf{x}_j, -)/2\sigma^2)$$

$$\times B_{j+1}(x_j(2), \ldots, x_j(K-1), -)$$

$$+ \exp(\lambda_j(\mathbf{x}_j, +)/2\sigma^2)$$

$$\times B_{j+1}(x_j(2), \ldots, x_j(K-1), +). \tag{4.43}$$

The value $B_j(\mathbf{x}_j)$ is the sum of the lengths of the subpaths that start at state $\mathbf{x}_j$ at the jth stage and end at the destination node. One possible reason we may want to compute (4.41) is that for all j

$$L = L_j(1) + L_j(-1).$$

So once the value of L is known we can simply focus on computing $L_j(1)$ for all j. Let us fix a given value of j. The computation of $L_j(1)$ resembles that of L except that now we are summing over half of the paths in the trellis, as we are discarding any $\mathbf{b}$ for which $b_j = -1$. Let us delete the 2^{K-1} arcs connecting the jth stage with the $(j+1)$th stage for which $b_j = -1$, and consider the remaining 2^{K-1} arcs one at a time. Each of those arcs corresponds to a different choice of the state at the jth stage $\mathbf{x}_j$, leading to state

$$\mathbf{x}_{j+1} = (x_j(2), \ldots, x_j(K-1), +).$$

The sum of all the paths that go through arc $(\mathbf{x}, b)$ is denoted by

$$L_j(\mathbf{x}, b) = \sum_{\mathbf{b}:b_j = b, \mathbf{x}_j = \mathbf{x}} \exp(\Omega(\mathbf{b})/2\sigma^2). \tag{4.44}$$

Obviously,

$$L_j(1) = \sum_{\mathbf{x}\in\{-1,1\}^{K-1}} L_j(\mathbf{x}, 1),$$

so if the values taken by the function in (4.44) are available, we can then compute $L_j(1)$ with 2^{K-1} operations. Because the sum is distributive over products, the sum of all the lengths of the paths that go through the arc that connects state $\mathbf{x}_j$ with state $\mathbf{x}_{j+1} = f(\mathbf{x}_j, b)$ can be expressed as

$$L_j(\mathbf{x}_j, b) = F_j(\mathbf{x}_j) \, \exp(\lambda_j(\mathbf{x}_j, b_j)/2\sigma^2) \, B_{j+1}(\mathbf{x}_{j+1}). \qquad (4.45)$$

The conclusion is that minimum bit-error-rate multiuser detection can be implemented by a *backward–forward* dynamic programming algorithm that carries out two independent iterations, each of which requires a computational effort proportional to the number of stages and to the number of states per stage 2^{K-1}. Once those backward and forward iterations are completed, the generation of the a posteriori probabilities of each bit (cf. (4.18)) requires a further computational effort proportional to 2^{K-1}. Therefore, the time complexity per bit is $O(2^K)$. In real-time implementation, the same kind of finite-horizon approximation we discussed above can be applied for the forward iteration. Analogously, we cannot afford to run the backward iteration from the end of the frame, and a sliding horizon can be chosen with negligible performance loss.

Comparing the optimum detectors to the conventional single-user matched-filter detector of Section 3, we see that aside from the increased computational complexity, the maximum-likelihood multiuser detector requires knowledge of the received amplitudes and crosscorrelations. The latter parameters can be generated by the receiver by cross-correlating the normalized replicas of the signature waveforms stored at the receiver with the delays and phases supplied by the synchronization system. The minimum bit-error-rate multiuser detector requires, in addition, knowledge of the noise spectral level. An advantage of the single-user matched-filter detector arises in those situations where only a subset of active users needs to be demodulated by a particular receiver. Then, unlike the optimum detector, the bank of single-user matched filters can simply ignore those users in which it is not interested. Judicious practical implementations of the optimum detector would favor neglecting those unwanted users that are comparatively weak with respect to the user(s) of interest.

4.3 MINIMUM ERROR PROBABILITY IN THE SYNCHRONOUS CHANNEL

The fundamental quantity of interest is the minimum error probability $P_k(\sigma)$, achievable by a receiver that observes (4.10) or, equivalently, the sufficient statistic (4.12). $P_k(\sigma)$ will serve as a baseline of comparison for the bit-error-rate of any multiuser detector.

In the single-user case, recall that the minimum bit-error-rate is given by (3.30)

$$P_1(\sigma) = Q\left(\frac{A_1}{\sigma}\right).$$

In the K-user channel, the same conclusion is obtained from (4.1) if $\rho_{1k} = 0$ for $k \neq 1$.

As long as different transmitted information vectors $(b_1, \ldots, b_K)$ lead to different received signals (a sufficient but not necessary condition is that the signature waveforms be linearly independent), we have

$$P_k(0) = 0. \tag{4.46}$$

Recall that the conventional single-user matched-filter receiver does not have this property and it may make errors even in the hypothetical scenario of complete absence of background noise.

4.3.1 TWO-USER CHANNEL

We study the two-user case first, where, as usual, we can find many of the key ingredients of the general solution. Referring to the decision regions of the minimum bit-error-rate detector in Figure 4.5, we see that in order to compute the minimum error probability for, say, user 1, we can condition on each of the four possible transmitted bit pairs and compute the probability that the received vector will cross the smooth boundary in Figure 4.5. This involves integrating a two-dimensional Gaussian distribution on a subset of the plane that has no particular structure, and therefore, we cannot expect that a closed-form solution will exist for the minimum error probability (even in terms of the Q-function as in the single-user matched-filter receiver).

In view of the inability to find an expression for the minimum error probability even in the simplest possible case, we will attempt to upper and lower bound it with computable quantities. Indeed, we will show tight bounds that result in explicit expressions for the asymptotic multiuser efficiency.

Let us first consider the problem of upper bounding $P_1(\sigma)$. We have already obtained one such upper bound, the error probability of the conventional detector (3.77):

$$P_1(\sigma) \le \frac{1}{2}Q\left(\frac{A_1 - A_2\rho}{\sigma}\right) + \frac{1}{2}Q\left(\frac{A_1 + A_2\rho}{\sigma}\right). \qquad (4.47)$$

This bound becomes tight when either of the following asymptotics are satisfied:

- $\sigma \to \infty$,
- $A_2\rho/A_1 \to 0$.

Otherwise (4.47) is a rather loose bound and it is important to derive better bounds.

Let us analyze the jointly optimum detector. Although it does not result in minimum bit-error-rate decisions, the jointly optimum detector is an excellent candidate to upper bound the minimum error probability because its decisions differ from those of the minimum bit-error-rate detector with negligible probability (unless the signal-to-noise ratio is very low).

Referring to the minimum distance decision regions in Figure 4.1, we can express the probability that the minimum distance detector makes an error in b_1 as

$$P_1^m(\sigma) = \frac{1}{4}P[++ \to -+] + \frac{1}{4}P[++ \to --]$$

$$+ \frac{1}{4}P[-- \to +-] + \frac{1}{4}P[-- \to ++]$$

$$+ \frac{1}{4}P[-+ \to +-] + \frac{1}{4}P[-+ \to ++]$$

$$+ \frac{1}{4}P[+- \to -+] + \frac{1}{4}P[+- \to --], \qquad (4.48)$$

where we have employed the notation

$$P[b_1 b_2 \to \hat{b}_1 \hat{b}_2] \qquad (4.49)$$

to denote the probability that the observations fall in the minimum distance region of $(\hat{b}_1, \hat{b}_2)$ conditioned on (b_1, b_2) being transmitted. We can exploit the inherent symmetry of the problem to simplify (4.48) by noticing that

$$P[++ \to --] = P[-- \to ++],$$

$$P[++ \to -+] = P[-- \to +-],$$

$$P[+- \to -+] = P[-+ \to +-],$$

$$P[+- \to --] = P[-+ \to ++].$$

Therefore,

$$P_1^m(\sigma) = \frac{1}{2}P[-- \to +-] + \frac{1}{2}P[-- \to ++]$$

$$+ \frac{1}{2}P[-+ \to +-] + \frac{1}{2}P[-+ \to ++]. \qquad (4.50)$$

Each of the terms in (4.50) is equal to the integral of a two-dimensional Gaussian distribution on a polytope, which cannot be written either in closed form or in terms of the Q-function. So the question is whether we have gained anything by analyzing the minimum distance detector in lieu of the minimum probability of error detector. Even though there is no hope of evaluating (4.50) in closed form, the four quantities therein can be upper bounded quite easily. Recall from (4.6) that the minimum distance detector outputs the pair $(\hat{d}_1, \hat{d}_2) \in \{-1, +1\}^2$ that maximizes

$$\Omega_2(d_1, d_2) = d_1 A_1 y_1 + d_2 A_2 y_2 - d_1 d_2 A_1 A_2 \rho, \qquad (4.51)$$

where

$$y_1 = A_1 b_1 + A_2 b_2 \rho + n_1, \qquad (4.52)$$

$$y_2 = A_2 b_2 + A_1 b_1 \rho + n_2, \qquad (4.53)$$

(b_1, b_2) are the transmitted bits, and n_1 and n_2 are jointly Gaussian with zero mean, variance σ^2, and correlation $\sigma^2 \rho$.

Let us focus on the first term in (4.50), that is, the probability that the observed vector due to $(--)$ is closest to $(+-)$. Clearly this can only happen if the observations are closer to $(+-)$ than to $(--)$. Accordingly,

$$P[-- \to +-] \le P[\Omega_2(--) < \Omega_2(+-)|(--) \text{ transmitted}]. \qquad (4.54)$$

Note that equality does not hold in (4.54) because the observations could be closer to $(+-)$ than to $(--)$, but yet fall in the minimum distance region of either $(++)$ or $(-+)$ (Figure 4.1). Another way to view the inequality in (4.54) is by noticing in Figure 4.1 that the region

$$\Omega_2(--) < \Omega_2(+-)$$

is the half-plane to the right of the line that bisects the segment between $(--)$ and $(+-)$. This half-plane contains the whole minimum distance region of $(+-)$. Now it is easy to evaluate the right side of (4.54) exactly; this probability is equal to the probability of error of a binary problem, and therefore it involves the integration of a one-dimensional Gaussian distribution on a semi-infinite interval, and, thus, can be written in terms of the Q-function:

$$P[\Omega_2(--) < \Omega_2(+-)|(--) \text{ transmitted}]$$

$$= P[-A_1 y_1 - A_2 y_2 - A_1 A_2 \rho < A_1 y_1 - A_2 y_2 + A_1 A_2 \rho$$

$$| y_1 = -A_1 - A_2 \rho + n_1; \; y_2 = -A_2 - A_1 \rho + n_2]$$

$$= P[n_1 > A_1]$$

$$= Q\left(\frac{A_1}{\sigma}\right). \tag{4.55}$$

Proceeding analogously with the other three terms in (4.50) we obtain

$$P[-- \to ++] \le P[\Omega_2(--) < \Omega_2(++)|(--) \text{ transmitted}]$$

$$= P[-A_1 y_1 - A_2 y_2 - A_1 A_2 \rho < A_1 y_1$$

$$+ A_2 y_2 - A_1 A_2 \rho \mid (--) \text{ transmitted}]$$

$$= P[A_1 y_1 + A_2 y_2 > 0 \mid y_1 = -A_1 - A_2 \rho$$

$$+ n_1; \; y_2 = -A_2 - A_1 \rho + n_2]$$

$$= P[A_1 n_1 + A_2 n_2 > A_1^2 + A_2^2 + 2A_1 A_2 \rho]$$

$$= Q\left(\frac{\sqrt{A_1^2 + A_2^2 + 2A_1 A_2 \rho}}{\sigma}\right), \tag{4.56}$$

where (4.56) follows because the random variable $A_1 n_1 + A_2 n_2$ is a zero-mean Gaussian with variance $\sigma^2(A_1^2 + A_2^2 + 2A_1 A_2 \rho)$ (guaranteed to be nonnegative because $|\rho| \le 1$). Analogously, we obtain

$$P[- + \to +-] \le P[\Omega_2(-+) < \Omega_2(+-)|(-+) \text{ transmitted}]$$

$$= P[-A_1 y_1 + A_2 y_2 + A_1 A_2 \rho < A_1 y_1 - A_2 y_2$$

$$+ A_1 A_2 \rho|(-+) \text{ transmitted}]$$

$$= P[A_1 y_1 - A_2 y_2 > 0 \mid y_1 = -A_1 + A_2 \rho$$

$$+ n_1; \; y_2 = A_2 - A_1 \rho + n_2]$$

$$= P[A_1 n_1 - A_2 n_2 > A_1^2 + A_2^2 - 2A_1 A_2 \rho]$$

$$= Q\left(\frac{\sqrt{A_1^2 + A_2^2 - 2A_1 A_2 \rho}}{\sigma}\right). \tag{4.57}$$

And, finally,

$$P[- + \rightarrow ++] \le P[\Omega_2(-+) < \Omega_2(++) \mid (-+) \text{ transmitted}]$$

$$= P[-A_1 y_1 + A_2 y_2 + A_1 A_2 \rho < A_1 y_1$$

$$+ A_2 y_2 - A_1 A_2 \rho \mid (-+) \text{transmitted}]$$

$$= P[y_1 > A_2 \rho | y_1 = -A_1 + A_2 \rho + n_1]$$

$$= P[n_1 > A_1]$$

$$= Q\left(\frac{A_1}{\sigma}\right). \tag{4.58}$$

Putting together (4.55)–(4.58), we obtain

$$P_1^m(\sigma) \le Q\left(\frac{A_1}{\sigma}\right) + \frac{1}{2} Q\left(\frac{\sqrt{A_1^2 + A_2^2 - 2A_1 A_2 \rho}}{\sigma}\right)$$

$$+ \frac{1}{2} Q\left(\frac{\sqrt{A_1^2 + A_2^2 + 2A_1 A_2 \rho}}{\sigma}\right). \tag{4.59}$$

We will now make the important observation that the last term in (4.59) is superfluous if $\rho \ge 0$. (If $\rho < 0$, then the second term is superfluous.) To see this, let us consider Figure 4.1 (where $\rho > 0$). It is easy to see that if $(--)$ is transmitted and the received vector is closest to $(++)$, then the received vector must be closer to $(+-)$ than to $(--)$. This is because the minimum distance region of $(++)$ is contained in the half-space of all points that are closer to $(+-)$ than to $(--)$. This implies that

$$P[-- \rightarrow ++] + P[-- \rightarrow +-]$$

$$= P[-- \rightarrow ++ \text{ or } -- \rightarrow +-]$$

$$\le P[\Omega_2(--) < \Omega_2(+-)|(--) \text{ transmitted}] \tag{4.60}$$

and, therefore, we can simply drop the third term from (4.59). Reasoning analogously in the case $\rho < 0$, we get with full generality

$$P_1(\sigma) \le P_1^m(\sigma) \le Q\left(\frac{A_1}{\sigma}\right) + \frac{1}{2} Q\left(\frac{\sqrt{A_1^2 + A_2^2 - 2A_1 A_2 |\rho|}}{\sigma}\right). \tag{4.61}$$

Let us now turn to the derivation of a *lower bound* on the minimum error probability for user 1. We can accomplish this by evaluating the exact minimum error probability of a hypothetical genie-aided receiver that has some side information about the transmitted bits not available to the original receiver. We will consider three different genies leading to different lower bounds.

First, suppose that the genie informs the receiver of the value of b_2, the bit transmitted by user 2. Clearly the best strategy (since b_1, b_2 and the noise are independent) is to subtract $A_2 b_2 s_2$ from the received waveform y, in which case we obtain a single-user channel

$$y(t) - A_2 b_2 s_2(t) = A_1 b_1 s_1(t) + \sigma n(t),$$

whose minimum error probability is $Q\left(\frac{A_1}{\sigma}\right)$.

Alternatively, we could consider a seemingly less forthright genie who tells the receiver whether the transmitted bits are equal or not, and if they are equal, it even tells the receiver whether $(b_1, b_2) = (+1, +1)$ or $(-1, -1)$. The receiver either gets full information or has to solve a binary hypothesis-testing problem with equiprobable hypotheses:

$$x_1 = A_1 s_1 - A_2 s_2,$$

$$x_2 = -x_1.$$

The probability of error of an optimum receiver for b_1 with this side information equals

$$\frac{1}{2} Q\left(\frac{\sqrt{A_1^2 + A_2^2 - 2A_1 A_2 \rho}}{\sigma}\right), \tag{4.62}$$

where the factor of $\frac{1}{2}$ accounts for the probability that the transmitted bits are different and the argument of the Q-function in (4.62) is equal to $\|x_1\|/\sigma$.

The symmetric counterpart to the genie in the previous paragraph reveals the true transmitted bits in case they are different. In that case, the minimum probability of error becomes

$$\frac{1}{2} Q\left(\frac{\sqrt{A_1^2 + A_2^2 + 2A_1 A_2 \rho}}{\sigma}\right).$$

These lower bounds to the minimum bit-error-rate together with the upper bound in (4.61) result in

$$\max\left\{ Q\left(\frac{A_1}{\sigma}\right), \frac{1}{2} Q\left(\frac{\sqrt{A_1^2 + A_2^2 - 2A_1 A_2 |\rho|}}{\sigma}\right)\right\}$$

$$\leq P_1(\sigma) \tag{4.63}$$

$$\leq Q\left(\frac{A_1}{\sigma}\right) + \frac{1}{2} Q\left(\frac{\sqrt{A_1^2 + A_2^2 - 2A_1 A_2 |\rho|}}{\sigma}\right). \tag{4.64}$$

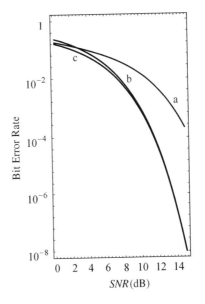

FIGURE 4.11.
Bit-error-rate in a two-user channel with $\rho = 0.4$, $A_1 = A_2$: (a) single-user matched filter, (b) maximum-likelihood upper bound to minimum bit-error-rate, (c) genie lower bound to minimum bit-error-rate.

Figure 4.11 shows the upper and lower bounds to the minimum error probability in the case $\rho = 0.4$ along with the error probability of the single-user matched-filter detector. We can see that the bounds (4.63) and (4.64) are very close. Furthermore, at very low signal-to-background noise ratios the single-user matched filter error probability is a better upper bound than (4.64) – the upper bound to the error probability of the jointly optimum (maximum-likelihood) detector. This does not imply that the conventional receiver exhibits lower bit-error-rate than the maximum-likelihood detector at very low signal-to-noise ratios, although such a phenomenon does occur under certain circumstances (Problem 4.28).

Since both arguments of the Q-functions in (4.63) are identical to those in (4.64), both expressions are dominated by the same term as $\sigma \to 0$. It follows that the ratio of the upper bound (4.64) to the lower bound (4.63) converges to 1 unless both arguments are identical:

$$A_2 = 2A_1|\rho|,$$

in which case, the ratio converges to 1.5.

We can compute the following expression for the optimum asymptotic multiuser efficiency of user 1:

$$\eta_1 = \min\left\{1,\ 1 + \frac{A_2^2}{A_1^2} - 2|\rho|\frac{A_2}{A_1}\right\}. \tag{4.65}$$

To check (4.65) we can use (3.41) to show that if

$$A_1 > \sqrt{A_1^2 + A_2^2 - 2A_1 A_2 |\rho|},$$

then the first term in the upper bound (4.64) is negligible with respect to the second term. Thus, the bounds (4.63) and (4.64) imply that

$$\lim_{\sigma \to 0} \frac{P_1(\sigma)}{Q\left(\frac{\sqrt{r}A_1}{\sigma}\right)} = 0, \tag{4.66}$$

for any

$$r < 1 + \frac{A_2^2}{A_1^2} - 2|\rho|\frac{A_2}{A_1}.$$

Analogously, if

$$A_1 < \sqrt{A_1^2 + A_2^2 - 2A_1 A_2 |\rho|},$$

then the second term in the upper bound (4.64) is negligible with respect to the first term, and (4.66) holds for any $r < 1$. Finally, if

$$A_1 = \sqrt{A_1^2 + A_2^2 - 2A_1 A_2 |\rho|},$$

then

$$P_1(\sigma) = c\left(\frac{A_1}{\sigma}\right) Q\left(\frac{A_1}{\sigma}\right),$$

where

$$1 \le c\left(\frac{A_1}{\sigma}\right) \le 1.5;$$

thus, (4.66) holds for any $r < 1$.

It is useful to compare (4.65) to the asymptotic multiuser efficiency of the conventional single-user matched filter (Figure 4.12),

$$\eta_1^c = \max{}^2\left\{0, 1 - \frac{A_2}{A_1}|\rho|\right\}.$$

As we would expect, if $A_2 \ll A_1$ then both the asymptotic multiuser efficiency of the conventional detector and the optimum asymptotic efficiency are close to unity. However, (4.65) is not monotonic in A_2/A_1. Actually, if

$$\frac{A_2}{A_1} \ge 2|\rho| \tag{4.67}$$

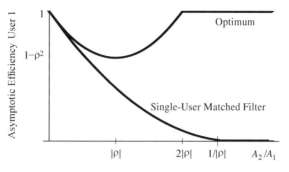

FIGURE 4.12.
Optimum and
single-user
asymptotic
multiuser
efficiencies for
two synchronous
users.

then $\eta_1 = 1$. Therefore, as long as the energy of user 2 exceeds the threshold given by (4.67) the asymptotic bit-error-rate of user 1 is equivalent to the single-user case where user 2 is not active. The explanation of this behavior of the optimum receiver is that if the interfering user is sufficiently powerful, then the primary source of errors committed in the optimum demodulation of user 1 is the background Gaussian noise, rather than the randomness of the information carried by the interfering signal. Note that according to the threshold in (4.67), an interferer (user 2) who is 3 dB weaker than the user of interest (user 1) has no appreciable effect on the bit-error-rate as long as the crosscorrelation is below 0.35 (a mild condition on the signal design). Interestingly, the same is true if the relative power of the interferer is *higher* than -3 dB for that crosscorrelation. If $A_1 \ll A_2$, then it is possible to explain the fact that $\eta_1 = 1$ using the technique of successive decoding in Figure 1.8: first, user 2 is decoded by its single-user matched filter (thus neglecting the presence of the comparatively weak user 1) with very low error probability (much lower than the error probability of user 1 if it were the only user of the channel); the remodulated signal of user 2 is then subtracted from the received waveform; the resulting channel is almost a single-user channel for user 1, whose main interference is the additive background noise rather than the uncancelled (actually boosted) interference from user 2 due to the occasional errors of the conventional receiver of user 2. This approach fails to explain why $\eta_1 = 1$ in the whole range given by (4.67); in fact, as we will see in Chapter 7, the successive decoder achieves neither minimum bit-error-rate nor maximum asymptotic multiuser efficiency. Figure 4.13 exhibits the asymptotic multiuser efficiency as a function of $|\rho|$ and the relative amplitude of the interferer.

The near–far resistance is obtained by minimizing (4.65) over $A_2/A_1 \geq 0$. The least-favorable relative amplitude of user 2 is

$$\frac{A_2}{A_1} = |\rho|,$$

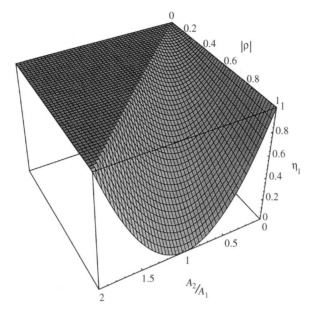

FIGURE 4.13.
Optimum
asymptotic
multiuser
efficiency as a
function of $|\rho|$ and
relative amplitude
of interferer.

which yields the near–far resistance for either user:

$$\bar{\eta}_k = 1 - \rho^2. \tag{4.68}$$

Figure 4.14 shows the two-user power-tradeoff region so that the optimum bit-error-rate of both users is not higher than 3×10^{-5}, for $|\rho| = 0.8, 0.9$, and 0.95. Figure 4.14 should be compared with Figure 3.9, which shows the same kind of plot for the conventional receiver, although

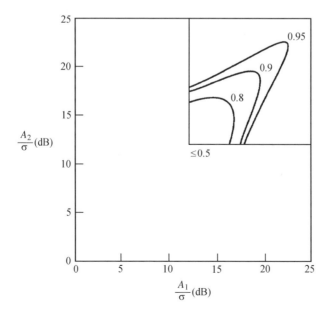

FIGURE 4.14.
Signal-to-noise
ratios necessary
to achieve
optimum
bit-error-rate not
higher than
3×10^{-5} for both
users.

185

in Figure 3.9, lower values of crosscorrelations were shown. For the (jointly or individually) optimum receiver, the permissible signal-to-noise ratios are indistinguishable from the single-user signal-to-noise ratios as long as the crosscorrelation satisfies $|\rho| \leq 0.5$ (Problem 4.26). For the very high cross-correlation values shown in Figure 4.14, we see the detrimental effect of insisting on equal powers. Indeed in those cases it is preferable to let one user transmit at single-user signal-to-noise ratio, a strategy that minimizes the sum of the powers necessary to achieve the prescribed bit-error-rate. The reason is clear: if both signature waveforms are very much alike, then similar amplitudes complicate the task of the optimum receiver.

4.3.2   K-USER SYNCHRONOUS CHANNEL

We now turn attention to the analysis of the minimum bit-error-rate of the K-user synchronous channel. At first sight, it may not be immediately obvious how to generalize the ideas that lead to the tight bounds on the minimum error probability in the two-user case. To that end we need to introduce several definitions.

The normalized difference between any pair of distinct transmitted vectors is referred to as an *error vector*. The set of error vectors that affects the kth user is

$$E_k = \{\epsilon \in \{-1, 0, 1\}^K, \epsilon_k \neq 0\}$$

and the set of (nonzero) error vectors is denoted by

$$E = \bigcup_{k=1}^{K} E_k.$$

As an example, consider the following transmitted $\mathbf{b}$ and demodulated $\hat{\mathbf{b}}$ bit vectors along with the corresponding error vector:

$$\mathbf{b} = [+1 \quad -1 \quad -1 \quad +1 \quad +1]^T,$$
$$\hat{\mathbf{b}} = [-1 \quad -1 \quad +1 \quad +1 \quad +1]^T,$$
$$\epsilon = [+1 \quad 0 \quad -1 \quad 0 \quad 0]^T.$$

Simply put, the error vector component is zero if no error occurs at that component or equal to the transmitted vector component otherwise. Note that for the $\mathbf{b}$ in this example, the error vector

$$\epsilon = [+1 \quad 0 \quad +1 \quad 0 \quad 0]^T$$

is not compatible with the transmitted vector. The set of error vectors that are compatible (or admissible) with a given $\mathbf{b} \in \{-1, 1\}^K$ being transmitted is denoted by

$$A(\mathbf{b}) = \{\epsilon \in E, \; \epsilon_i = b_i \text{ or } 0\} \tag{4.69}$$

$$= \{\epsilon \in E, \; 2\epsilon - \mathbf{b} \in \{-1, 1\}^K\}. \tag{4.70}$$

The admissible error vectors that affect the kth user are

$$A_k(\mathbf{b}) = A(\mathbf{b}) \cap E_k.$$

The number of nonzero components (weight) of an error vector and the energy of a hypothetical multiuser signal modulated by ϵ are denoted, respectively, by

$$w(\epsilon) = \sum_{k=1}^{K} |\epsilon_k|$$

and

$$\|S(\epsilon)\|^2 = \int_0^T \left(\sum_{k=1}^{K} \epsilon_k A_k s_k(t) \right)^2 dt$$

$$= \epsilon^T \mathbf{H} \epsilon, \tag{4.71}$$

where

$$\mathbf{H} = \mathbf{ARA}.$$

An error vector $\epsilon \in E$ is *decomposable* into $\epsilon' \in E$ and $\epsilon'' \in E$ if

1) $\epsilon = \epsilon' + \epsilon''$;
2) if $\epsilon_k = 0$, then $\epsilon'_k = \epsilon''_k = 0$;
3) $\langle S(\epsilon'), S(\epsilon'') \rangle \geq 0$.

If ϵ is decomposable into ϵ' and ϵ'' we write

$$\epsilon \overset{\text{dec}}{=} \epsilon' + \epsilon''.$$

The subset of *indecomposable* vectors in E_k is denoted by F_k. Note that

$$\langle S(\epsilon'), S(\epsilon'') \rangle = \int_0^T \sum_{k=1}^{K} \sum_{j=1}^{K} \epsilon'_k \epsilon''_j A_k A_j s_k(t) s_j(t) \, dt$$

$$= \epsilon'^T \mathbf{H} \epsilon''. \tag{4.72}$$

To familiarize ourselves with the concept of indecomposability, let us consider a few examples:

- If $\rho_{1k} = 0, k \neq 1$, then the indecomposable vectors for user 1 are

$$F_1 = \{[+1 \quad 0 \cdots 0], \quad [-1 \quad 0 \cdots 0]\}.$$

- In the two-user case with $\rho \neq 0$, it is immediate to check that

$$F_1 = \{[+1 \quad 0], \quad [-1 \quad 0], \quad [+1 \quad -\text{sgn}(\rho)], \quad [-1 \quad +\text{sgn}(\rho)]\}. \tag{4.73}$$

- If the signature waveforms are such that all crosscorrelations are identical and positive:

$$\rho_{kl} = \rho > 0,$$

and

$$A_1 = \cdots = A_K,$$

then it is straightforward (Problem 4.57) to show that the only indecomposable error vectors (for user 1) have either weight one or two:

$$
\begin{array}{cccccc}
[+1, & 0, & 0, & \cdots & 0, & 0], \\
[+1, & -1, & 0, & \cdots & 0, & 0], \\
[+1, & 0, & -1, & \cdots & 0, & 0], \\
& & \cdots & & & \\
[+1, & 0, & 0, & \cdots & -1, & 0], \\
[+1, & 0, & 0, & \cdots & 0, & -1],
\end{array}
$$

plus their antipodal images. This means that in this special case of a total of 3^{K-1} error sequences with $\epsilon_1 = 1$, only K are indecomposable.

The main result in the analysis of the optimum multiuser detector is the following upper bound on the minimum probability of error of the kth user:

Proposition 4.1

$$P_k(\sigma) \leq \sum_{\epsilon \in F_k} 2^{-w(\epsilon)} Q\left(\frac{\|S(\epsilon)\|}{\sigma}\right). \tag{4.74}$$

Before we proceed to justify (4.74), we point out that it lends itself to further simplification. Since $\epsilon \in F_k \Leftrightarrow -\epsilon \in F_k$ and $w(\epsilon) = w(-\epsilon)$, $\|S(\epsilon)\| = \|S(-\epsilon)\|$, we can eliminate from the sum in (4.74) all those vectors with $\epsilon_k = -1$, by replacing $w(\epsilon)$ with $w(\epsilon) - 1$ therein.

It follows from (4.73) that (4.74) reduces to (4.61) if $K = 2$.

188

As in the two-user case, in order to show Proposition 4.1 we bound the error probability of a 2^K-ary hypothesis test in terms of the error probabilities of binary tests. If $\hat{\mathbf{b}} = [\hat{b}_1, \ldots, \hat{b}_K]$ is the vector put out by the detector and $\mathbf{b}$ is the true transmitted vector, we can upper bound the error probability by

$$P_k(\sigma) \leq \sum_{\epsilon \in E_k} P[\epsilon \in A(\mathbf{b}); \ \Omega(\mathbf{b} - 2\epsilon) = \max_{\mathbf{d}} \Omega(\mathbf{d})] \tag{4.75}$$

$$\leq \sum_{\epsilon \in E_k} P[\epsilon \in A(\mathbf{b}); \ \Omega(\mathbf{b} - 2\epsilon) \geq \Omega(\mathbf{b})]. \tag{4.76}$$

The inequality in (4.75) is not necessarily an equality because the minimum distance detector does not result in minimum bit-error-rate; the inequality in (4.75) is a consequence of the obvious fact that if $\mathbf{b} - 2\epsilon$ is the most likely vector, then it must be more likely than $\mathbf{b}$. For each ϵ, the event $\epsilon \in A(\mathbf{b})$ depends only on the transmitted vector. As all transmitted bits are equiprobable and equally likely,

$$P[\epsilon \in A(\mathbf{b})] = \prod_{k=1}^{K} P[(b_k - \epsilon_k)\epsilon_k = 0]$$

$$= 2^{-w(\epsilon)}. \tag{4.77}$$

Because of symmetry, the event $\Omega(\mathbf{b} - 2\epsilon) \geq \Omega(\mathbf{b})$ turns out to be independent of the transmitted vector $\mathbf{b}$. To see this, recall from (4.11) that $\Omega(\mathbf{b}) = 2\mathbf{b}^T \mathbf{A}\mathbf{y} - \mathbf{b}^T \mathbf{H}\mathbf{b}$. Thus,

$$\Omega(\mathbf{b} - 2\epsilon) - \Omega(\mathbf{b}) = 2(\mathbf{b} - 2\epsilon)^T \mathbf{A}\mathbf{y} - (\mathbf{b} - 2\epsilon)^T \mathbf{H}(\mathbf{b} - 2\epsilon)$$

$$-2\mathbf{b}^T \mathbf{A}\mathbf{y} + \mathbf{b}^T \mathbf{H}\mathbf{b}$$

$$= -4\epsilon^T \mathbf{A}\mathbf{y} - 4\epsilon^T \mathbf{H}\epsilon + 2\mathbf{b}^T \mathbf{H}\epsilon + 2\epsilon^T \mathbf{H}\mathbf{b}$$

$$= -4\epsilon^T \mathbf{A}\mathbf{n} - 4\epsilon^T \mathbf{H}\epsilon, \tag{4.78}$$

where we have used the discrete-time model in (2.78), $\mathbf{y} = \mathbf{R}\mathbf{A}\mathbf{b} + \mathbf{n}$. Since $\epsilon^T \mathbf{A}\mathbf{n}$ is a Gaussian random variable with zero mean and variance

$$E[\epsilon^T \mathbf{A}\mathbf{n}\mathbf{n}^T \mathbf{A}\epsilon] = \sigma^2 \epsilon^T \mathbf{H}\epsilon = \sigma^2 \|S(\epsilon)\|^2,$$

we can conclude from (4.78) that

$$P[\Omega(\mathbf{b} - 2\epsilon) \geq \Omega(\mathbf{b})] = Q\left(\frac{\|S(\epsilon)\|}{\sigma}\right). \tag{4.79}$$

We have seen that the event $\{\epsilon \in A(\mathbf{b})\}$ depends on $\mathbf{b}$ but not on $\mathbf{n}$ and the event $\{\Omega(\mathbf{b} - 2\epsilon) \geq \Omega(\mathbf{b})\}$ depends on $\mathbf{n}$ but not on $\mathbf{b}$; therefore, they are independent events. Accordingly, the right side of (4.76) is (using (4.77) and (4.79))

$$P_k(\sigma) \leq \sum_{\epsilon \in E_k} 2^{-w(\epsilon)} Q\left(\frac{\|S(\epsilon)\|}{\sigma}\right). \qquad (4.80)$$

This bound gives reasonably tight results if the signal-to-noise ratio is high. However, its main shortcoming is that its generalization to the asynchronous case is completely useless: in that case, error vectors become *error sequences* and the sum in (4.80) becomes a divergent infinite series as $M \to \infty$. The reason for this behavior is that many of the terms in the summation in (4.80) are redundant; once those terms are omitted, then the bounding series not only converges but approximates the error probability closely. Recall that in the two-user case one of the terms in (4.59) was redundant and could be omitted from the bound. The second inequality in (4.76) can be strengthened by removing all decomposable error sequences from the right side. This simplification relies on the following result:

Proposition 4.2 *(Verdú [490]) For every* $\mathbf{b} \in \{-1, +1\}^K$ *and* $\epsilon \in A_k(\mathbf{b})$, *there exists* $\epsilon' \in A_k(\mathbf{b}) \cap F_k$ *such that if* $\mathbf{b} - 2\epsilon$ *is the most likely vector, then* $\mathbf{b} - 2\epsilon'$ *is more likely than* $\mathbf{b}$.

Proof. If $\epsilon \in F_k$, then the result is trivially satisfied with $\epsilon' = \epsilon$. If ϵ is decomposable, then we prove by induction on $w(\epsilon)$ that it can be decomposed directly into an indecomposable vector. More precisely, we will show the existence of $\epsilon^* \in F_k$ such that $\epsilon \stackrel{\text{dec}}{=} \epsilon^* + (\epsilon - \epsilon^*)$.

If a sequence of weight two is decomposable, then it is decomposable into its two nonzero components, both of which are indecomposable since they have unit weight. So in this case the sought-after ϵ^* is all-zero except in the kth component.

Now suppose that the claim is true for any sequence whose weight is strictly less than $w(\epsilon)$. Find the (not necessarily unique) decomposition $\epsilon \stackrel{\text{dec}}{=} \epsilon^1 + \epsilon^2$, $\epsilon^1 \in A_k(\mathbf{b})$, with largest inner product, that is,

$$\langle S(\epsilon^1), S(\epsilon^2) \rangle \geq \langle S(\epsilon^a), S(\epsilon^b) \rangle \geq 0$$

for any decomposition $\epsilon \stackrel{\text{dec}}{=} \epsilon^a + \epsilon^b$. If $\epsilon^1 \in F_k$, we have found the sought-after decomposition of ϵ. Otherwise, using the induction hypothesis we can find $\epsilon^3 \in F_k$ such that

$$\epsilon^1 \stackrel{\text{dec}}{=} \epsilon^3 + \epsilon^4.$$

The vector ϵ can be written in various ways, none of which involve cancellations of $+1$ and -1:

$$\epsilon = \epsilon^1 + \epsilon^2 \tag{4.81}$$

$$= (\epsilon^2 + \epsilon^3) + \epsilon^4 \tag{4.82}$$

$$= \epsilon^3 + (\epsilon^2 + \epsilon^4). \tag{4.83}$$

Now we will check that

$$\epsilon \overset{\text{dec}}{=} \epsilon^3 + (\epsilon^2 + \epsilon^4),$$

and thus ϵ^3 is the sought-after ϵ^*. We just need to verify that the inner product of the signals modulated by ϵ^3 and $(\epsilon^2 + \epsilon^4)$ is nonnegative. To do so, we observe that if this were not true, then the right-hand side of the identity

$$\langle S(\epsilon^2) + S(\epsilon^3), S(\epsilon^4) \rangle - \langle S(\epsilon^1), S(\epsilon^2) \rangle$$

$$= 2\langle S(\epsilon^3), S(\epsilon^4) \rangle - \langle S(\epsilon^3), S(\epsilon^4) + S(\epsilon^2) \rangle$$

would be strictly positive, contradicting the choice of $\epsilon^1 + \epsilon^2$ as the largest-inner-product decomposition of ϵ.

Now, we show that $\epsilon' = \epsilon^*$ satisfies the conditions of Proposition 4.2. Since $\epsilon_i = 0$ implies $\epsilon_i' = 0$, we have $\epsilon' \in A_k(\mathbf{b})$. To show that $\mathbf{b} - 2\epsilon'$ is more likely than $\mathbf{b}$, let $\epsilon'' = \epsilon - \epsilon'$ and use (4.11) to show

$$\Omega(\mathbf{b} - 2\epsilon') - \Omega(\mathbf{b}) = [\Omega(\mathbf{b} - 2\epsilon) - \Omega(\mathbf{b} - 2\epsilon'')] + 8\langle S(\epsilon'), S(\epsilon'') \rangle, \tag{4.84}$$

where both terms in the right side of (4.83) are nonnegative because $\mathbf{b} - 2\epsilon$ is the most likely vector and $\epsilon \overset{\text{dec}}{=} \epsilon' + \epsilon''$, respectively. This shows that $\mathbf{b} - 2\epsilon'$ is more likely than $\mathbf{b}$ and the proof of Proposition 4.2 concludes.

In Figure 4.15 we see a graphical illustration of the fact that if ϵ is decomposable into ϵ' and ϵ'' and $\mathbf{b} - 2\epsilon$ is the most likely vector, then $\mathbf{b} - 2\epsilon'$ (and $\mathbf{b} - 2\epsilon''$) is more likely than $\mathbf{b}$.

As $\sigma \to 0$, the sum in (4.74) will be dominated by the term or terms with the smallest argument of the Q-function. That corresponds to the error vector that achieves

$$d_{k,\min} = \min_{\epsilon \in F_k} \|S(\epsilon)\| = \min_{\substack{\epsilon \in \{-1,0,1\}^K \\ \epsilon_k = 1}} \|S(\epsilon)\|, \tag{4.85}$$

where we have used the fact that if an error vector is decomposable into $\epsilon \overset{\text{dec}}{=} \epsilon^1 + \epsilon^2$, then $\|S(\epsilon)\| \geq \max\{\|S(\epsilon^1)\|, \|S(\epsilon^2)\|\}$.

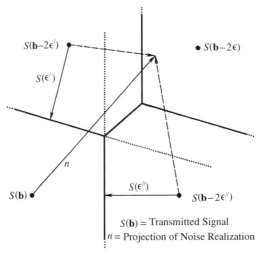

$S(\mathbf{b}) = $ Transmitted Signal
$n = $ Projection of Noise Realization

FIGURE 4.15.
If ϵ is decomposable into ϵ' and ϵ'', and $\mathbf{b} - 2\epsilon$ is the most likely vector, then both $\mathbf{b} - 2\epsilon'$ and $\mathbf{b} - 2\epsilon''$ are more likely than $\mathbf{b}$.

The distance between the multiuser signals modulated by arbitrary $\mathbf{b}^1$ and $\mathbf{b}^2$ is equal to

$$\|S(\mathbf{b}^1) - S(\mathbf{b}^2)\| = 2\|S(\epsilon)\|$$

with $\epsilon = \frac{1}{2}(\mathbf{b}^1 - \mathbf{b}^2)$. Therefore, we can interpret $d_{k,\min}$ in (4.85) as one half of the minimum distance between two multiuser signals that differ in the kth bit. We will henceforth refer to this quantity as the *minimum distance of the kth user*.

The coefficient multiplying $Q(d_{k,\min}/\sigma)$ in (4.74) is equal to

$$C_k^U = \sum_{\substack{\epsilon \in F_k s.t. \\ \|S(\epsilon)\| = d_{k,\min}}} 2^{-w(\epsilon)}. \tag{4.86}$$

As $\sigma \to 0$, the bound in (4.74) is dominated by the largest term in the sum, that is,

$$\lim_{\sigma \to 0} \sum_{\epsilon \in F_k} 2^{-w(\epsilon)} \frac{Q(\|S(\epsilon)\|/\sigma)}{Q(d_{k,\min}/\sigma)} = C_k^U. \tag{4.87}$$

The main conclusion that we can draw from (4.87) is that the upper bound on the kth user bit-error-rate behaves for high signal-to-noise ratio as a single-user system with energy $d_{k,\min}^2$. The question now is whether the minimum bit-error-rate actually behaves that way. To answer this question, we will generalize the reasoning employed to obtain the two-user lower bound in (4.63).

Let us choose (in an arbitrary fashion, for now) one of the error vectors in F_k that achieve $d_{k,\min}$, and let us denote this vector by ϵ^*. We construct the following genie:

- If the transmitted vector **b** is compatible with ϵ^*, that is,

$$\epsilon^* \in A(\mathbf{b}),$$

  then the genie tells the receiver that the true transmitted vector belongs to the pair

$$\{\mathbf{b}, \mathbf{b} - 2\epsilon^*\}.$$

- If the transmitted vector **b** is compatible with $-\epsilon^*$, that is,

$$\epsilon^* \in A(-\mathbf{b}),$$

  then the genie tells the receiver that the true transmitted vector belongs to the pair

$$\{\mathbf{b}, \mathbf{b} + 2\epsilon^*\}.$$

- If the transmitted vector is such that

$$\epsilon^* \notin A(\mathbf{b}) \cup A(-\mathbf{b}),$$

  then the genie goes ahead and reveals **b** to the receiver.

Note that there is no ambiguity in the description of the genie because $A(\mathbf{b})$ and $A(-\mathbf{b})$ are disjoint.

Let us evaluate the probability that a receiver that processes optimally the available information (observed waveform plus the side information) makes an error in the kth component. An error will occur if and only if the transmitted **b** and the noise realization are such that either of the following events occurs:

$$\{\epsilon^* \in A(\mathbf{b})\} \cap \{\Omega(\mathbf{b} - 2\epsilon^*) \geq \Omega(\mathbf{b})\},$$

$$\{\epsilon^* \in A(-\mathbf{b})\} \cap \{\Omega(\mathbf{b} + 2\epsilon^*) \geq \Omega(\mathbf{b})\}.$$

These events are nonoverlapping and have identical probabilities. Following the analysis in (4.77)–(4.79), we obtain the bound

$$P_k(\sigma) \geq 2^{1-w(\epsilon^*)} Q\left(\frac{d_{k,\min}}{\sigma}\right). \tag{4.88}$$

Since we have the freedom to choose ϵ^* from those vectors in F_k that achieve the lowest $\|S(\epsilon)\|$, we maximize the lower bound in (4.87) by choosing ϵ^*

with the smallest possible weight. Let

$$w_{k,\min} = \min_{\substack{\epsilon \in F_k \\ \|S(\epsilon)\|=d_{k,\min}}} w(\epsilon).$$

We have shown that the minimum bit-error-rate of the kth user is lower bounded by

$$P_k(\sigma) \geq 2^{1-w_{k,\min}} Q\left(\frac{d_{k,\min}}{\sigma}\right). \tag{4.89}$$

A lower bound that is, in general, tighter is given in Problem 4.23. It is tempting to tighten (4.89) by replacing the genie with a less generous one. For example, we could let the genie reveal the true transmitted vector only when the transmitted vector is not compatible with any of the error vectors that achieve $d_{k,\min}$; otherwise, the genie gives the true transmitted vector and one of its nearest neighbors. It would seem that this strategy would lead to the stronger lower bound where the coefficient in (4.89) is replaced by the probability that the transmitted vector is compatible with any one of the error vectors with minimum energy that affect the kth user. Yet, for any pair of data vectors supplied by the genie, the probability of each vector must be $1/2$, for, otherwise, the probability of error of the binary test would be strictly less than $Q(d_{k,\min}/\sigma)$. This condition may fail to hold if the choice of nearest neighbors is such that some data vectors are more likely to be chosen by the genie than others (Problem 4.24).

Figure 4.16 shows the upper and lower bounds to the minimum bit-error-rate in the case of fifteen equal-power users with equicorrelated signature

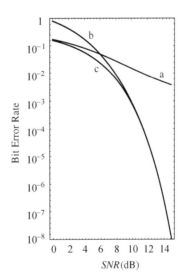

FIGURE 4.16.
Bit-error-rate in a fifteen-user channel with equal-power users and $\rho_{kl} = 0.09$; (a) conventional, (b) maximum-likelihood upper bound to minimum bit-error-rate, (c) lower bound to minimum bit-error-rate.

waveforms $\rho_{kl} = 0.09$, as well as the error probability achieved by the single-user matched filter. The upper and lower bounds are very tight (and indistinguishable from single-user error probability) except when the bit-error-rate is very high. For very low signal-to-noise ratios, the error probability of the conventional single-user matched filter is a better upper bound to the minimum bit-error-rate than (4.74).

4.4 K-USER OPTIMUM ASYMPTOTIC EFFICIENCY AND NEAR–FAR RESISTANCE

The bounds in (4.74) and (4.89) are dominated as $\sigma \to 0$ by a term equal to (modulo a factor independent of σ)

$$Q\left(\frac{d_{k,\min}}{\sigma}\right).$$

Therefore they are asymptotically tight in the sense that their ratio converges to a nonzero constant:

$$\sum_{\substack{\epsilon:\epsilon_k=1 \\ \|S(\epsilon)\|=d_{k,\min}}} 2^{w_{k,\min}-w(\epsilon)}.$$

This is sufficient to obtain the optimum asymptotic multiuser efficiency. From (4.87) and (4.89) we get

$$\eta_k = \frac{d_{k,\min}^2}{A_k^2} \tag{4.90}$$

$$= \min_{\substack{\epsilon \in \{-1,0,1\}^K \\ \epsilon_k=1}} \frac{1}{A_k^2} \epsilon^T \mathbf{ARA}\epsilon, \tag{4.91}$$

which in the two-user case is easily seen to reduce to (4.65). According to (4.72), the computation of optimum asymptotic multiuser efficiency using (4.91) entails solving for the minimum-energy error vector, a combinatorial optimization problem for which no polynomial-in-K algorithm is known. Furthermore, it was shown in Verdú [496] that (4.91) belongs to the class of hard combinatorial problems discussed in Section 4.1 (which includes optimum multiuser detection, traveling salesman, integer linear programming, etc.). In contrast to the asymptotic multiuser efficiency of the single-user matched filter, the optimum multiuser efficiency of the kth user depends not only on $\{\rho_{kj}, j \neq k\}$, but on $\{\rho_{lj}, l \neq k, j \neq k\}$.

The optimum near–far resistance is obtained by minimizing (4.91) with respect to the amplitudes:

$$\bar{\eta}_k = \min_{\substack{A_i \geq 0 \\ i \neq k}} \min_{\substack{\epsilon \in \{-1,0,1\}^K \\ \epsilon_k = 1}} \frac{1}{A_k^2} \epsilon^T \mathbf{A} \mathbf{R} \mathbf{A} \epsilon$$

$$= \min_{\substack{\mathbf{w} \in R^K \\ w_k = 1}} \mathbf{w}^T \mathbf{R} \mathbf{w} \tag{4.92}$$

$$= \min_{\substack{a_l \in R \\ l \neq k}} \left\| s_k + \sum_{\substack{j=1 \\ j \neq k}}^K a_j s_j \right\|^2. \tag{4.93}$$

Now, to solve the optimization problem in (4.92), we will assume first that the crosscorrelation matrix $\mathbf{R}$ is nonsingular. Let us denote by $\mathbf{R}_k$ the $(K-1) \times (K-1)$ symmetric matrix that results by striking out the kth column and the kth row from $\mathbf{R}$; moreover, let $\mathbf{a}_k$ be the $(K-1)$-vector that results from eliminating the kk component (equal to 1) from the kth column of $\mathbf{R}$. It is easy to check that

$$\min_{\substack{\mathbf{w} \in R^K \\ w_k = 1}} \mathbf{w}^T \mathbf{R} \mathbf{w} = \min_{\mathbf{z} \in R^{K-1}} (1 + 2\mathbf{z}^T \mathbf{a}_k + \mathbf{z}^T \mathbf{R}_k \mathbf{z}). \tag{4.94}$$

If $\mathbf{z} = \mathbf{z}_0 - \mathbf{R}_k^{-1} \mathbf{a}_k$ for an arbitrary $\mathbf{z}_0$, it is immediate to verify that the positive definiteness of $\mathbf{R}_k$ implies that $\mathbf{z}_0 = \mathbf{0}$ minimizes the function in (4.94). Thus, the minimum in (4.94) is achieved by

$$\mathbf{z} = -\mathbf{R}_k^{-1} \mathbf{a}_k, \tag{4.95}$$

yielding a value for the near–far resistance equal to

$$\bar{\eta}_k = 1 - \mathbf{a}_k^T \mathbf{R}_k^{-1} \mathbf{a}_k \tag{4.96}$$

$$= \frac{1}{(\mathbf{R}^{-1})_{kk}}. \tag{4.97}$$

We have obtained (4.97) using the following standard formula for the inverse of a block-partitioned matrix:

Proposition 4.3 *If $\mathbf{A}$ and $\mathbf{B}$ are nonsingular, then*

$$\begin{bmatrix} \mathbf{A} & \mathbf{C}^T \\ \mathbf{C} & \mathbf{B} \end{bmatrix}^{-1} = \begin{bmatrix} [\mathbf{A} - \mathbf{C}^T \mathbf{B}^{-1} \mathbf{C}]^{-1} & -\mathbf{A}^{-1} \mathbf{C}^T \Delta^{-1} \\ -\Delta^{-1} \mathbf{C} \mathbf{A}^{-1} & \Delta^{-1} \end{bmatrix}, \tag{4.98}$$

where Δ, the Schur complement of $\mathbf{A}$, is defined by

$$\Delta = \mathbf{B} - \mathbf{C} \mathbf{A}^{-1} \mathbf{C}^T,$$

and the northwest block in the right side of (4.98) can be expressed, alter-natively, as[2]

$$\left[\mathbf{A} - \mathbf{C}^T \mathbf{B}^{-1} \mathbf{C}\right]^{-1} = \mathbf{A}^{-1} + \mathbf{A}^{-1} \mathbf{C}^T \mathbf{\Delta}^{-1} \mathbf{C} \mathbf{A}^{-1}. \tag{4.99}$$

An alternative expression for the optimum near–far resistance in terms of the signature vectors of (2.98) is

$$\bar{\eta}_k = \mathbf{s}_k^T \left[\mathbf{I} - \mathbf{S}_k \left(\mathbf{S}_k^T \mathbf{S}_k\right)^{-1} \mathbf{S}_k^T\right] \mathbf{s}_k, \tag{4.100}$$

where $\mathbf{s}_k$ is the signature vector of the kth user and $\mathbf{S}_k$ is the $L \times K$ matrix that results from striking out the kth column from $\mathbf{S}$. To verify (4.100), we can use (4.96), $\|\mathbf{s}_k\| = 1$,

$$\mathbf{a}_k = \mathbf{S}_k^T \mathbf{s}_k,$$

and (2.101),

$$\mathbf{R}_k = \mathbf{S}_k^T \mathbf{S}_k.$$

Yet another representation for the optimum near–far resistance can be obtained from the Cholesky factorization of the crosscorrelation matrix (Proposition 2.3) and the result of Problem 2.51:

$$\bar{\eta}_1 = F_{11}^2. \tag{4.101}$$

Let us now examine the optimum near–far resistance of user k when $\mathbf{R}$ is singular. It follows from (4.93) that $\bar{\eta}_k = 0$ when s_k belongs to the subspace spanned by all the other signature waveforms; otherwise, the dimensionality, L, of the space spanned by the interferers is less than $K - 1$, $\mathbf{R}_k$ is not invertible, and direct use of the expressions (4.96) and (4.97) is not possible to compute the near–far resistance of the kth user. However, as far as the computation of that quantity is concerned, it is equivalent to consider another channel with $L + 1$ users: the original kth signature waveform and a subset of L interfering signature waveforms that span the space spanned by the $K - 1$ interfering signature waveforms $\{s_j, j \neq k\}$ (cf. Problem 4.48).

A general formula for near–far resistance that applies regardless of whether $\mathbf{R}$ is nonsingular involves replacing the inverse of $\mathbf{R}$ by the so-called Moore–Penrose generalized inverse discussed in Chapter 5.

---

[2] Equation (4.99) is known as the *matrix inversion lemma*.

Let us now analyze the optimum near–far resistance of direct-sequence spread-spectrum signature waveforms. For a given set of signature waveforms one can compute its crosscorrelation matrix and $\bar{\eta}_k$ via (4.97). However, we would like to get insight into the level of near–far resistance achievable by direct-sequence spread-spectrum, by analyzing the statistical behavior of the near–far resistance when the direct-sequence spread-spectrum waveforms are randomly chosen, in the sense that the polarity of each chip for each user is independently equally likely to be $+1$ or -1. The distribution of the near–far resistance of such a random constellation depends only on the number of users and on the number of chips per bit, N. The expected near–far resistance for random signature waveforms is a useful lower bound on what can be achieved by careful design of a constellation with low crosscorrelation properties. It is also relevant to spread-spectrum systems where the length of the pseudonoise sequence is much longer than the symbol interval.

For any given set of K signature waveforms, suppose that the dimensionality of the space spanned by the $K - 1$ users interfering with user 1 is equal to L. This means that the rank of $\mathbf{R}_1$ is equal to L. As we discussed, as far as computing $\bar{\eta}_1$, we can eliminate from consideration all but a subset of L interferers whose signature waveforms span the whole space spanned by the original interferers. For convenience, we will label these L interferers as users $2, \ldots, L + 1$. From (4.96) we get

$$\bar{\eta}_1 = 1 - \sum_{k=2}^{L+1} \sum_{l=2}^{L+1} \rho_{1k} \rho_{1l} \left(\mathbf{R}_1^{-1}\right)_{kl}. \tag{4.102}$$

To find the expectation of (4.102) with respect to all the signature waveforms, we will condition first on the signature waveforms of the L interferers. In order to take the average of (4.102) with respect to the signature waveform of user 1, we just need to compute $E_1[\rho_{1k}\rho_{1l}]$, where the subscript in the expectation denotes the signature waveform with respect to which we are averaging.

Let us denote by $d_{ki} \in \{+1, -1\}$ the polarity of the ith chip of the kth user's signature sequence. Since the crosscorrelation between two N-chip signature waveforms is equal to the number of coincident polarities minus the number of discrepancies divided by N, we have (cf. Problem 2.21)

$$\rho_{lk} = \frac{1}{N} \sum_{i=1}^{N} d_{li} d_{ki}, \tag{4.103}$$

and

$$E_1[\rho_{1k}\rho_{1l}] = \frac{1}{N^2} \sum_{i=1}^{N} \sum_{j=1}^{N} E_1[d_{1i}d_{ki}d_{1j}d_{lj}]$$

$$= \frac{1}{N^2} \sum_{i=1}^{N} d_{ki}d_{li}$$

$$= \frac{\rho_{kl}}{N}. \tag{4.104}$$

In particular, note that the crosscorrelation between two randomly chosen N-chip direct-sequence signature waveforms has zero mean according to (4.102), and variance equal to $1/N$ according to (4.103) with $k = l$.

Substituting (4.104) into (4.102) we obtain

$$E_1[\bar{\eta}_1] = 1 - \sum_{k=2}^{L+1} \sum_{l=2}^{L+1} E_1[\rho_{1k}\rho_{1l}](\mathbf{R}_1^{-1})_{kl}$$

$$= 1 - \sum_{k=2}^{L+1} \sum_{l=2}^{L+1} \frac{1}{N}(\mathbf{R}_1)_{lk}(\mathbf{R}_1^{-1})_{kl}$$

$$= 1 - \frac{1}{N} \sum_{l=2}^{L+1} (\mathbf{R}_1\mathbf{R}_1^{-1})_{ll}$$

$$= 1 - \frac{L}{N}, \tag{4.105}$$

which depends on the signature waveforms of the $K - 1$ interferers only through their rank, L. Note for future use that the derivation leading to (4.105) holds verbatim as long as

$$\sum_{i=1}^{N} d_{ki}^2 = N$$

for $k = 2, \ldots, K$ even if the interfering signature waveforms are not bipolar ($d_{ki} \notin \{+1, -1\}$).

If $K = 2$, the rank of the interfering space is $L = 1$. Thus, the expected near–far resistance averaged with respect to the spreading code of the desired user is $1 - 1/N$, regardless of the spreading code of the interferer. When $K > 2$, we can proceed with the averaging of L with respect to the signature waveforms of the interferers.

Let $\overline{[J, N]}$ denote the expected dimensionality of J synchronous random $\pm$ signature waveforms with N chips per bit. Naturally,

$$\overline{[J, N]} \leq \min\{J, N\}. \tag{4.106}$$

Taking the expectation of (4.105) with respect to the signature waveforms of the interferers, we get that the average optimum near–far resistance with synchronous random direct-sequence spread-spectrum waveforms is

$$E[\bar{\eta}_1] = 1 - \frac{\overline{[K-1, N]}}{N}$$

$$\geq 1 - \frac{\min\{K-1, N\}}{N}$$

$$= \left[1 - \frac{K-1}{N}\right]^+. \tag{4.107}$$

Equation (4.107) gives a simple lower bound on optimum performance as a function of the number of users and the spreading factor of the CDMA system. This bound is asymptotically tight as the number of users grows:

$$\lim_{K \to \infty} \left(E[\bar{\eta}_1] - \left[1 - \frac{K-1}{N}\right]^+ \right) = 0, \tag{4.108}$$

where N is allowed to grow as K grows. To verify (4.108) first note that if N does not go to infinity, then obviously

$$\lim_{K \to \infty} \overline{[K-1, N]} = N,$$

and both sides of (4.107) vanish as $K \to \infty$. If N does go to infinity with K, notice that we can lower bound the expected rank by

$$\overline{[K-1, N]} \geq \min\{K-1, N\}\left(1 - \Gamma_{\min\{K-1,N\}}\right), \tag{4.109}$$

where Γ_n is the probability that an $n \times n$ matrix with random independent ± 1 elements is singular. Therefore,

$$E[\bar{\eta}_1] = 1 - \frac{\overline{[K-1, N]}}{N}$$

$$\leq 1 - \frac{\min\{K-1, N\}}{N}\left(1 - \Gamma_{\min\{K-1,N\}}\right)$$

$$= \left[1 - \frac{K-1}{N}\right]^+ + \frac{\min\{K-1, N\}}{N}\Gamma_{\min\{K-1,N\}}$$

$$\leq \left[1 - \frac{K-1}{N}\right]^+ + \Gamma_{\min\{K-1,N\}}.$$

To finish the proof of (4.108) all we need is

$$\Gamma_{\min\{K-1,N\}} \to 0$$

when both N and K grow without bound. This is given by the following combinatorial result:

Proposition 4.4 (*Komlós [222]*) *The probability that an $n \times n$ matrix with random independent ± 1 elements is singular goes to 0 as $n \to \infty$.*

As a matter of fact, when both $N \to \infty$ and $K \to \infty$,

$$\min\{K - 1, N\} - \overline{[K - 1, N]}$$

vanishes exponentially fast in $\min\{K - 1, N\}$ (Kahn et al. [191]), and therefore so does

$$E[\bar{\eta}_1] - \left[1 - \frac{K - 1}{N}\right]^+.$$

If

$$\lim_{K \to \infty} \frac{K}{N} = \beta < 1,$$

then not only

$$\lim_{K \to \infty} E[\bar{\eta}_1] = 1 - \beta, \qquad (4.110)$$

but in fact

$$\lim_{K \to \infty} \bar{\eta}_1 = 1 - \beta, \qquad (4.111)$$

where convergence is in mean-square sense. In other words, the variance of the near–far resistance due to the random choice of signature sequences vanishes with the number of users. In order to show (4.111), we will first change the distribution under which $\mathbf{R}$ is chosen by conditioning on the event $\{\mathbf{R}$ is invertible$\}$. Because the probability of this event goes to 1 and $0 \le \bar{\eta}_1 \le 1$, such a change of distribution has no effect on the asymptotic variance of $\bar{\eta}_1$. Consider the normalized variance:

$$\frac{\text{var}(\bar{\eta}_1)}{E^2[\bar{\eta}_1]} \le E\left[\frac{\bar{\eta}_1^2 + E^2[\bar{\eta}_1] - 2\bar{\eta}_1 E[\bar{\eta}_1]}{\bar{\eta}_1 E^2[\bar{\eta}_1]}\right] \qquad (4.112)$$

$$= E\left[\frac{1}{\bar{\eta}_1}\right] - \frac{1}{E[\bar{\eta}_1]}, \qquad (4.113)$$

where $\bar{\eta}_1 \le 1$ leads to (4.112). To conclude the proof of the mean-square convergence result in (4.111) all we need to show now is that the average near–far resistance is asymptotically equal to its harmonic mean:

$$\lim_{K \to \infty} E\left[\frac{1}{\bar{\eta}_1}\right] = \lim_{K \to \infty} \frac{1}{E[\bar{\eta}_1]}. \qquad (4.114)$$

This follows by invoking the asymptotic distribution of the eigenvalues of the crosscorrelation matrix (Proposition 2.1):

$$E\left[\frac{1}{\bar{\eta}_1}\right] = E[(\mathbf{R}^{-1})_{11}] \tag{4.115}$$

$$= \frac{1}{K}\sum_{k=1}^{K} E[(\mathbf{R}^{-1})_{kk}] \tag{4.116}$$

$$= \frac{1}{K}E[\text{trace}(\mathbf{R}^{-1})] \tag{4.117}$$

$$= \frac{1}{K}\sum_{k=1}^{K} E[\lambda_k(\mathbf{R}^{-1})] \tag{4.118}$$

$$= E\left[\frac{1}{\lambda_k(\mathbf{R})}\right] \tag{4.119}$$

$$\rightarrow \int \frac{1}{x}f_\beta(x)\,dx \tag{4.120}$$

$$= \frac{1}{1-\beta}. \tag{4.121}$$

Equation (4.115) comes from (4.97); the symmetry in the assignment of signature sequences to the users results in (4.116); Equation (4.118) reflects the fact that the trace of a square matrix is equal to the sum of its eigenvalues (denoted here by λ_k); Proposition 2.1 is used to write (4.120); finally the integral in (4.121) can be verified in Gradshteyn and Ryzhik [125].

4.5 MINIMUM ERROR PROBABILITY IN THE ASYNCHRONOUS CHANNEL

In Section 4.2 we found implementations of optimum multiuser detection for the asynchronous CDMA channel. In this section, we find bounds on the bit-error-rate achieved by those detectors. Here it will be preferable not to take into account the dynamic-programming-based optimum detection structure obtained in Section 4.2.

Recall from Section 2.9 that to obtain the minimum error probability in the asynchronous channel, we may analyze the discrete-time model of sufficient statistics in (2.106). Each bit in the frame of the kth user (Section 2.2),

$$\{b_k[-M], \ldots, b_k[0], \ldots, b_k[M]\}, \tag{4.122}$$

will, in general, have a different probability of error. For example, we would expect that the probability of error will decrease as we get closer to one of the boundaries of the frame. But we are interested in the bit-error-rate as $M \to \infty$, which can be expressed as

$$P_k(\sigma) = \lim_{M \to \infty} P_k^M(\sigma), \tag{4.123}$$

where we have denoted the minimum error probability of bit 0 within (4.122) by

$$P_k^M(\sigma) = P[\hat{b}_k[0] \neq b_k[0]]. \tag{4.124}$$

A channel with an M-frame will have the same error probabilities as a channel with an $(M + 1)$-frame and perfect side information of $b_j[\pm(M + 1)]$, $j = 1, \ldots, K$. Therefore, $P_k^M(\sigma)$ is monotonically increasing with M and, thus, the limit in (4.123) is guaranteed to exist.

As usual (cf. Chapter 2), we can view the asynchronous channel with K users and frame-length equal to $2M + 1$ as a "synchronous" channel with $(2M + 1)K$ "users" (one per bit), such that the "user" of interest is the one that transmits $b_k[0]$. This enables us to find upper and lower bounds on $P_k^M(\sigma)$ using the results in Section 4.3 on the synchronous minimum error probability. Now all we need to do is view the error vectors ϵ as being $(2M + 1)K$ dimensional; F_k is now the set of those indecomposable vectors such that $\epsilon_k[0] \neq 0$; $d_{k,\min}^2$ is the minimum energy among all vectors in F_k (or all error vectors for that matter).

We are not done yet since we have to take the limit as $M \to \infty$ in (4.123). Now error vectors become doubly infinite *error sequences*.

Let us examine first the lower bound in (4.89). We would end up with a trivial bound if all error sequences achieving $d_{k,\min}$ had an infinite number of nonzero components. Fortunately, it can be shown (Problem 4.45) that $d_{k,\min}$ is always achieved by a finite-length error sequence whose weight is upper bounded by a function of K.

Regarding the upper bound,

$$P_k(\sigma) \leq \sum_{\epsilon \in F_k} 2^{-w(\epsilon)} Q\left(\frac{\|S(\epsilon)\|}{\sigma}\right), \tag{4.125}$$

we now have an infinite series, so we have to elucidate whether it converges, and if it does, whether it gives tight results for high signal-to-noise ratio as in the synchronous case. The weaker bound (4.80), which includes all error sequences (not just indecomposable ones), is useless in the asynchronous case because it diverges for all σ (Problem 4.46). For a given set of signature waveforms and offsets, the series in (4.124) will behave according to one of the following alternatives:

A. It converges for all $\sigma > 0$.
B. It converges for $0 < \sigma < \sigma_0$.
C. It diverges for all $\sigma > 0$.

We refer the reader to Verdú [490] for a proof of the results on the convergence analysis of the series in (4.125). The main conclusions that can be drawn from that analysis are:

1. Two-user channels belong to type A.
2. For any set of K signature waveforms, if the offsets are uniformly distributed and independent, then type C happens with zero probability.
3. Type C may only occur for some signature waveforms with very heavy crosscorrelations and where subsets of users are synchronized (Problem 4.39).

Provided that the signature waveforms and offsets are such that they belong to either type A or B, it is shown in Verdú [490] that (4.87) holds, and in that case C_k^U is finite. This fact and the aforementioned behavior of the lower bound enable us to conclude that as long as the bounding series is not everywhere divergent (type C) (which as we mentioned occurs with zero probability), the optimum asymptotic efficiency is given by

$$\eta_k = \frac{d_{k,\min}^2}{A_k^2}. \tag{4.126}$$

In the two-user case, it can be shown (Problem 4.44) that (4.126) particularizes to (Figure 4.17)

$$\eta_1 = 1 - \frac{A_2}{A_1}\left[\left(2|\rho_{12}| - \frac{A_2}{A_1}\right)^+ + \left(2|\rho_{21}| - \frac{A_2}{A_1}\right)^+\right]. \tag{4.127}$$

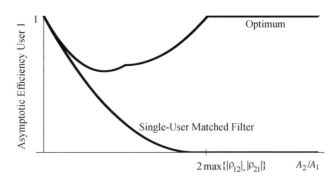

FIGURE 4.17.
Optimum and single-user matched-filter asymptotic multiuser efficiencies as functions of the amplitude of the interferer.

The minimization of (4.127) with respect to $A_2/A_1 \geq 0$ yields

$$\bar{\eta}_1 = 1 - \max^2 \left\{ |\rho_{12}|, |\rho_{21}|, \frac{|\rho_{12}| + |\rho_{21}|}{\sqrt{2}} \right\}. \tag{4.128}$$

This corresponds to the version of near–far resistance where the received amplitudes are held constant for all bits (cf. (3.117)).

We will not attempt to generalize (4.128) to the K-user case. Instead, as we anticipated in Chapter 3, a more meaningful and tractable measure of near–far resistance can be obtained by letting the amplitudes vary over time in an arbitrary fashion (cf. (3.118)). We need to follow parallel lines to the synchronous case analysis that lead us from (4.91) to (4.97). In this case, however, we are working with what amounts to an infinite crosscorrelation matrix, and it is desirable to give an expression for near–far resistance directly in terms of the asynchronous crosscorrelation matrices $\mathbf{R}[0]$ and $\mathbf{R}[1]$. As justified in Chapter 5, the expression for the optimum asynchronous near–far resistance is

$$\bar{\eta}_k = \left[\frac{1}{\pi} \int_0^\pi [\mathbf{R}^T[1]e^{j\omega} + \mathbf{R}[0] + \mathbf{R}[1]e^{-j\omega}]_{kk}^{-1} \, d\omega \right]^{-1}. \tag{4.129}$$

In the two-user case, the first step to evaluate (4.129) is to take the inverse of the Hermitian matrix therein:

$$[\mathbf{R}^T[1]e^{j\omega} + \mathbf{R}[0] + \mathbf{R}[1]e^{-j\omega}]^{-1}$$

$$= \frac{1}{1 - [\rho_{12}^2 + \rho_{21}^2 + 2\rho_{12}\rho_{21}\cos\omega]}$$

$$\times \begin{bmatrix} 1 & -\rho_{12} - \rho_{21}e^{-j\omega} \\ -\rho_{12} - \rho_{21}e^{j\omega} & 1 \end{bmatrix}. \tag{4.130}$$

Substituting (4.130) into (4.129) and using a table of integrals (e.g., (Gradshteyn and Ryzhik [125])), we can evaluate

$$\bar{\eta}_k^{-1} = \frac{1}{2\pi} \int_{-\pi}^\pi \frac{1}{1 - [\rho_{12}^2 + \rho_{21}^2 + 2\rho_{12}\rho_{21}\cos\omega]} \, d\omega$$

$$= \frac{1}{\sqrt{\left(1 - \rho_{12}^2 - \rho_{21}^2\right)^2 - 4\rho_{21}^2\rho_{12}^2}}$$

$$= \frac{1}{\sqrt{[1 - (\rho_{12} + \rho_{21})^2][1 - (\rho_{12} - \rho_{21})^2]}}, \tag{4.131}$$

and therefore,

$$\bar{\eta}_k = \sqrt{[1 - (\rho_{12} + \rho_{21})^2][1 - (\rho_{12} - \rho_{21})^2]}. \tag{4.132}$$

Note that (4.132) is the geometric mean of the synchronous near–far resistances obtained with the even and odd periodic crosscorrelations $\rho_{12} + \rho_{21}$ and $\rho_{12} - \rho_{21}$.

We can use the results in Section 4.4 on the near–far resistance of random direct-sequence spread spectrum to obtain a lower bound on the expected optimum near–far resistance in the asynchronous chip-synchronous case (i.e., when the relative offsets are fractions of the chip duration). To that end, we first lower bound optimum near–far resistance by that achieved by a demodulator that discards the received signal outside the interval of the bit of the desired user. In that event, the problem is equivalent to one with two synchronous interfering pseudo-users per interfering user, that is, $2K - 2$ interfering pseudo-users. The signature waveforms of interfering pseudo-users that overlap with the desired signal on the left [resp. right] are assumed to have $d_{ki} = 0$ for i sufficiently high [resp. low]. As we pointed out, the above analysis is general enough to accommodate interfering pseudo-users whose direct-sequence waveforms take values in $\{-1, 0, 1\}$. We can conclude that in a symbol-asynchronous chip-synchronous CDMA channel the optimum near–far resistance satisfies

$$E[\bar{\eta}_1] \geq \left[1 - \frac{2K - 2}{N} \right]^+ . \tag{4.133}$$

In the case of a suboptimal receiver whose observables are the chip matched-filter outputs of the desired user, a similar analysis can be shown to hold (Problem 4.60).

4.6 PERFORMANCE ANALYSIS IN RAYLEIGH FADING

The generalization of the optimum detector to coherent demodulation in the presence of frequency-flat or frequency-selective fading is conceptually trivial. Thanks to the assumption that the magnitudes and phases of $\{A_k, k = 1, \ldots, K\}$ are known at the receiver, the derivation of the optimum detectors is entirely similar to the one we saw for the real-valued channel and it will not be repeated here. Instead, this section is devoted to the analysis of the minimum error probability when the received amplitudes are Rayleigh-distributed and known to the receiver. The major conclusion of this analysis is that the asymptotic multiuser efficiency is equal to 1 as long as the signature waveforms are linearly independent.

The relevant model here is the complex-valued channel in (2.41),

$$y(t) = \sum_{k=1}^{K} \mathsf{A}_k b_k \mathsf{s}_k(t) + \sigma \mathsf{n}(t), \tag{4.134}$$

where

$$\mathsf{A}_k = A_k \mathsf{U}_k$$

and U_k are independent complex Gaussian random variables with independent real and imaginary parts, each with unit second moment.

The method of indecomposable vectors can also be shown to apply in this case (Zvonar and Brady [570]). Here, we will concentrate on showing that in the synchronous channel the asymptotic multiuser efficiency of the probability of error averaged with respect to the fading coefficients $\{\mathsf{A}_k, k = 1, \ldots, K\}$ is equal to 1. For this purpose, it will be sufficient to average the upper bound in (4.80). This requires the average

$$E\left[Q\left(\frac{\|S(\epsilon)\|}{\sigma} \right) \right], \tag{4.135}$$

where according to (4.71)

$$\|S(\epsilon)\|^2 = \epsilon^T \mathbf{u}^* \mathbf{H} \mathbf{u} \epsilon$$

$$= \mathbf{u}^* \operatorname{diag}\{\epsilon\} \mathbf{H} \operatorname{diag}\{\epsilon\} \mathbf{u} \tag{4.136}$$

with

$$\mathbf{u} = \begin{bmatrix} \mathsf{U}_1 \\ \vdots \\ \mathsf{U}_K \end{bmatrix}.$$

The unnormalized crosscorrelation matrix $\mathbf{H}$ (assumed to be nonsingular) admits the factorization

$$\mathbf{H} = \mathbf{F}^T \mathbf{F},$$

with which we can define the zero-mean Gaussian complex K-vector

$$\mathbf{X} = \frac{1}{\sigma} \mathbf{F} \operatorname{diag}\{\epsilon\} \mathbf{u}.$$

The covariance matrix of $\mathbf{X}$ is

$$E[\mathbf{X}\mathbf{X}^T] = \sigma^{-2} \mathbf{F} \operatorname{diag}\{\epsilon\} E[\mathbf{u}\mathbf{u}^*] \operatorname{diag}\{\epsilon\} \mathbf{F}^T$$

$$= 2\sigma^{-2} \mathbf{F} \operatorname{diag}^2\{\epsilon\} \mathbf{F}^T, \tag{4.137}$$

which has rank $w(\epsilon)$. The rationale for introducing $\mathbf{X}$ is that the argument in the Q-function of (4.135) is $\|\mathbf{X}\|$. To apply the averaging formula (3.64), we need to investigate the eigenvalues of (4.137). Denote by $\mathbf{F}_\epsilon$ the $w(\epsilon) \times w(\epsilon)$ matrix that results from striking out the rows and columns of $\mathbf{F}$ corresponding to indices for which $\epsilon_i = 0$. The eigenvalues of $\mathbf{F}_\epsilon \mathbf{F}_\epsilon^T$ can be shown to be equal to the eigenvalues of the positive definite $w(\epsilon) \times w(\epsilon)$ matrix that results from striking out all the rows and columns from $\mathbf{H}$ corresponding to indices for which $\epsilon_i = 0$. Those eigenvalues will be denoted by $\lambda_i(\epsilon)$, $i = 1, \ldots, w(\epsilon)$, and their product (equal to the determinant of the surviving submatrix of $\mathbf{H}$) will be denoted by $\Delta(\epsilon)$. Assuming that those eigenvalues are distinct we can apply the formula in (3.64) to obtain

$$P_k^F(\sigma) \leq \sum_{\epsilon \in E_k} 2^{-w(\epsilon)} \sum_{i=1}^{w(\epsilon)} \frac{\alpha_i(\epsilon)}{2} \left(1 - \frac{1}{\sqrt{\frac{\sigma^2}{\lambda_i(\epsilon)} + 1}} \right), \qquad (4.138)$$

where

$$\alpha_j(\epsilon) = \prod_{i=1, i \neq j}^{w(\epsilon)} \frac{\lambda_j(\epsilon)}{\lambda_j(\epsilon) - \lambda_i(\epsilon)}.$$

For the purposes of an asymptotic analysis of $P_k^F(\sigma)$, $\sigma \to 0$, it is convenient to use the Taylor series expansion

$$1 - \frac{1}{\sqrt{1+x^2}} = \sum_{n=1}^{\infty} (-1)^{n+1} \frac{(2n-1)(2n-3) \cdots 1}{n! \, 2^n} x^n, \qquad (4.139)$$

which, when substituted in (4.138), indicates that performance depends on the crosscorrelation matrix only through the quantities

$$\sum_{i=1}^{w(\epsilon)} \frac{\alpha_i(\epsilon)}{\lambda_i^n(\epsilon)} = \begin{cases} 0, & n = 1, \ldots, w(\epsilon) - 1; \\ \frac{(-1)^{w(\epsilon)-1}}{\Delta(\epsilon)}, & n = w(\epsilon), \end{cases} \qquad (4.140)$$

an equation justified in Zvonar and Brady [570]. Putting together (4.138), (4.139), and (4.140) we obtain the following series for the upper bound on error probability:

$$P_k^F(\sigma) \leq \sum_{\epsilon \in E_k} \frac{2^{-2w(\epsilon)-1}}{w(\epsilon)!} \frac{(2w(\epsilon) - 1)(2w(\epsilon) - 3) \cdots 1}{\Delta(\epsilon)} \sigma^{2w(\epsilon)} + o(\sigma^{2w(\epsilon)}). \qquad (4.141)$$

For $\sigma \to 0$, the right side of (4.141) is dominated by the two terms with $w(\epsilon) = 1$, for which $\Delta(\epsilon) = A_k^2$. Consequently,

$$P_k^F(\sigma) \leq \frac{\sigma^2}{4A_k^2} + o(\sigma^2). \qquad (4.142)$$

Comparing (4.142) with the asymptotic result for the single-user channel (3.131), we conclude that the asymptotic multiuser efficiency is equal to 1. We recall that the assumptions used to reach this result are that the signature waveforms of all K users are linearly independent and that the eigenvalues of the unnormalized crosscorrelation matrix of any subset of users are distinct.

The analysis in the asynchronous case is more involved. For example, the upper bound that includes all sequences in E_k is useless; however, a coarser notion of decomposability that does not depend on the actual amplitudes can be developed so that the bounding series excludes any sequence including a run of $K - 1$ or more consecutive zeros. Furthermore, performance in the asynchronous case depends on the autocorrelation function of the fading coefficients. Nevertheless, the property of unit asymptotic multiuser efficiency is shown in Zvonar and Brady [570] to remain true in the asynchronous channel unless the users are assigned identical signature waveforms.

4.7 OPTIMUM NONCOHERENT MULTIUSER DETECTION

This section briefly considers the problem of noncoherent multiuser demodulation. The channel of interest is (3.170)

$$y(t) = \sum_{k=1}^{K} |A_k| e^{j\theta_k} s_k(t; b_k) + \sigma n(t), \qquad (4.143)$$

where $b_k \in \mathcal{A}$ and θ_k are independent and uniformly distributed and unknown to the receiver. The single-user receiver in Section 3.8 correlates with each of the signal hypotheses of the desired user and selects that corresponding to the largest magnitude output. Such a receiver does not need to know either the magnitudes $|A_k|$ or their distribution. In contrast, the way the maximum-likelihood K-user receiver processes the $|\mathcal{A}|K$ sufficient statistics (Problem 4.63) does depend on those distributions. Maximum-likelihood K-user detection for (4.143) is equivalent to a (single-user) non-orthogonal m-ary noncoherent hypothesis testing problem, with $m = |\mathcal{A}|^K$.

The structure of (4.143) can be exploited more fully in the particular case of Rayleigh fading where the received magnitudes are unknown to the receiver and $|A_k| e^{j\theta_k}$ are independent, zero-mean, Gaussian, complex random variables. In that case, the $|\mathcal{A}|K$ observables are zero-mean complex Gaussian when conditioned on each of the $|\mathcal{A}|^K$ hypotheses. Thus, the information is carried in the covariance matrix of the observables under each of the hypotheses. The optimum decision rule can be found in Problem 3.11.

209

4.8   BIBLIOGRAPHICAL NOTES

The derivation and analysis of the optimum multiuser receiver date back to the early eighties (Verdú [485, 486, 487]). The chief reason why multiuser detection did not develop until then was the belief shared by many spread-spectrum practitioners that the conventional matched-filter detector is essentially optimum (cf. Verdú [502]). Techniques from both minimax robustness and non-Gaussian signal detection to improve the performance of the conventional single-user detector in multiuser channels were proposed in Poor [334] to deal with the non-Gaussian nature of multiaccess interference. Several earlier works had already investigated receivers that used the knowledge of the interfering signature waveforms to improve the performance of CDMA systems; Timor [445, 446] showed that it was possible to double the maximum number of simultaneous users achievable with the conventional noncoherent demodulator of frequency-hopped FSK systems; Horwood and Gagliardi [160] and Schneider [392, 393] considered multiuser receivers for synchronous systems. Those ideas were closely related to earlier work to combat cross-talk in multichannel communication systems in Shnidman [409], Kaye and George [206], Etten [80, 81], among others. Those works dealt with Pulse Amplitude Modulation signals transmitted through multi-input multi-output dispersive channels, which can be viewed as a vector generalization of the conventional scalar intersymbol interference model. The maximum-likelihood sequence detector for single-user intersymbol interference is due to Forney [96], with contemporaneous discoveries in Omura [308] and Kobayashi [212] and an earlier hint in Viterbi [508]. The Viterbi algorithm was originally introduced in the context of decoding of convolutional codes in Viterbi [507].

The optimum detection algorithm for asynchronous channels presented in this chapter appeared in Verdú [485, 487, 490]. The metric in that algorithm can be viewed as a periodically time-varying generalization of that proposed in Ungerboeck [459] for single-user channels. The applicability of the Viterbi algorithm for the demodulation of asynchronous multiuser channels had been hinted earlier in Schneider [392] and Kohno et al. [215]. The vector versions of the Viterbi algorithm proposed in Etten [81] and Verdú [486] do not exploit fully the structure of the asynchronous channel and (for the same complexity) can accommodate only a fraction of the number of users of the algorithm in Verdú [490]. The analysis of the computational complexity and the proof that the optimum multiuser detection problem is combinatorially hard appeared in Verdú ([487] and [496]). Optimum synchronous multiuser detection was shown to be solvable in polynomial time for nonpositive crosscorrelations in Sankaran and Ephremides [380]. Other signature waveform designs that lead to polynomial-in-K jointly optimum detection are developed in Learned et al. [230].

The minimum bit-error-rate multiuser receiver based on a backward–forward dynamic programming algorithm is due to Verdú [487] and Verdú and Poor [503], which rediscovered the backward–forward paradigm earlier used in data demodulation problems in Chang and Hancock [50], Abend and Fritchman [7], and Bahl et al. [18]. In the case of single-user intersymbol interference, the backward–forward dynamic programming algorithm of Verdú [487] provided significant computational savings over the algorithm proposed in Hayes et al. [138], which has been adapted for soft-decision multiuser detection in Hafeez and Stark [132].

Various tree-search algorithms that provide suboptimal demodulation with reduced computational complexity have been presented in Xie et al. [548] and Wei et al. [526].

The complexity of optimum multiuser detection for hybrid FDMA/CDMA with random assignment of frequency band is studied in Liu et al. [249].

Combined maximum-likelihood multiuser demodulation and decoding of trellis codes has been studied in Giallorenzi and Wilson [113], Fawer and Aazhang [88], Giallorenzi and Wilson [114], Giallorenzi and Wilson [115], and Shama and Vojcic [405]. Reference Vojcic et al. [517] consider maximum-likelihood decoding of the error-correcting code using as observables the a posteriori probabilities (4.18) supplied by a backward–forward multiuser detector.

The upper-bound analysis based on indecomposable sequences was first given in Verdú [490] in the context of asynchronous channels. The same method was applied in Verdú [495] to the single-user intersymbol interference problem to yield a bound that is tighter than the Forney bound (Forney [96]). A branch-and-bound algorithm to compute the indecomposable-sequence bound is given in Verdú [488]. The genie-based lower-bound analysis elaborates on the original idea in Forney [95]. Neither that reference nor Verdú [490] give full justification for the constant in their claimed lower bounds. The optimum near–far resistance was obtained in Verdú [489] and Lupas and Verdú [254] in the synchronous case and in Lupas and Verdú [253], and Lupas and Verdú [255] in the asynchronous case; Visotsky and Madhow [506] examine the optimum near–far resistance for nonlinearly modulated channels (Problem 4.64). The lower bound on the expected maximum near–far resistance with random binary direct-sequence signature waveforms is due to Madhow and Hoing [262]; the asymptotic analysis is new. Optimum near–far resistance with random binary direct-sequence signature waveforms and rectangular chips is averaged with respect to offsets and carrier phases in Acar [11] and Acar and Tantaratana [12]. The (nonasymptotic) distribution of the near–far resistance with random Gaussian direct-sequence signatures is found in Müller et al. [292]. The asymptotic analysis of the optimum asymptotic efficiency achieved by random signature sequences is an open problem.

The derivation and analysis of the optimum multiuser detector in the presence of frequency-flat Rayleigh fading (Section 4.6) is due to Zvonar and Brady [570], and a system that incorporates trellis-coded modulation is analyzed in Caire et al. [41]. A modified detector that takes into account imperfect estimation of the fading coefficients is proposed in Vasudevan and Varanasi [480]. Frequency-flat and frequency-selective Rician fading is considered in Varanasi and Vasudevan [478] and Vasudevan and Varanasi [479]. Optimum noncoherent detection in the presence of Rayleigh fading is studied in Russ and Varanasi [370, 371].

In environments where the positions of the transmitters relative to the receiver are not fixed or where the received amplitudes are not known a priori for whatever reasons, it is of interest to investigate the problem of multiuser amplitude estimation from the received waveform, as those amplitudes are required by the optimum detector and other detectors reviewed in future chapters. This problem was treated first in Verdú [487] under the assumption that the transmitted information sequence is known (i.e., when a start-up training sequence is sent). This assumption was dropped in Poor [335] and Steinberg and Poor [336]. Further work on this problem and the problem of phase estimation was undertaken in Miller [276] and Strom and Miller [428] including the derivation of Crámer–Rao bounds. The case of unknown information sequences is treated in Moon et al. [286] and Miller [276], when the amplitudes are modeled as arbitrarily varying, and in Xie et al. [549], which gives a tree-search algorithm for joint data detection and amplitude estimation in the basic CDMA asynchronous channel. Joint optimum multiuser demodulation and estimation of multipath fading coefficients is developed in Wang and Poor [521] assuming a rational spectrum for the fading coefficients. The solution consists of a dynamic programming algorithm where each surviving path is assigned a different Kalman filter for the estimation of the fading coefficients. The effect of receiver mismatch on optimum performance is treated in Gray et al. [126].

Joint amplitude and delay estimation is studied in Iltis and Mailaender [171] and Lim and Ramussen [241]. Although conventional code acquisition and synchronization systems for CDMA have neglected the presence of multiaccess interference, a number of works have exploited the structure of the multiaccess interference to increase robustness, near–far resistance, and acquisition/synchronization speed (Iltis and Mailaender [172]; Madhow [258]; Bensley and Aazhang [271]; Strom et al. [431]; Strom et al. [430]; Zheng et al. [561]; and Srinivasan et al. [419]). Methods to determine which users are active at any given time are explored in Halford and Brandt-Pearce [136].

Optimum multiuser detection with array observations is pursued in Miller [276], Miller and Schwartz [277], Kohno et al. [219], and Lee and Pickholtz [233].

Reference Verdú [492] derives and analyzes an optimal detector for the optical multiple-access point-process channel. The noncoherence and nonlinearity of this channel model do not prevent the existence of dynamic-programming based demodulators that exploit the sequential structure of the likelihood function; however, they do render the method of indecomposable sequences less powerful.

4.9   PROBLEMS

PROBLEM 4.1. Let $\hat{\mathbf{b}}$ be the vector in the set $\{-1, +1\}^K$ that maximizes

$$\Omega(\mathbf{b}) = 2\mathbf{b}^T \mathbf{A} \mathbf{y} - \mathbf{b}^T \mathbf{H} \mathbf{b}.$$

Show that

$$\hat{b}_k = \text{sgn}\left(y_k - \sum_{j \neq k} \hat{b}_j A_j \rho_{jk} \right). \tag{4.144}$$

PROBLEM 4.2. Let $\hat{\mathbf{b}}$ be the vector in the set $\{-1, +1\}^K$ that maximizes

$$\Omega(\mathbf{b}) = 2\mathbf{b}^T \mathbf{A} \mathbf{y} - \mathbf{b}^T \mathbf{H} \mathbf{b}.$$

(a) Show that $\hat{\mathbf{b}}$ minimizes

$$\|\mathbf{r} - \mathbf{S} \mathbf{A} \mathbf{b}\|,$$

where $\mathbf{r}$ are L-dimensional observations in an orthonormal space (2.97) and **S** is the matrix of signature vectors (2.98) in that space.

(b) Show that $\hat{\mathbf{b}}$ maximizes

$$\sum_{k=1}^{K} A_k b_k \mathbf{s}_k^T \left(\mathbf{r} - \sum_{j=k+1}^{K} A_j b_j \mathbf{s}_j \right).$$

PROBLEM 4.3. Let $\hat{b}_1 \in \{-1, +1\}$ and $\hat{b}_2 \in \{-1, +1\}$ be the bits that maximize the function

$$A_1 y_1 b_1 + A_2 y_2 b_2 - A_1 A_2 \rho b_1 b_2.$$

Show that $\hat{b}_1$ and $\hat{b}_2$ satisfy

(a) If $\min\{A_1|y_1|, A_2|y_2|\} \geq A_1 A_2 |\rho|$, then

$$\hat{b}_1 = \text{sgn}(y_1),$$

$$\hat{b}_2 = \text{sgn}(y_2).$$

Otherwise,

$$\hat{b}_1 = \text{sgn}(A_1 y_1 - \text{sgn}(\rho) A_2 y_2),$$

$$\hat{b}_2 = \text{sgn}(A_2 y_2 - \text{sgn}(\rho) A_1 y_1).$$

(b)

$$\hat{b}_1 = \text{sgn}\left(A_1 y_1 + \frac{1}{2}|A_2 y_2 - A_1 A_2 \rho| - \frac{1}{2}|A_2 y_2 + A_1 A_2 \rho| \right)$$

$$\hat{b}_2 = \text{sgn}\left(A_2 y_2 + \frac{1}{2}|A_1 y_1 - A_1 A_2 \rho| - \frac{1}{2}|A_1 y_1 + A_1 A_2 \rho| \right).$$

PROBLEM 4.4. Show that the scheme in Figure 4.18 outputs

$$\log\left(\frac{1}{P[b_1 = +1|y]} - 1 \right)$$

for the two-user synchronous channel.

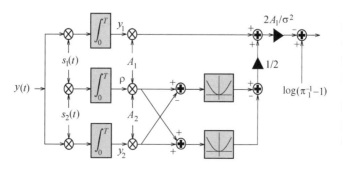

FIGURE 4.18.
Computation of the a posteriori probability $P[b_1 = +1|y]$ from the a priori probability $\pi_1 = P[b_1 = +1]$.

PROBLEM 4.5. Consider a K-user synchronous CDMA system demodulated by a bank of single-user matched-filter receivers.

(a) Show that at least one of the K bits selected by the maximum-likelihood (jointly optimum) multiuser detector is in agreement with the corresponding output of the bank of single-user matched filters.

(b) Consider the special case $\rho_{ij} = \rho > 0$ for all $i \neq j$. Let user l be such that

$$A_l|y_l| \geq A_j|y_j|,$$

for $j \neq l$. Show that sgn(y_l) is the jointly optimal decision for the lth user.

(c) Show that if the crosscorrelations are identical but negative the result of (b) remains true for $K = 2$ but it is false for $K > 2$ (cf. Kempf [207]).

PROBLEM 4.6. Consider a receiver for user 1 that uses y_1, the output of the filter matched to the signature waveform of user 1 as its only observable. Derive the decision rule on y_1 that minimizes the probability of error of user 1 for a synchronous K-user CDMA system, in the following cases:

(a) The open-eye condition (3.92) is satisfied.
(b) $K = 3$, $A_1 = A_2 = A_3$, $\rho_{12} = -0.8$, $\rho_{13} = 0.3$, $A_1/\sigma \to \infty$.

PROBLEM 4.7. Consider a two-user receiver for synchronous CDMA that outputs $(\hat{b}_1, \hat{b}_2)$ that maximizes

$$\langle y, A_1 b_1 s_1 + A_2 b_2 s_2 \rangle.$$

Depict the decision regions for $\rho = 0.2$ in the cases $A_1 = A_2$ and $A_1 = 6A_2$. Compare with Figures 4.1 and 4.2.

PROBLEM 4.8. Find an algorithm to find the argument that maximizes (4.11) with time complexity per bit equal to $O(2^K/K)$.

PROBLEM 4.9. Consider a synchronous three-user channel with $A_1 = A_2 = A_3$,

$$s_3 = \frac{s_1 + s_2}{\sqrt{2}},$$

and orthogonal s_1 and s_2.

(a) Depict the decision regions for jointly optimal detection.
(b) Depict the decision regions if $b_3 = b_1 b_2$.

PROBLEM 4.10. Show that the maximization of

$$\Omega(\mathbf{b}) = 2\mathbf{b}^T \mathbf{A} \mathbf{y} - \mathbf{b}^T \mathbf{H} \mathbf{b}$$

over $\mathbf{b} \in \{-1, 1\}^K$ is equivalent to the minimization of

$$\mathbf{x}^T \mathbf{G} \mathbf{x} + \mathbf{z}^T \mathbf{x}$$

over $\mathbf{x} \in \{0, 1\}^K$, with

$$\mathbf{G} = \mathbf{H} - \operatorname{diag} \phi,$$

$$\mathbf{z} = \phi - \mathbf{H}[1 \cdots 1]^T - \mathbf{A} \mathbf{y},$$

and ϕ an arbitrary K-vector. [*Note*: This transformation has been used in Shi et al. [408] and Wu and Wang [544] to obtain suboptimal multiuser detectors.]

PROBLEM 4.11. Consider the K-user, D-channel synchronous diversity model in (2.75):

$$y_1(t) = \sum_{k=1}^K A_{1k} b_k s_{1k}(t) + \sigma n_1(t),$$

$$\vdots \qquad\qquad\qquad (4.145)$$

$$y_D(t) = \sum_{k=1}^K A_{Dk} b_k s_{Dk}(t) + \sigma n_D(t),$$

where the noise processes are independent.

(a) Show that the set of observables

$$\{y_{dk}, d = 1, \dots, K; k = 1, \dots K\}$$

is a sufficient statistic for $b_1, \dots, b_K$, with

$$y_{dk} = \int_0^T y_d(t) s_{dk}^*(t) \, dt. \qquad (4.146)$$

(b) The observable

$$y_{2k1} = \int_0^T y_1(t) s_{2k}^*(t) \, dt$$

depends on $b_1, \dots, b_K$ unless s_{2k} is orthogonal to $\{s_{1j}, j = 1, \dots, K\}$. Explain why y_{2k1} need not be included in the sufficient statistics found in (a).

PROBLEM 4.12. Give a backward dynamic programming algorithm to maximize (4.34).

PROBLEM 4.13. Consider the general setup of (nonlinear) non-antipodal modulation in which user k transmits

$$s_k(t; u)$$

in order to send symbol $u \in \{u_1, \ldots, u_{n_k}\}$ (cf. Section 2.5.2).

(a) Show that the collection of observables

$$y_{j,u} = \int_{-\infty}^{\infty} v_j(t; u) y(t)\, dt \qquad (4.147)$$

with (cf. (4.21))

$$v_{k+iK}(t; u) = s_k(t - iT - \tau_k; u)$$

are sufficient statistics.

(b) (Verdú [490]) Generalize the metric (4.34) for a dynamic programming solution to maximum-likelihood, asynchronous, nonantipodal multiuser detection. The new metric should depend on the received amplitudes A_k, the matched filter outputs in (4.147), and the crosscorrelations ($k < l$)

$$\rho_{kl}(u_k, u_l) = \int_{\tau_l - \tau_k}^{T} s_k(t; u_k) s_l(t - \tau_l + \tau_k; u_l)\, dt, \quad (4.148a)$$

$$\rho_{lk}(u_l, u_k) = \int_{0}^{\tau_l - \tau_k} s_k(t; u_k) s_l(t + T - \tau_l + \tau_k; u_l)\, dt.$$

$$(4.148b)$$

(c) Find the time complexity per bit of the algorithm in (b) assuming $n_k = m$.

(d) Does the time complexity per bit found in (c) decrease if the modulation is linear?

PROBLEM 4.14. (Verdú [487]) Suppose that the K users can be divided in S groups such that within each group users are synchronized but users in different groups are not mutually synchronized. Find a maximum-likelihood multiuser detector with time complexity per bit equal to $O(S2^K/K)$.

PROBLEM 4.15. (Verdú [486])

(a) Show that the asynchronous likelihood function (4.31) can be decomposed as

$$\Omega(\mathbf{b}) = \sum_{i=-M}^{M} \bar{\lambda}_i(\mathbf{x}_i, \mathbf{b}[i]), \qquad (4.149)$$

217

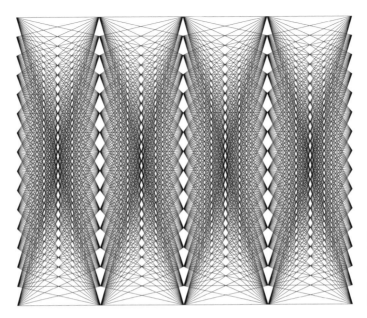

FIGURE 4.19.
Trellis for a 4-user channel using the decomposition in Problem 4.15.

where $\mathbf{x}_M = 0$,

$$\mathbf{x}_{i+1} = \mathbf{b}[i] = [b_1[i], \dots, b_K[i]]^T,$$

and

$$\bar{\lambda}_i(\mathbf{x}, \mathbf{b}) = 2\mathbf{y}^T[i]\mathbf{A}\mathbf{b} - \mathbf{b}^T\mathbf{A}\mathbf{R}[0]\mathbf{A}\mathbf{b} - 2\mathbf{b}^T\mathbf{A}\mathbf{R}[1]\mathbf{A}\mathbf{x}.$$

$$(4.150)$$

(b) Note that the trellis corresponding to the decomposition in (4.150) has the property that each state is connected to all previous states (Figure 4.19). Show that the time complexity per bit of the dynamic programming algorithm derived from the decomposition in (4.150) is $O(4^K/K)$.

PROBLEM 4.16. (Duel-Hallen [74]) Derive a dynamic programming algorithm for maximum-likelihood detection in a K-user asynchronous channel using the whitened observables in (2.117).

PROBLEM 4.17. Consider a four-user synchronous channel with equal received powers:

$$y(t) = Ab_1s_1(t) + Ab_2s_2(t) + Ab_3s_3(t) + Ab_4s_4(t) + \sigma n(t).$$

Assume that the signature waveforms s_1, s_2, s_3, s_4 are the following direct-sequence spread-spectrum waveforms of length $N = 3$: $[+++], [+-+], [+--], [--+]$ with duration-$\frac{1}{3}$ chip waveform $p_{1/3}$. In this problem we explore two alternative implementations of maximum-likelihood (jointly optimal) detection.

(a) Count the number of arithmetic operations (sums and comparisons of real numbers) that it takes to select the maximum-likelihood $(\hat{b}_1, \hat{b}_2, \hat{b}_3, \hat{b}_4)$ by exhaustive computation of the sixteen possible values of $\Omega(\mathbf{b})$.

(b) Define the chip matched filter outputs

$$r_1 = \frac{1}{A} \int_0^{1/3} p_{1/3}(t) y(t)\, dt, \tag{4.151}$$

$$r_2 = \frac{1}{A} \int_{1/3}^{2/3} p_{1/3}\left(t - \frac{1}{3}\right) y(t)\, dt, \tag{4.152}$$

$$r_3 = \frac{1}{A} \int_{2/3}^{1} p_{1/3}\left(t - \frac{2}{3}\right) y(t)\, dt. \tag{4.153}$$

Let $\xi_i = \text{sgn}(r_i)$ and

$$\zeta_1 = 2 \max|r_i| - \frac{4}{3}, \tag{4.154}$$

$$\zeta_2 = |r_1| + |r_2| + |r_3| - 1. \tag{4.155}$$

Consider the following decision rule:
If $\zeta_1 \geq \max\{\zeta_2, 0\}$ then

$(\hat{b}_1, \hat{b}_2, \hat{b}_3, \hat{b}_4)$

$$= \begin{cases} (\xi_1, \xi_1, \xi_1, -\xi_1), & \text{if } |r_1| > \max\{|r_2|, |r_3|\}, \\ (\xi_2, -\xi_2, -\xi_2, -\xi_2), & \text{if } |r_2| > \max\{|r_1|, |r_3|\}, \\ (\xi_3, \xi_3, -\xi_3, \xi_3), & \text{if } |r_3| > \max\{|r_1|, |r_2|\}. \end{cases}$$

If $\zeta_2 > \max\{\zeta_1, 0\}$ then

ξ_1	ξ_2	ξ_3	$\hat{b}_1$	$\hat{b}_2$	$\hat{b}_3$	$\hat{b}_4$
+	+	+	+	+	−	−
+	+	−	+	−	+	−
+	−	+	+	+	+	+
+	−	−	−	+	+	−
−	+	+	+	−	−	+
−	+	−	−	−	−	−
−	−	+	−	+	−	+
−	−	−	−	−	+	+

If $\zeta_1 < 0$ and $\zeta_2 < 0$, then

$$(\hat{b}_1, \hat{b}_2, \hat{b}_3, \hat{b}_4) = (+1, -1, +1, +1).$$

Show that this decision rule is maximum likelihood.

(c) Count the number of operations of the decision rule in (b).

PROBLEM 4.18. (Sendonaris and Aazhang [400]) Assume that the amplitudes $A_1, \ldots, A_K$ in a synchronous CDMA channel are unknown to the receiver, and they are independent, with probability density functions denoted by p_{A_k}, $k = 1, \ldots, K$. Show that the jointly optimum decisions satisfy for some $\mathbf{x} \in R^K$:

$$\hat{\mathbf{b}} = \text{sgn}(\mathbf{x}),$$

$$\mathbf{R}\mathbf{x} - \mathbf{y} - \sigma^2 \mathbf{v}(\mathbf{x}) = 0,$$

where

$$\mathbf{v}(\mathbf{x}) = \left[\frac{p'_{A_1}(x_1)}{p_{A_1}(x_1)} \hat{b}_1, \ldots, \frac{p'_{A_K}(x_K)}{p_{A_K}(x_K)} \hat{b}_K \right]^T.$$

PROBLEM 4.19. Consider a general additive multiuser channel

$$y(t) = \sum_{k=1}^{K} A_k s_k(t; b_k) + \sigma n(t),$$

where both A_k and $s_k(t; b)$ are allowed to be random, unknown to the receiver. Assuming that b_k are independent from user to user and equally likely to take any of its permissible values, show that the minimum error probability for the first user is lower bounded by the minimum error probability of the single-user problem:

$$y(t) = A_1 s_1(t; b_1) + \sigma n(t).$$

PROBLEM 4.20. (*Beneficial Multiuser Interference*) Consider the following multiuser channel:

$$y(t) = (b_1 s_1(t) + A_2 b_2 s_2(t))^2 + \sigma n(t),$$

where $b_1 \in \{0, 1\}$, $b_2 \in \{0, 1\}$, $n(t)$ is additive white Gaussian noise, and $s_1(t) = s_2(t)$. Denote the minimum probability of error for user 1 by $P_1(A_2, \sigma)$. Show that for any σ,

$$\lim_{A_2 \to \infty} \frac{P_1(A_2, \sigma)}{P_1(0, \sigma)} = \frac{1}{2}.$$

PROBLEM 4.21. Find upper and lower bounds on the minimum bit-error-rate for user 1 in the four-user equal amplitude channel of Problem 4.17.

PROBLEM 4.22. Assume $\sigma = 0$ and $K = 3$.

(a) Find a set of signature waveforms and received amplitudes such that in a synchronous CDMA channel, the minimum bit-error-rate satisfies

$$P_k(0) > 0, \quad k = 1, 2, 3.$$

(b) With the same set of signature waveforms you chose in (a), find a set of received amplitudes such that

$$P_k(0) = 0, \quad k = 1, 2, 3.$$

(c) Find a set of signature waveforms and received amplitudes such that in an asynchronous CDMA channel with $\tau_1 = 0$, $\tau_2 = T/\sqrt{2}$, and $\tau_3 = T/\sqrt{3}$

$$P_k^M(0) > 0, \quad k = 1, 2, 3.$$

(d) Show that for any set of received nonzero amplitudes and signature waveforms, there exists a set of offsets such that

$$P_k^M(0) = 0, \quad k = 1, 2, 3.$$

PROBLEM 4.23. (Mazo [269]) Let E_k^+ be the set of error vectors such that $b_k = 1$, and let E_k^- be the set of error vectors such that $b_k = -1$. Choose an arbitrary one-to-one correspondence $\phi : E_k^+ \rightarrow E_k^-$.

(a) Prove that

$$P_k(\sigma) \geq 2^{1-K} \sum_{\mathbf{b} \in E_k^+} Q\left(\frac{\|S(\mathbf{b} - \phi(\mathbf{b}))\|}{2\sigma}\right) \tag{4.156}$$

[*Hint*: $P_k(\sigma)$ is the error probability of a binary equiprobable test (Problem 3.1) between the distributions

$$2^{1-K} \sum_{\mathbf{b} \in E_k^+} P[y \mid \mathbf{b}]$$

and

$$2^{1-K} \sum_{\mathbf{b} \in E_k^+} P[y \mid \phi(\mathbf{b})].]$$

(b) Conclude that

$$P_k(\sigma) \geq \alpha_k 2^{1-K} Q\left(\frac{d_{k,\min}}{\sigma}\right), \tag{4.157}$$

where α_k is the maximum number of pairs $(\mathbf{b}, \mathbf{d})$ such that every $\mathbf{b} \in E_k^+$ belongs to at most one such pair, every $\mathbf{d} \in E_k^-$ belongs to at most one such pair, and

$$S(\mathbf{b} - \mathbf{d}) = 2d_{k,\min}.$$

(c) Derive lower bound (4.89) as a consequence of (4.157).

PROBLEM 4.24. Consider a three-user synchronous CDMA system with $A_1 = A_2 = A_3 = 1$ and crosscorrelation matrix

$$\mathbf{R} = \begin{bmatrix} 1 & -0.6 & -0.6 \\ -0.6 & 1 & 0.2 \\ -0.6 & 0.2 & 1 \end{bmatrix}.$$

(a) Show that $d_{1,\min}^2 = 0.8$.
(b) Show that $P_1(\sigma) \geq \frac{1}{2}Q\left(\frac{0.8}{\sigma}\right)$.
(c) Show that the probability that the transmitted vector lies at minimum distance $d_{1,\min}$ from a vector that differs in the first component (and possibly others) is $3/4$.
(d) Consider the following "improved" genie. List all pairs $(\mathbf{b}, \mathbf{d})$ with $\mathbf{b} \in E_1^+$, $\mathbf{d} \in E_1^-$, and $S(\mathbf{b}, \mathbf{d}) = 2d_{1,\min}$ (cf. Problem 4.23). If the true transmitted vector does not appear in that list (i.e., its closest neighbor with a different first component is not at minimum distance), then the genie reveals the true transmitted vector to the receiver. If the true transmitted vector appears m times in that list, the genie reveals one of those m pairs chosen equiprobably. Now consider the situation when the receiver is informed by the genie of the pair $\{[+1, +1, +1], [-1, -1, +1]\}$. Show that, given that side information, the probability that $[+1, +1, +1]$ was transmitted is equal to $1/3$ (before the received waveform is observed). Conclude that the reasoning we gave to derive the lower bound (4.89) breaks down for this genie.

PROBLEM 4.25. Throughout this text expressions and bounds on error probabilities are sought in terms of the Q-function, equivalent to the one-dimensional Gaussian cumulative distribution function (cdf), which can be readily computed. However, numerical algorithms exist to compute higher-order Gaussian cdfs as well (Drezner [71]). Define

$$\Phi(\alpha_1, \alpha_2, \gamma) = P[X_1 \geq \alpha_1, X_2 \geq \alpha_2], \tag{4.158}$$

where X_1 and X_2 are jointly normal random variables with zero mean, unit variance, and correlation coefficient $-1 \leq \gamma \leq 1$.

Consider a two-user synchronous channel with equal received signal-to-noise ratios and crosscorrelation ρ and the jointly optimum receiver.

(a) Give an exact expression for the error probability for user 1 in terms of the Φ-function.

(b) Use the result in (a) to show the upper bound in (4.61).

(c) Express the probability that the jointly optimum receiver decodes both bits correctly in terms of the Φ-function (Varanasi and Aazhang [476]).

(d) Show that if $\gamma > 0$, then

$$\Phi(\alpha, \alpha, \gamma) > Q^2(\alpha). \qquad (4.159)$$

PROBLEM 4.26. Consider a two-user synchronous CDMA channel with crosscorrelation ρ.

(a) Find the worst asymptotic effective energy of the optimum detector $\omega(A_1, A_2)$ (cf. (3.115)).

(b) Verify that, if $|\rho| \leq 0.5$, then the equal-power penalty satisfies

$$\frac{1}{A^2}\omega(A, A) = 0 \text{ dB}.$$

(c) Use (b) to explain the shape of the optimal power-tradeoff region in Figure 4.14 when $|\rho| \leq 0.5$.

PROBLEM 4.27. A two-user system employs the same signature waveforms and both users are received with equal power:

$$y(t) = A\, b_1\, s(t) + A\, b_2\, s(t) + \sigma n(t).$$

(a) Show that the minimum probability of error is equal to

$$P_k(\sigma) = \frac{1}{4} + \frac{1}{2}Q\left(\frac{2A}{\sigma}\right)$$

and is achieved by

$$\hat{b}_1 = \hat{b}_2 = \text{sgn}\left(\int_0^T y(t)s(t)\,dt\right).$$

(b) Show that the bit-error-rate of the jointly optimum (maximum-likelihood) detector is equal to

$$P_k^m(\sigma) = \frac{1}{4} + \frac{1}{4}Q\left(\frac{A}{\sigma}\right) + \frac{1}{4}Q\left(\frac{3A}{\sigma}\right).$$

PROBLEM 4.28. Consider a two-user synchronous CDMA channel where the open-eye condition is satisfied with equality, that is,

$$A_1 = A_2|\rho|.$$

(a) Show the following *exact* expression for the bit-error-rate of user 1 achieved by the jointly optimum (maximum-likelihood) detector:

$$P_1^m(\sigma) = Q\left(\frac{A_1}{\sigma}\right) + \frac{1}{2}Q\left(\xi(\rho)\frac{A_1}{\sigma}\right)$$

$$- \frac{3}{2}Q\left(\frac{A_1}{\sigma}\right)Q\left(\xi(\rho)\frac{A_1}{\sigma}\right)$$

$$+ \frac{1}{2}Q\left(\frac{3A_1}{\sigma}\right)Q\left(\xi(\rho)\frac{A_1}{\sigma}\right), \qquad (4.160)$$

with

$$\xi(\rho) = \frac{\sqrt{1-\rho^2}}{|\rho|}.$$

(b) Show that the bit-error-rate of the single-user matched-filter detector is given by

$$P_1^c(\sigma) = \frac{1}{4} + \frac{1}{2}Q\left(\frac{2A_1}{\sigma}\right).$$

(c) Conclude that the conventional single-user matched filter has *lower* bit-error-rate than the jointly optimum (maximum-likelihood) detector in the special case where

$$|\rho| = 0.9,$$

$$\frac{A_1}{\sigma} = -3\,\text{dB},$$

$$A_2 = \frac{A_1}{|\rho|}.$$

PROBLEM 4.29. Consider two synchronous users with amplitudes A_1 and A_2 and signature waveforms with crosscorrelation ρ. Let

$$P_{1,2}(\sigma) = \min P[(b_1, b_2) \neq (\hat{b}_1, \hat{b}_2)],$$

where the minimum is over all decision rules.

(a) Find an exact expression for $P_{1,2}(\sigma)$ in the special case where $\rho = 0$.

(b) Find

$$\eta_{1,2} = \frac{2}{\min\{A_1^2, A_2^2\}} \lim_{\sigma \to 0} \sigma^2 \log 1/P_{1,2}(\sigma).$$

(c) Under what condition on (A_1, A_2, ρ) is $\eta_{1,2} = 1$?

PROBLEM 4.30. (*Nonlinear Modulation*) Assume that user k transmits

$$\begin{cases} A_k s_k^0(t), & t \in [0, T] \quad \text{to send '0'}, \\ A_k s_k^1(t), & t \in [0, T] \quad \text{to send '1'}. \end{cases}$$

(a) Generalize the two-user lower bound (4.63) on the minimum bit-error-rate.

(b) Generalize the two-user upper bound (4.59) on the minimum bit-error-rate.

PROBLEM 4.31. Explain why $\eta_1 = 1$ when two synchronous users are assigned the same signature waveform and $2A_1 = A_2$ (Figure 4.13).

PROBLEM 4.32. Find a crosscorrelation matrix $\mathbf{R}$ such that

$$\|\mathbf{a}_1\|^2 < \min\{\mathbf{a}_1^T \mathbf{R}_1 \mathbf{a}_1, \mathbf{a}_1^T \mathbf{R}_1^{-1} \mathbf{a}_1\},$$

where $\mathbf{a}_1 = [\rho_{12}, \ldots, \rho_{1K}]$ and $\mathbf{R}_1$ is the matrix that results from $\mathbf{R}$ by striking out row and column 1 (cf. Acar and Tantaratana [12]).

PROBLEM 4.33. Consider a three-user synchronous CDMA channel, where user 2 belongs to the same cell as user 1 and user 3 belongs to a neighboring cell. Since user 3 cannot be arbitrarily near the base station of user 1, it is of interest to obtain a more general version of the near–far resistance of user 1 incorporating a bound on the relative power of the out-of-cell interferer:

$$\bar{\eta}_1(\alpha_3) = \min_{0 \le A_3/A_1 \le \alpha_3} \min_{0 \le A_2/A_1} \eta_1. \qquad (4.161)$$

(a) Show that

$$\bar{\eta}_1 = \frac{1 - \rho_{12}^2 - \rho_{13}^2 - \rho_{23}^2 + 2\rho_{12}\rho_{13}\rho_{23}}{1 - \rho_{23}^2}.$$

(b) Show that

$$\bar{\eta}_1(\alpha_3) = \begin{cases} \bar{\eta}_1, & \text{if } \alpha_3 \ge \frac{|\rho_{23}\rho_{12} - \rho_{13}|}{1 - \rho_{23}^2}, \\ 1 - \rho_{12}^2 - 2|\rho_{23}\rho_{12} - \rho_{13}|\alpha_3 \\ \quad + (1 - \rho_{23}^2)\alpha_3^2, & \text{otherwise.} \end{cases}$$

PROBLEM 4.34. Assuming all users have the same amplitude, find the set of indecomposable error vectors for user 1 for a three-user synchronous channel where the signature waveforms assigned to users 1, 2, and 3 are $N = 3$ direct-sequence spread-spectrum waveforms equal to $[+ - +]$, $[+ + +]$, and $[- - +]$, respectively.

PROBLEM 4.35. For the signature waveforms in Problem 4.34 and $A_1 = 1$, find the lowest energy achieved by any indecomposable error vector, over all possible values of A_2 and A_3.

PROBLEM 4.36. Consider a minimum bit-error-rate receiver for a two-user synchronous channel with crosscorrelation ρ. The received powers A_1^2 and A_2^2 have to be set to meet the specifications:

$$\lim_{\sigma \to 0} \frac{P_1(\sigma)}{Q\left(\frac{A}{\sigma}\right)} \le 1,$$

$$\lim_{\sigma \to 0} \frac{P_2(\sigma)}{Q\left(\frac{A}{\sigma}\right)} \le 1.$$

(a) Suppose that $A_1 = A$. For what values of A_2 will the design specifications be met?

(b) Suppose that $A_1 = A_2$. For what values of A_1 will the design specifications be met?

(c) Compare the minimum values of $A_1^2 + A_2^2$ achieved in cases (a) and (b).

[*Hint*: See Figure 4.14.]

PROBLEM 4.37. Consider a three-user synchronous CDMA channel with crosscorrelation matrix ($|\delta| < 1/2$)

$$\mathbf{R} = \begin{bmatrix} 1 & \delta & -\delta \\ \delta & 1 & \delta \\ -\delta & \delta & 1 \end{bmatrix}.$$

(a) Find the near–far resistance of all three users as a function of δ.

(b) Explain why the near–far resistance is identical for all three users.

PROBLEM 4.38. Consider a K-user asynchronous Gaussian channel.

(a) Show that if the signature waveforms and offsets are such that the bounding series (4.74) converges for at least small enough σ (Type A or B), then C_k^U defined in (4.86) is finite.

(b) Give an example of an asynchronous multiuser channel where $C_k^U = +\infty$.

PROBLEM 4.39. Consider a six-user channel where all users are received with the same power and all signature waveforms are equal to a rectangular pulse on $[0, T]$. All users except user 1 are synchronized (i.e., $\tau_k = \tau_1 + T/2$, $k = 2, \ldots, 6$).

(a) (Verdú [490]) Show that the bounding series (4.74) on $P_1(\sigma)$ diverges for all $\sigma > 0$.

(b) Find a nontrivial upper bound to the error probability of user 1.

226

PROBLEM 4.40. Consider a K-user asynchronous channel. Show that if the noise realization is such that

$$|y_k| > \sum_{\substack{j=1 \\ j \neq k}}^{K} A_j \left(|\rho_{jk}| + |\rho_{kj}| \right),$$

then the jointly optimum detector outputs

$$\hat{b}_k = \text{sgn}(y_k).$$

PROBLEM 4.41. Consider a three-user synchronous channel with crosscorrelation matrix

$$\mathbf{R} = \begin{bmatrix} 1 & 0.6 & 0.6 \\ 0.6 & 1 & 0.6 \\ 0.6 & 0.6 & 1 \end{bmatrix}.$$

Let $A_1 = A_2 = 1$. Find η_1 as a function of A_3.

PROBLEM 4.42. Consider a three-user synchronous channel with crosscorrelation matrix

$$\mathbf{R} = \begin{bmatrix} 1 & 0.5 & 0.5 \\ 0.5 & 1 & \rho_{23} \\ 0.5 & \rho_{23} & 1 \end{bmatrix}.$$

Let $A_1 = A_2 = A_3 = 1$. Find η_1 as a function of $-0.5 \leq \rho_{23} \leq 1$.

PROBLEM 4.43. The *branch-and-bound* algorithm is a well-known combinatorial optimization method (e.g., Papadimitrou and Steiglitz [314]) that can be used to obtain the minimum of a function:

$$d(e_1, \ldots, e_K) = \sum_{i=1}^{K} d_i(e_1, \ldots, e_i),$$

where the functions d_i are assumed to be nonnegative and the variables e_i belong to a finite alphabet. The principle is simple: we can construct a K-level tree where nodes at the i level are labeled by $(e_1, \ldots, e_i)$ and the edge connecting $(e_1, \ldots, e_{i-1})$ to $(e_1, \ldots, e_{i-1}, e_i)$ has weight $d_i(e_1, \ldots, e_i)$ (where the root is denoted as e_0). The sum of the weights from the root to the leaf $(e_1, \ldots, e_K)$ is equal to the cost function $d(e_1, \ldots, e_K)$. If we know an upper bound to the minimum cost,[3]

$$\min_{e_1, \ldots, e_K} d(e_1, \ldots, e_K) \leq \delta,$$

---

[3] Such an upper bound may be improved with the minimum cost found by the algorithm so far.

then it is not necessary to conduct an exhaustive search of the tree to find the leaf with minimum cost. As soon as an intermediate node is found to have weight greater than δ all its descendants can be removed from further consideration.

(a) Give a branch-and-bound algorithm to compute $d_{1,\min}^2$ in the case of K synchronous users. *Hint*: (Schlegel and Wei [389]) Using the Cholesky factorization of the crosscorrelation matrix in Proposition 2.2, we can write

$$\epsilon^T H \epsilon = \| FA \epsilon \|^2$$

$$= \sum_{i=1}^{K} \left(\sum_{j=1}^{i} F_{ji} A_j \epsilon_j \right)^2 .$$

(b) What is the worst-case complexity of the algorithm as a function of K?

(c) Is it possible to find a branch-and-bound algorithm for jointly optimum multiuser detection of K synchronous users?

PROBLEM 4.44. Derive the expression in (4.127) for the two-user asynchronous asymptotic efficiency. [*Hint*: (1) Find the minimum distance achieved by sequences whose only nonzero components (in addition to $\epsilon_1[0]$) can be, but need not be, $\epsilon_2[-1]$ and $\epsilon_2[0]$; (2) show that the minimum distance cannot be achieved by sequences that contain a string of zeros amidst nonzeros; (3) represent the energy of an error sequence with the aid of a two-state trellis (Figure 4.8) and show that to achieve minimum distance an error sequence must have $\epsilon_1[1] = \epsilon_1[-1] = 0$.]

PROBLEM 4.45. Show that for any set of K signature waveforms and offsets, the asynchronous minimum distance $d_{k,\min}$ is always achieved by an error sequence with $w(\epsilon) \leq 2K \, 3^{K-1}$.

PROBLEM 4.46. Show in the special case of two asynchronous users that the upper bound in (4.80) diverges for all σ.

PROBLEM 4.47. Show that in any K-user synchronous or asynchronous channel the conventional single-user matched filter is asymptotically optimal (as $\sigma \to \infty$) in the sense that

$$\lim_{\sigma \to \infty} \frac{\sigma}{A_k} \left(P_k^c(\sigma) - P_k(\sigma) \right) = 0. \tag{4.162}$$

[*Hint*: Use (3.54).]

PROBLEM 4.48. Consider a K-user synchronous channel with signature waveforms

$$\{s_1, \ldots, s_K\}$$

and an $(L+1)$-user synchronous channel with signature waveforms

$$\{s_1, u_1, \ldots, u_L\}.$$

Show that if $\{u_1, \ldots, u_L\}$ and $\{s_2, \ldots, s_K\}$ span the same space, then the near–far resistance of user 1 is the same in both channels. [*Hint*: Use (4.93)].

PROBLEM 4.49.

(a) Verify Proposition 4.3.
(b) Suppose that $\mathbf{A}$ is a $K \times K$ invertible matrix and that $\mathbf{x}$ and $\mathbf{y}$ are column K-vectors such that $\mathbf{A} + \mathbf{xy}^T$ is invertible. Verify the Sherman–Morrison formula:

$$[\mathbf{A} + \mathbf{xy}^T]^{-1} = \mathbf{A}^{-1} - \frac{(\mathbf{A}^{-1}\mathbf{x})(\mathbf{y}^T\mathbf{A}^{-1})}{1 + \mathbf{y}^T\mathbf{A}^{-1}\mathbf{x}}.$$

PROBLEM 4.50. Show that (4.104) need not hold if users $2, \ldots, K$ employ direct-sequence spread-spectrum waveforms whose spreading factor is different from that of the direct-sequence signature waveform assigned to user 1.

PROBLEM 4.51. Consider a random-sequence direct-sequence spread-spectrum system with spreading factor equal to N (cf. Problem 2.21). Show the following results where the expectations are with respect to an equiprobable choice of signatures and the indices k, l, m, n are assumed to be different:

(a) $E[\rho_{kl}^4] = \frac{3}{N^2} - \frac{2}{N^3}$.
(b) $E[\rho_{kl}^2\rho_{km}^2] = \frac{1}{N^2}$.
(c) $E[\rho_{km}\rho_{lm}\rho_{kn}\rho_{ln}] = \frac{1}{N^3}$.
(d) $E[\rho_{kj}\rho_{jl}\rho_{lk}] = \frac{1}{N^2}$.

PROBLEM 4.52. A four-user synchronous CDMA channel employs direct-sequence spread-spectrum waveforms with spreading factor $N = 3$.

(a) Find the expected dimensionality $\overline{[3, 3]}$ obtained when all the chips are independent equally likely to be $+1$ or -1.
(b) What is the expected optimal near–far resistance?

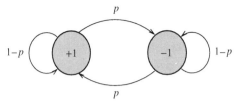

FIGURE 4.20.
Markov chain modeling data stream in Problem 4.53.

(c) Show a choice for the four signature waveforms that maximizes the near–far resistance of user 1. What is that maximum near–far resistance?

PROBLEM 4.53. Describe the jointly optimum detector for two synchronous users assigned CDMA waveforms with crosscorrelation ρ whose data streams are binary Markov chains with transition rates given in Figure 4.20.

PROBLEM 4.54. Consider a single binary data source that is transmitted via a two-user synchronous CDMA channel with crosscorrelation ρ. The "second user" can be considered a helper that operates in two different modes:

(a) $b_2[i] = b_1[i]$.
(b) $b_2[i] = b_1[i]b_1[i-1]$.

The asymptotic efficiency (in this case greater than unity) quantifies the improvement in signal-to-noise ratio (for low background noise) due to the helper. Find the optimum asymptotic efficiency as a function of A_2/A_1 and ρ under both modes of operation.

PROBLEM 4.55. Show that (4.104) need not hold when the users are chip asynchronous.

PROBLEM 4.56. Consider a K-user synchronous channel where the signature waveforms are such that $\rho_{jk} = \rho$ for all j and k (cf. Problem 3.41).

(a) Find the optimum near–far resistance as a function of ρ and K.
(b) Particularize the solution to (a) to a direct-sequence spread spectrum system where the users are assigned distinct circularly shifted versions of the same maximal-length shift-register sequence of period N (cf. (2.16)).
(c) Take the limit of the result in (b) as $K = N \to \infty$. Compare with the asymptotic average result in (4.108).

PROBLEM 4.57. Consider a K-user synchronous channel where the signature waveforms are such that $\rho_{jk} = \rho > 0$ for all j and k and

$$A_1 = \cdots = A_K.$$

(a) Show that there are no indecomposable error vectors of weight greater than or equal to 3.
(b) Particularize the bounds (4.74) and (4.88) to this case.
(c) Find the asymptotic efficiency of the optimum receiver as a function of ρ and K.
(d) Can you find a nontrivial upper bound on the minimum bit-error-rate as $K \to \infty$?

PROBLEM 4.58. Consider a K-user synchronous crosscorrelation matrix such that

$$\rho_{1k} > 0, \quad k = 2, \ldots, K,$$

$$\rho_{kj} = 0, \quad j = 2, \ldots, K, \quad k = 2, \ldots, K.$$

Assume that all users have identical received signal-to-noise ratios.

(a) Find F_1, the set of indecomposable vectors for user 1.
(b) Find F_2, the set of indecomposable vectors for user 2.
(c) Find $d_{1,\min}^2$ and $d_{2,\min}^2$.
(d) Find the optimum near–far resistance for user 1. Under what conditions on ρ_{1k}, $k = 2, \ldots, K$ is the crosscorrelation matrix positive definite?
(e) Describe the jointly optimal decision rule for users $2, \ldots, K$.
(f) Find the best upper and lower bounds you can find for the minimum error probability of user 1 when $K = 3$, $\rho_{12} = 0.6$, and $\rho_{13} = 0.4$.

PROBLEM 4.59. As a benchmark to estimate computation times of optimum detectors in this problem, use the figure of 5×10^6 floating-point operations per second (which is achieved by a typical personal computer in 1997).

(a) Estimate the number of 8 kilobit-per-second CDMA-multiplexed streams that can be demodulated optimally if the streams are (1) synchronous, (2) asynchronous.
(b) Consider a two-user asynchronous CDMA channel used to transmit the ASCII binary version (7 bits per character including punctuation and spaces) of the following messages (cf. Epstein et al. [78]) (started slightly off-phase):

USER 1: The problems of two users do not amount to a hill of beans in this crazy world.

USER 2: I remember every detail. The interferers wore gray; the desired user wore blue.

Estimate the time it will take to demodulate those messages optimally with

- the fastest known algorithm for optimum multiuser detection (Section 4.2),
- a brute-force optimum receiver (which does not use a Viterbi algorithm).

PROBLEM 4.60. Consider a two-user asynchronous CDMA channel where the signature waveforms are direct-sequence spread-spectrum waveforms with $N > 1$ chips per symbol. Suppose that the observables employed by the receiver for user 1 are the N chip matched-filter outputs corresponding to the user of interest, that is, the observables used to demodulate $b_1[0]$ are

$$\int y(t) p_{T_c} \left(t - \frac{lT}{N} \right) dt, \quad l = 0, \ldots, N-1.$$

Suppose further that those observables are processed in the optimal way to minimize the bit-error-rate for user 1.

(a) Show by means of an example that the resulting bit-error-rate is strictly higher than the minimum achievable with a receiver that processes the incoming waveform $y(t)$ optimally. [*Hint:* Consider the noiseless case.]

(b) Assume that the signature waveform of user 1 is selected randomly and equiprobably among all 2^N choices. Show that the expected near–far resistance for user 1 achieved by the receiver considered in this problem is equal to

$$1 - \frac{2}{N}$$

for all offsets $\tau \in (0, T)$.

PROBLEM 4.61. Show that if direct-sequence spread-spectrum signatures with N chips per symbol are assigned randomly, then the expected log near–far resistance of any multiuser detector satisfies

$$E[\log \bar{\eta}_k] = -\infty$$

for all N and $K \geq 2$. (*Note:* This is notwithstanding the fact that $\log \bar{\eta}_k$ converges in probability to

$$\lim_{K \to \infty} \log \left(1 - \frac{K}{N} \right),$$

because of (4.111).)

PROBLEM 4.62. Particularize the bound in (4.138) to the two-user case.

PROBLEM 4.63. In a noncoherent detection problem, the receiver obtains

$$y(t) = \sum_{k=1}^{K} |A_k| e^{j\theta_k} s_k(t; b_k) + \sigma n(t) \qquad (4.163)$$

and has no knowledge of the received phases so it assumes they are independent and uniformly distributed. Show that

$$\{y_k(b_k), k = 1, \ldots, K, b_k \in \mathcal{A}\}$$

is a sufficient statistic with

$$y_k(b_k) = \int_0^T y(t) s_k^*(t; b_k) \, dt,$$

regardless of the distribution of the received magnitudes $|A_k|$.

PROBLEM 4.64. (Visotsky and Madhow [506]) Consider a two-user nonlinearly modulated channel

$$y(t) = A_1 s_1(t, b_1) + A_2 s_1(t, b_2) + \sigma n(t)$$

with binary modulation $b_k \in \{+, -\}$ and such that the receiver knows A_1 and A_2.

(a) Show that the near–far resistance of user 1 is equal to zero if and only if

$$s_1(t, +) - s_1(t, -) = \alpha s_2(t, +) - \alpha s_2(t, -)$$

for some α.

(b) Generalize the necessary and sufficient condition found in (a) to the K-user case.

CHAPTER FIVE

DECORRELATING DETECTOR

The large gaps in performance and complexity between the conventional single-user matched filter and the optimum multiuser detector encourage the search for other multiuser detectors that exhibit good performance/complexity tradeoffs. Chapters 5 and 6 examine various *linear* multiuser detectors. The decorrelating detector not only is a simple and natural strategy but it is optimal according to three different criteria: least-squares, near–far resistance, and maximum-likelihood when the received amplitudes are unknown.

5.1 THE DECORRELATING DETECTOR IN THE SYNCHRONOUS CHANNEL

The output vector of the bank of K matched filter outputs can be written as (Section 2.9)

$$\mathbf{y} = \mathbf{RAb} + \mathbf{n}, \qquad (5.1)$$

where $\mathbf{n}$ is a Gaussian random vector with zero mean and covariance matrix $\sigma^2 \mathbf{R}$. In order to motivate the derivation of the decorrelating detector it is useful to recall from Section 3.4 that the conventional receiver may make errors even in the absence of noise ($\sigma = 0$), that is, it may happen that

$$\hat{b}_k = \text{sgn}((\mathbf{RAb})_k) \neq b_k.$$

In contrast, the optimum receiver demodulates the data error-free in the absence of noise (cf. (4.46)). Of course, this is a desirable feature for any multiuser detector, even though the situation of absence of background noise

234

is completely hypothetical. A simple receiver suffices to recover the transmitted bits error-free in the absence of noise, without requiring knowledge of the received amplitudes. Let us assume that the crosscorrelation matrix $\mathbf{R}$ is invertible. If we premultiply the vector of matched filter outputs by $\mathbf{R}^{-1}$, then

$$\mathbf{R}^{-1}\mathbf{y} = \mathbf{R}^{-1}\mathbf{RAb} = \mathbf{Ab}, \tag{5.2}$$

if $\sigma = 0$. So we can simply take the sign of each of the components in (5.2) to recover the transmitted data:

$$\hat{b}_k = \operatorname{sgn}((\mathbf{R}^{-1}\mathbf{y})_k) \tag{5.3}$$

$$= \operatorname{sgn}((\mathbf{Ab})_k) \tag{5.4}$$

$$= b_k. \tag{5.5}$$

We conclude that if the signature waveforms are linearly independent the detector in (5.3) achieves perfect demodulation for every active user.

Let us bring in the noise now. Processing the matched filter bank outputs (5.1) with $\mathbf{R}^{-1}$ results in

$$\mathbf{R}^{-1}\mathbf{y} = \mathbf{Ab} + \mathbf{R}^{-1}\mathbf{n}. \tag{5.6}$$

Notice that the kth component of (5.6) is still free from interference caused by any of the other users, that is, it is independent of all $\{b_j\}$, $j \neq k$. The only source of interference is the background noise. This is why the detector that performs (5.3) is called the *decorrelating detector* (Figure 5.1).

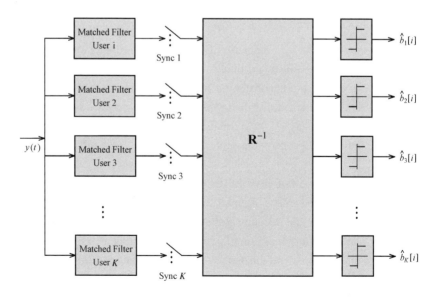

FIGURE 5.1.
Decorrelating
detector for the
synchronous
channel.

235

From the viewpoint of implementation, two desirable features of this multiuser detector are:

- It does not require knowledge of the received amplitudes.
- It can readily be decentralized, in the sense that the demodulation of each user can be implemented completely independently.

To see the second property, use R_{kj}^+ as a shorthand for $(\mathbf{R}^{-1})_{kj}$ and note that the kth output of the linear transformation $\mathbf{R}^{-1}$ equals

$$(\mathbf{R}^{-1}\mathbf{y})_k = \sum_{j=1}^{K} R_{kj}^+ y_j$$

$$= \sum_{j=1}^{K} R_{kj}^+ \langle y, s_j \rangle$$

$$= \langle y, \sum_{j=1}^{K} R_{jk}^+ s_j \rangle$$

$$= \langle y, \tilde{s}_k \rangle, \tag{5.7}$$

where we have introduced the notation

$$\tilde{s}_k(t) = \sum_{j=1}^{K} R_{jk}^+ s_j(t). \tag{5.8}$$

The signal in (5.8) has unit inner product with its corresponding signature waveform:

$$\langle \tilde{s}_k, s_k \rangle = \int_0^T \sum_{j=1}^{K} R_{jk}^+ s_j(t) s_k(t) \, dt = [\mathbf{R}^{-1}\mathbf{R}]_{kk} = 1, \tag{5.9}$$

and, consequently, $||\tilde{s}_k|| \geq 1$ because of the Cauchy–Schwarz inequality.

We will refer to any linear combination of $\{s_1, \ldots, s_K\}$ that is orthogonal to all of them but s_k as a decorrelating linear transformation for s_k against $\{s_j\}_{j \neq k}$. Obviously, this transformation does not exist if s_k is a linear combination of $\{s_j\}_{j \neq k}$. If $\{s_1, \ldots, s_K\}$ are linearly independent, then $\tilde{s}_k$ in (5.8) is the unique (up to a nonzero constant) decorrelating transformation for s_k against $\{s_j\}_{j \neq k}$.

Because of (5.7) we can view the decorrelating detector for the kth user as a modified matched filter (Figure 5.2). This alternative implementation is equivalent because the decision statistics fed to the zero-thresholds are identical. The computational complexity of the receiver in Figure 5.2 is identical to the complexity of the bank of single-user matched-filters, except that if the crosscorrelations are not known a priori (e.g., due to channel

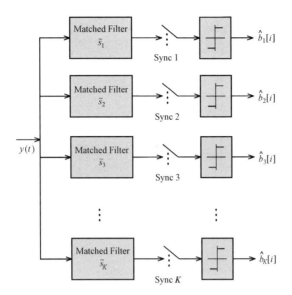

FIGURE 5.2.
Modified matched
filter bank for
decorrelation.

distortion) not only do they have to be generated from the replicas of the received signature waveforms, but the coefficients $R_{kj}^+, j = 1, \ldots, K$ have to be computed.

The decision statistic of the decorrelating detector $(\mathbf{R}^{-1}\mathbf{y})_k$ (or the output of the filter matched to (5.8)) contains no trace of the signals modulated by the interfering users. Indeed, for any vector $(a_1, \ldots, a_K) \in R^K$,

$$\int_0^T \left[\sum_{i \neq k} a_i s_i(t) \right] \tilde{s}_k(t) \, dt = \int_0^T \left[\sum_{i \neq k} a_i s_i(t) \right] \left[\sum_{j=1}^K R_{jk}^+ s_j(t) \right] dt$$

$$= \sum_{i \neq k} a_i [\mathbf{R}^{-1}\mathbf{R}]_{ik}$$

$$= 0. \tag{5.10}$$

Another way to state this key property is to cast it in the geometry of the linear vector space spanned by the K signature waveforms: the decorrelating detector correlates with the projection (scaled version of $\tilde{s}_k$) of $s_k(t)$ on the subspace orthogonal to the subspace spanned by the interfering signature waveforms $\{s_j, j \neq k\}$.

In the two-user case,

$$\mathbf{R}^{-1} = \begin{bmatrix} 1 & \rho \\ \rho & 1 \end{bmatrix}^{-1} = \frac{1}{1 - \rho^2} \begin{bmatrix} 1 & -\rho \\ -\rho & 1 \end{bmatrix}. \tag{5.11}$$

Since positive multiplication factors are irrelevant when taking signs, we see that in a two-user channel the decorrelator for user 1 resembles the conventional single-user matched filter except that it replaces s_1 by $s_1 - \rho s_2$,

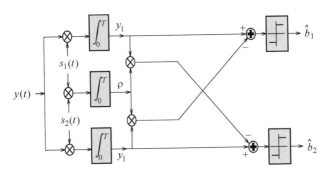

FIGURE 5.3.
Decorrelating
receiver for two
synchronous
users.

or equivalently it processes the single-user matched filter outputs as (Figure 5.3)

$$\hat{b}_1 = \text{sgn}(y_1 - \rho y_2).$$

The decision regions of the two-user decorrelating detector are depicted in Figure 5.4 for the case $A_1 = A_2$. Note that the location of the four multiuser signals, $A_1 b_1 s_1 + A_2 b_2 s_2$, varies with the amplitudes whereas the decision regions are invariant to A_1 and A_2. Figure 5.4 should be compared to the decision regions of (a) the single-user matched filter detector (Figure 3.12), (b) jointly optimum detector (Figure 4.1), and (c) minimum bit-error-rate detector (Figure 4.5). We see that in both Figures 3.12 and 5.4 the decision regions are delimited by straight lines going through the origin. For $K > 2$, they generalize to hyperplanes containing the origin. This is true of any *linear* multiuser detector, that is, any detector that takes the sign of the correlation of the received signal and a given waveform.

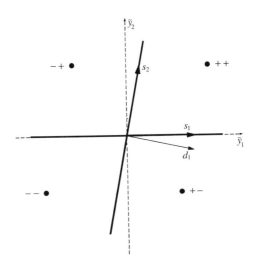

FIGURE 5.4.
Decision regions
of the two-user
decorrelating
detector;
$A_1 = A_2$.

238

The decorrelating linear transformation is related to the whitened matched filter model in (2.87). Recall from Proposition 2.2 that

$$\mathbf{R} = \mathbf{F}^T \mathbf{F},$$

where $\mathbf{F}$ is a lower triangular matrix. The signal

$$\bar{s}_1(t) = \sum_{j=1}^{K} (\mathbf{F}^{-T})_{1j} s_j(t) \tag{5.12}$$

is a decorrelating transformation for s_1 against $\{s_2, \ldots, s_K\}$, because

$$\langle \bar{s}_1, s_k \rangle = \sum_{j=1}^{K} (\mathbf{F}^{-T})_{1j} \rho_{jk}$$

$$= (\mathbf{F}^{-T} \mathbf{R})_{1k}$$

$$= (\mathbf{F}^{-T} \mathbf{F}^T \mathbf{F})_{1k}$$

$$= \mathbf{F}_{1k}, \tag{5.13}$$

which is nonzero only for $k = 1$. From (2.86) and (5.12), it follows that the first whitened matched filter is actually the decorrelating linear transformation for user 1. This property does not hold for the other users; however, the kth whitened matched filter (cf. (2.86) and (5.12)),

$$\bar{s}_k(t) = \sum_{j=1}^{K} (\mathbf{F}^{-T})_{kj} s_j(t), \tag{5.14}$$

is a decorrelating linear transformation for s_k against $\{s_{k+1}, \ldots, s_K\}$. To see this, analogously to (5.13) we have

$$\langle \bar{s}_k, s_j \rangle = F_{kj},$$

which is positive for $j = k$ and zero for $j > k$. Furthermore, $\mathbf{F}^{-T}$ is an upper triangular matrix whose kth row depends only on $\{s_k, \ldots, s_K\}$ (Problem 2.54).

Up to now we have introduced the decorrelating detector in a rather ad hoc way: simply by noticing that it results in error-free demodulation in the absence of noise. Yet, the decorrelating detector can be obtained as the solution to various optimization problems. In this chapter we will see three criteria according to which the decorrelating detector is optimum and then, in Chapter 6, we will see that it is the limit of another linear detector that maximizes output signal-to-noise ratio.

In contrast to what we assumed in the derivation of the optimum detector, suppose now that the detector knows neither the received amplitudes nor any prior distributions for them. It is then natural to consider joint maximum-likelihood estimation of amplitudes and transmitted bits. Because the noise is white and Gaussian, the most likely bits and amplitudes are those that best explain the received waveform in a mean-square sense, that is, the arguments that achieve

$$\min_{\substack{b\in\{-1,1\}^K}} \min_{\substack{A_k\geq0\\k=1,\cdots K}} \int_0^T \left[y(t) - \sum_{k=1}^K A_k b_k s_k(t) \right]^2 dt. \qquad (5.15)$$

It should be emphasized that the one-shot maximum-likelihood problem in (5.15) arises in the hypothetical situation in which there is neither a priori information on the received amplitudes nor any assumption on their time dependence. If the amplitudes are known to be constant (or otherwise correlated) over time, then a one-shot approach is suboptimal even in the synchronous case (Problem 5.10).

If we let $c_k = A_k b_k$ and, as usual, we denote the outputs of the filters matched to the signature waveforms by y_k, we see that the minimization in (5.15) is equivalent to the maximization

$$\max_{\mathbf{c}\in R^K} 2\mathbf{c}^T\mathbf{y} - \mathbf{c}^T\mathbf{R}\mathbf{c},$$

which is achieved by

$$\mathbf{c}^* = \mathbf{R}^{-1}\mathbf{y},$$

because

$$2\mathbf{c}^T\mathbf{y} - \mathbf{c}^T\mathbf{R}\mathbf{c} = \mathbf{y}^T\mathbf{R}^{-1}\mathbf{y} - (\mathbf{c} - \mathbf{c}^*)^T\mathbf{R}(\mathbf{c} - \mathbf{c}^*).$$

Now, the arguments that minimize (5.15) (i.e., the most likely bits and amplitudes) are simply

$$\hat{b}_k = \text{sgn}(c_k^*) = \text{sgn}((\mathbf{R}^{-1}\mathbf{y})_k)$$

and

$$\hat{A}_k = |c_k^*|.$$

Therefore, the decorrelating detector is seen to give the best joint estimate of the transmitted bits and amplitudes in the absence of any prior knowledge about the received amplitudes.

240

So far in this chapter, we have assumed that the crosscorrelation matrix is nonsingular. This assumption can be replaced by the milder condition that the kth user signature waveform is not spanned by the other signature waveforms; the decorrelating detector for user k can be found as above replacing $\{s_j\}$, $j \neq k$ by a linearly independent basis spanning the same space as $\{s_j\}$, $j \neq k$ (Problem 5.32). More generally, we can turn our attention to the L-dimensional discrete-time model in (2.97):

$$\mathbf{r} = \mathbf{SAb} + \sigma\mathbf{m}, \tag{5.16}$$

where $\mathbf{S}$ is the $L \times K$ matrix of signature vectors (2.98) and the covariance matrix of $\mathbf{m}$ is the $L \times L$ identity matrix. A reasonable way to make decisions is to let

$$\hat{b}_k = \mathrm{sgn}(x_k(\mathbf{r})),$$

where $\mathbf{x}(\mathbf{r})$ is an estimate of $\mathbf{Ab}$ defined as the K-vector that achieves the solution to the least-squares problem:

$$\min_{\mathbf{x} \in R^K} \|\mathbf{Sx} - \mathbf{r}\|. \tag{5.17}$$

A complete solution to this problem is given by the following well-known linear algebra result (e.g., Lancaster and Tismenetsky [227]):

Proposition 5.1 *The solution to (5.17) is*

$$\mathbf{x}(\mathbf{r}) = \mathbf{S}^+\mathbf{r},$$

where the $K \times L$ matrix $\mathbf{S}^+$ is the Moore–Penrose generalized inverse[1] of $\mathbf{S}$.

Whenever the underlying L-dimensional basis spans the continuous-time signature waveforms, the crosscorrelation matrix is equal to (2.101)

$$\mathbf{R} = \mathbf{S}^T\mathbf{S} \tag{5.18}$$

and the vector of matched filter outputs can be written as (2.103)

$$\mathbf{y} = \mathbf{S}^T\mathbf{r}. \tag{5.19}$$

---

[1] A generalized inverse $\mathbf{C}$ of a matrix $\mathbf{B}$ is any matrix that satisfies: $\mathbf{CBC} = \mathbf{C}$ and $\mathbf{BCB} = \mathbf{B}$. The Moore–Penrose generalized inverse is the unique inverse for which $\mathbf{CB}$ and $\mathbf{BC}$ are symmetric. It follows that if $\mathbf{B}$ is a (square) nonsingular matrix, then its Moore–Penrose generalized inverse is $\mathbf{B}^{-1}$.

Moreover, the solution given in Proposition 5.1 is equivalent to

$$\mathbf{S}^+\mathbf{r} = (\mathbf{S}^T\mathbf{S})^+\mathbf{S}^T\mathbf{r} \tag{5.20}$$

$$= (\mathbf{S}^T\mathbf{S})^+\mathbf{y}$$

$$= \mathbf{R}^+\mathbf{y}, \tag{5.21}$$

where (5.20) follows from Problem 5.37. Thus, by solving a least-squares problem we have generalized the decorrelator to the case where the cross-correlation matrix is singular.

How do we obtain the Moore–Penrose generalized inverse? We can always write $\mathbf{S}$ as the product of two full-rank matrices:

$$\mathbf{S} = \mathbf{E}\mathbf{G}^T,$$

where $\mathbf{E}$ is an $L \times r$ matrix, $\mathbf{G}$ is a $K \times r$ matrix, and r is the rank of $\mathbf{S}$. Then the Moore–Penrose generalized inverse of $\mathbf{S}$ becomes

$$\mathbf{S}^+ = \mathbf{G}(\mathbf{G}^T\mathbf{G})^{-1}(\mathbf{E}^T\mathbf{E})^{-1}\mathbf{E}^T. \tag{5.22}$$

If the matrix $\mathbf{S}$ has full rank, then (5.22) simplifies considerably:

$$\mathbf{S}^+ = \begin{cases} \mathbf{S}^T(\mathbf{SS}^T)^{-1}, & \text{if } L < K; \\ \mathbf{S}^{-1}, & \text{if } L = K; \\ (\mathbf{S}^T\mathbf{S})^{-1}\mathbf{S}^T, & \text{if } L > K. \end{cases} \tag{5.23}$$

A decorrelating transformation for $\mathbf{s}_1$ against $\{\mathbf{s}_2, \ldots, \mathbf{s}_K\}$ exists if and only if $\mathbf{s}_1$ is not spanned by $\{\mathbf{s}_2, \ldots, \mathbf{s}_K\}$. Denote

$$\mathbf{U} = [\,\mathbf{s}_2 \mid \cdots \mid \mathbf{s}_K\,]. \tag{5.24}$$

Then

$$\mathbf{d}_1 \stackrel{\text{def}}{=} [\mathbf{I} - \mathbf{U}\mathbf{U}^+]\mathbf{s}_1 \tag{5.25}$$

is a decorrelator for $\mathbf{s}_1$ against $\{\mathbf{s}_2, \ldots, \mathbf{s}_K\}$ provided $\mathbf{s}_1$ is not spanned by $\{\mathbf{s}_2, \ldots, \mathbf{s}_K\}$. To verify this property we show first that

$$\mathbf{d}_1^T\mathbf{s}_1 = \mathbf{s}_1^T[\mathbf{I} - \mathbf{U}\mathbf{U}^+]^T\mathbf{s}_1 \tag{5.26}$$

$$= \mathbf{s}_1^T[\mathbf{I} - \mathbf{U}\mathbf{U}^+]\mathbf{s}_1 \tag{5.27}$$

$$\neq 0, \tag{5.28}$$

where we used the fact that the product of a matrix with its Moore–Penrose inverse is symmetric, and (5.28) will follow by contradiction. Define the vector (cf. (5.27))

$$\mathbf{v} = \mathbf{U}\mathbf{U}^+\mathbf{s}_1.$$

Using the various properties in the definition of Moore–Penrose generalized inverse, we have

$$\|\mathbf{v}\|^2 = \mathbf{s}_1^T (\mathbf{U}\mathbf{U}^+)^T \mathbf{U}\mathbf{U}^+ \mathbf{s}_1$$

$$= \mathbf{s}_1^T \mathbf{U}\mathbf{U}^+ \mathbf{U}\mathbf{U}^+ \mathbf{s}_1$$

$$= \mathbf{s}_1^T \mathbf{U}\mathbf{U}^+ \mathbf{s}_1$$

$$= \mathbf{s}_1^T \mathbf{v}$$

$$= 1, \qquad (5.29)$$

where (5.29) holds if (5.28) is not true. As $\|\mathbf{s}_1\| = 1$, we must have $\mathbf{v} = \mathbf{s}_1$. It follows that we have found a linear combination of the columns of $\mathbf{U}$, namely $\mathbf{v}$, that is equal to $\mathbf{s}_1$. This contradicts our initial assumption.

Now it remains to verify that $\mathbf{d}_1$ is orthogonal to $\{\mathbf{s}_2, \ldots, \mathbf{s}_K\}$. To that end, denote $\mathbf{e}_k$ as the $(K\text{-}1)$-vector whose components are zero except for the $(k-1)$th component, which is equal to one. Then using (5.24) we get

$$\mathbf{d}_1^T \mathbf{s}_k = \mathbf{s}_1^T [\mathbf{I} - \mathbf{U}\mathbf{U}^+] \mathbf{s}_k$$

$$= \mathbf{s}_1^T [\mathbf{I} - \mathbf{U}\mathbf{U}^+] \mathbf{U}\mathbf{e}_k$$

$$= \mathbf{s}_1^T [\mathbf{U}\mathbf{e}_k - \mathbf{U}\mathbf{e}_k]$$

$$= 0. \qquad (5.30)$$

The expression for the decorrelator in (5.25) suggests a generalization of the decorrelator to situations where a group of more than one but less than K users needs to be demodulated: the unwanted users can be completely suppressed by a linear transformation $[\mathbf{I} - \mathbf{U}\mathbf{U}^+]$, where $\mathbf{U}$ is the matrix of their signature vectors. The resulting outputs can then be processed by a bank of matched filters for the desired users followed by any multiuser detection strategy (not necessarily decorrelation). See Problems 5.36, 5.45, and 5.46.

Another optimality criterion satisfied by the decorrelating detector will be studied in Section 5.5.

5.2 THE DECORRELATING DETECTOR IN THE ASYNCHRONOUS CHANNEL

The vector discrete-time model of the matched filter outputs for the asynchronous channel (2.106) is shown in Figure 2.16. Analogously to the synchronous case, we can look for a linear transformation that recovers

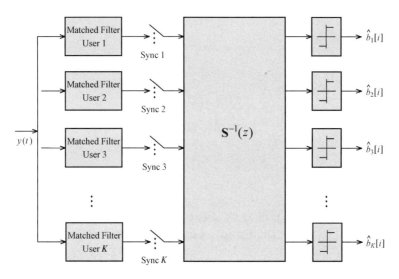

FIGURE 5.5.
Asynchronous
decorrelating
detector.

a scaled version of the input bits if $\sigma = 0$. An obvious choice is the inverse of the discrete-time channel transfer function (2.112)

$$\mathbf{S}^{-1}(z) = [\mathbf{R}^T[1]z + \mathbf{R}[0] + \mathbf{R}[1]z^{-1}]^{-1}.$$

This leads to the detector shown in Figure 5.5.

In the two-user case we have

$$\mathbf{S}^{-1}(z) = \begin{bmatrix} 1 & \rho_{12} + \rho_{21}z^{-1} \\ \rho_{12} + \rho_{21}z & 1 \end{bmatrix}^{-1}$$

$$= \frac{1}{1 - \rho_{12}^2 - \rho_{21}^2 - \rho_{12}\rho_{21}z - \rho_{12}\rho_{21}z^{-1}}$$

$$\times \begin{bmatrix} 1 & -\rho_{12} - \rho_{21}z^{-1} \\ -\rho_{12} - \rho_{21}z & 1 \end{bmatrix}. \qquad (5.31)$$

Equation (5.31) reflects the general structure of the matrix transfer function $\mathbf{S}^{-1}(z)$ as the product of a scalar transfer function $[\det \mathbf{S}(z)]^{-1}$ times the matrix transfer function adj $\mathbf{S}(z)$. The entries of adj $\mathbf{S}(z)$ do not contain autoregressive terms (denominators); they are polynomials in positive and negative powers of z. We may view the operation of multiplying the matched-filter output vectors by the matrix adj $\mathbf{S}(z)$ as the removal of the interference from other users. However, this introduces intersymbol interference among the previously noninterfering symbols of the same user. For example, in the two-user case premultiplying the matched filter outputs by

$$\text{adj } \mathbf{S}(z) = \begin{bmatrix} 1 & -\rho_{12} - \rho_{21}z^{-1} \\ -\rho_{12} - \rho_{21}z & 1 \end{bmatrix}$$

244

yields at the ith instant for both $k = 1$ and $k = 2$:

$$A_k \left[\left(1 - \rho_{12}^2 - \rho_{21}^2\right) b_k[i] - \rho_{12}\rho_{21}b_k[i+1] - \rho_{12}\rho_{21}b_k[i-1] \right].$$

Notice that the matched filter outputs of those bits that overlap with $b_1[i]$, (namely, $b_2[i-1]$ and $b_2[i]$) are contaminated by bits $b_1[i-1]$ and $b_1[i+1]$. Thus, the purpose of the scalar transfer function $[\det \mathbf{S}(z)]^{-1}$ is to act as a *zero-forcing equalizer*, that is, a scalar linear filter whose transfer function is the inverse of the equivalent single-user transfer function:

$$-\rho_{12}\rho_{21}z + \left(1 - \rho_{12}^2 - \rho_{21}^2\right) - \rho_{12}\rho_{21}z^{-1}.$$

A stable realization of $[\det \mathbf{S}(z)]^{-1}$ exists as long as $\det \mathbf{S}(z)$ has no zeros on the unit circle:

$$\det [\mathbf{R}^T[1]e^{j\omega} + \mathbf{R}[0] + \mathbf{R}[1]e^{-j\omega}] \neq 0 \quad \text{for all } \omega \in [0, 2\pi]. \quad (5.32)$$

In the two-user case (5.32) becomes

$$|\rho_{12}| + |\rho_{21}| < 1$$

(see Problem 5.17).

For example, consider two users who are assigned the same rectangular waveform. If their relative offset is τ, then (normalizing $T = 1$ without loss of generality) $\rho_{12} = 1 - \tau$ and $\rho_{21} = \tau$. In this case,

$$\det[\mathbf{R}^T[1]e^{j\omega} + \mathbf{R}[0] + \mathbf{R}[1]e^{-j\omega}] = 2(1 - \tau)\tau[1 + \cos\omega],$$

which is equal to 0 for any offset if $\omega = \pi$. So in this case (5.32) is not satisfied.

Although (5.32) is satisfied in most cases of interest, it should be noted that it is more stringent than the existence of a decorrelating detector for every finite frame-length M (cf. (2.6)). That case can be viewed as a synchronous problem as usual, where the nonsingularity of the equivalent cross-correlation matrix guarantees that the corresponding decorrelator exists. It can be checked that in the previous two-user example with identical signals, the crosscorrelation matrix is indeed nonsingular for all finite M.

Even if $\mathbf{S}^{-1}(z)$ admits a stable realization, it cannot be implemented causally, however, as $[\det \mathbf{S}(z)]^{-1}$ does not admit a causal stable inverse in any asynchronous problem. For example, in the two-user case, the inverse z transform of

$$\left[-\rho_{12}\rho_{21}z + \left(1 - \rho_{12}^2 - \rho_{21}^2\right) - \rho_{12}\rho_{21}z^{-1} \right]^{-1}$$

is equal to

$$h[n] = \frac{\xi^{|n|}}{\eta} \qquad (5.33)$$

with

$$\xi = \frac{1 - \rho_{12}^2 - \rho_{21}^2 - \eta}{2\rho_{12}\rho_{21}} \qquad (5.34)$$

and

$$\eta = \sqrt{[1 - (\rho_{12} + \rho_{21})^2][1 - (\rho_{12} - \rho_{21})^2]}, \qquad (5.35)$$

where (5.35) was encountered in (4.132). The impulse response in (5.33) is doubly infinite. Intuitively, this makes sense as all transmitted bits have to be taken into account in order to cancel the interference affecting any particular bit. To implement a real-time approximation of such an impulse response, it is customary to delay it and truncate it at a point where the tail of the impulse response has a negligible contribution.

The asynchronous decorrelating detector can also be implemented as a bank of independent correlators for each user, as in Figure 5.2. Note that the modified matched filters have infinite impulse responses in this case.

5.3 TRUNCATED-WINDOW DECORRELATING DETECTOR

The approach to the asynchronous decorrelator in Section 5.2 is not the only (amplitude-independent) linear strategy that results in error-free demodulation in the absence of background noise. There are other decorrelator-type detectors that achieve the same goal. Let us consider first the *one-shot decorrelating detector*. Here we take a one-shot approach where in order to demodulate every bit we discard all information outside its interval. For example, let us take this approach to demodulate user 1 in a two-user case; the situation is depicted in Figure 5.6.

Effectively, we have a three-user "synchronous" channel where the interferers have unit-energy signature waveforms:

$$s_2^L(t) = \begin{cases} \frac{1}{\sqrt{\vartheta_2}} s_2(t + T - \tau_2), & \text{if } 0 \le t \le \tau_2; \\ 0, & \text{if } \tau_2 < t \le T, \end{cases} \qquad (5.36)$$

$$s_2^R(t) = \begin{cases} 0, & \text{if } 0 \le t \le \tau_2; \\ \frac{1}{\sqrt{1-\vartheta_2}} s_2(t - \tau_2), & \text{if } \tau_2 < t \le T, \end{cases} \qquad (5.37)$$

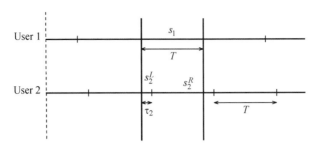

FIGURE 5.6.
One-shot
approach to
demodulation in
asynchronous
two-user channel.

where ϑ_2 is the partial energy of the interfering signal over the left over-lapping interval. To convert this problem into the standard form, we must first obtain a discrete-time model where the observables are the outputs of correlators with $s_1(t)$, $s_2^L(t)$, and $s_2^R(t)$ over $[0, T]$. Note that in contrast to all demodulators considered so far, the bank of filters matched to the signature waveforms cannot be used in this case. Nonetheless, the correlation with $s_2^L(t)$ and $s_2^R(t)$ can be easily implemented using the replica of $s_2(t)$ and the bit-epoch timing of user 2. The crosscorrelation matrix of the ensuing "three-user synchronous" channel is (assuming "user 2" modulates s_2^L and "user 3" modulates s_2^R)

$$\mathbf{R} = \begin{bmatrix} 1 & \rho_{21}/\sqrt{\vartheta_2} & \rho_{12}/\sqrt{1-\vartheta_2} \\ \rho_{21}/\sqrt{\vartheta_2} & 1 & 0 \\ \rho_{12}/\sqrt{1-\vartheta_2} & 0 & 1 \end{bmatrix}. \tag{5.38}$$

The first row of $\mathbf{R}^{-1}$ is a constant times the vector

$$[1 \quad -\rho_{21}/\sqrt{\vartheta_2} \quad -\rho_{12}/\sqrt{1-\vartheta_2}].$$

Therefore, the two-user one-shot decorrelator subtracts from the matched-filter output of the desired user the weighted outputs of the partial correlators.

It is conceptually straightforward to extend this strategy to any number of users. We emphasize that this approach also succeeds in demodulating the bit of the desired user error-free in the absence of background noise. Its disadvantages with respect to the asynchronous decorrelator in Section 5.2 include:

- Partial matched filtering of the interfering bits is required.
- The one-shot crosscorrelation matrix may be singular even if (5.32) is satisfied.
- The performance of the one-shot decorrelating detector is worse than the asynchronous decorrelator, in terms of both bit-error-rate and near–far resistance.

247

The strategy adopted in obtaining the one-shot decorrelating detector can be generalized (in order to improve performance) so as to extend the observation interval beyond $[0, T]$ while keeping a sliding observation window that spans several symbol lengths. Complexity increases with the number of fictitious interferers included in the window, and performance also improves up to a point where it is indistinguishable from that of the ideal asynchronous decorrelator that observes the whole transmitted sequence.

5.4 APPROXIMATE DECORRELATOR

The multiuser detector examined in this section belongs to the class of nondecorrelating linear detectors studied in Chapter 6. It is included here because of its close conceptual relationship with the decorrelating detector.

If the normalized crosscorrelations among all the signature waveforms are very small, $\mathbf{R}$ is strongly diagonal, and the decorrelating signals in (5.8) can be approximated by neglecting coefficients involving products of normalized crosscorrelations:

$$\tilde{s}_k(t) \approx s_k(t) - \sum_{j \neq k} \rho_{kj} s_j(t). \qquad (5.39)$$

This means that instead of processing the matched filter outputs with the matrix $\mathbf{R}^{-1}$, they are processed with $2\mathbf{I} - \mathbf{R}$.

The approximation in (5.39) is justified on the basis that

$$[\mathbf{I} + \delta\mathbf{M}]^{-1} = \mathbf{I} - \delta\mathbf{M} + o(\delta).$$

Analogously, in the asynchronous case, the decorrelating detector for the kth user can be approximated by a filter matched to

$$s_k(t) - \sum_{j \neq k} \rho_{jk} s_j(t - \tau_j) - \sum_{j \neq k} \rho_{kj} s_j(t - \tau_j + T). \qquad (5.40)$$

The same approach can be used to obtain an approximate one-shot decorrelator (Section 5.3), whereby the overlapping matched filter outputs are replaced by partial matched filter outputs.

Whenever the crosscorrelations are not known in advance and the detector coefficients have to be computed on-line, the approximation in (5.39) has the advantage that it does not need any processing of the crosscorrelations supplied by the crosscorrelators of the replicas of the signature waveforms. This detector is particularly advantageous in those asynchronous CDMA channels where the period of the signature waveform is much longer than the symbol period (cf. Section 2.3.4). In those cases, the crosscorrelations keep changing and it is cumbersome to perform the algebraic manipulations required to obtain the decorrelating transformation on-line. Although the approximate decorrelator has zero near–far resistance, its bit-error-rate performance has been shown to be quite superior to that of the conventional matched filter (cf. Problem 5.28).

5.5 PERFORMANCE ANALYSIS: SYNCHRONOUS CASE

We now return to the decorrelating detector for synchronous linearly independent signature waveforms. The decorrelating linear transformation is the projection of the signal of the desired user on the orthogonal space to the space spanned by the interfering signals, and, thus, its bit-error-rate is invariant to the amplitudes of the interfering signals. In fact, the bit-error-rate analysis of the decorrelating detector is the simplest among all multiuser detectors. The reason is that, thanks to property (5.10), the output of the filter matched to $\tilde{s}_k$ only has two components: one due to the signal of user k, which is equal to $A_k b_k$ (cf. (5.9)), and the other due to the background noise, which is Gaussian with zero mean and variance equal to the kk component of the covariance matrix

$$E[(\mathbf{R}^{-1}\mathbf{n})(\mathbf{R}^{-1}\mathbf{n})^T] = E[\mathbf{R}^{-1}\mathbf{n}\mathbf{n}^T\mathbf{R}^{-1}]$$

$$= \sigma^2 \mathbf{R}^{-1}\mathbf{R}\mathbf{R}^{-1}$$

$$= \sigma^2 \mathbf{R}^{-1}. \tag{5.41}$$

Consequently, the kth user bit-error-rate is simply

$$P_k^d(\sigma) = Q\left(\frac{A_k}{\sigma\sqrt{R_{kk}^+}}\right) \tag{5.42}$$

$$= Q\left(\frac{A_k}{\sigma}\sqrt{1 - \mathbf{a}_k^T \mathbf{R}_k^{-1}\mathbf{a}_k}\right), \tag{5.43}$$

which, as we would expect, is *independent of the interfering amplitudes*. The expression in the right side of (5.43) is obtained from (4.97). Recall that $\mathbf{a}_k$ is the kth column of $\mathbf{R}$ without the diagonal element, and $\mathbf{R}_k$ is the $(K-1) \times (K-1)$ matrix that results by striking out the kth row and column from $\mathbf{R}$. To obtain (5.42) we have assumed that the crosscorrelation matrix is nonsingular. The case of singular $\mathbf{R}$ is treated at the end of this section. Obviously, if the kth user is orthogonal to the other users, then the decorrelator coincides with the single-user matched filter and $R_{kk}^+ =$

1. Another expression for the argument of the Q-function in (5.43) was obtained in (4.100) in terms of the signature vectors:

$$\frac{1}{R_{kk}^+} = s_k^T[I - S_k (S_k^T S_k)^{-1} S_k^T]s_k. \tag{5.44}$$

An alternative expression to (5.42) can be given in terms of the unnormalized crosscorrelation matrix (2.80):

$$H = ARA.$$

Since

$$H^{-1} = A^{-1}R^{-1}A^{-1},$$

the diagonal elements of the inverse of the unnormalized crosscorrelation matrix do not depend on the energy of the interferers:

$$A_k^2 H_{kk}^+ = R_{kk}^+,$$

and (5.42) becomes

$$P_k^d(\sigma) = Q\left(\frac{1}{\sigma\sqrt{H_{kk}^+}}\right). \tag{5.45}$$

Yet another expression can be given for $P_1^d(\sigma)$ in terms of the Cholesky factor $R = F^T F$. Since $F_{11}^2 R_{11}^+ = 1$ (Problem 2.51), the bit-error-rate of user 1 is

$$P_1^d(\sigma) = Q\left(\frac{A_1 F_{11}}{\sigma}\right). \tag{5.46}$$

In general,

$$Q\left(\frac{A_k F_{kk}}{\sigma}\right)$$

is the bit-error-rate of the decorrelating linear transformation for s_k against $\{s_{k+1}, \ldots, s_K\}$ in the absence of users $1, \ldots, k-1$.

In the two-user case, we have (cf. (5.11)) $R_{kk}^+ = (1 - \rho^2)^{-1}$, and

$$P_k^d(\sigma) = Q\left(\frac{A_k\sqrt{1 - \rho^2}}{\sigma}\right). \tag{5.47}$$

Consequently, the performance penalty is identical for both users and the effective energy defined in Section 3.5 is independent of the noise level:

$$e_k^d(\sigma) = (1 - \rho^2)A_k^2,$$

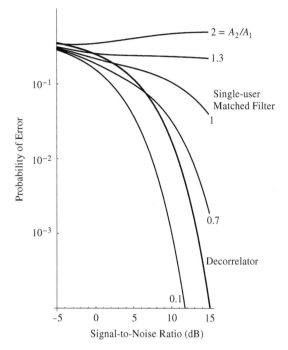

FIGURE 5.7.
Bit-error-rate
comparison of
decorrelator and
single-user
matched filter with
two users and
$\rho = 0.75$.

which translates into multiuser efficiency equal to

$$\eta_1 = \eta_2 = 1 - \rho^2.$$

It is interesting to contrast (5.47) with the error probability of the single-user matched filter detector (3.18):

$$P_1^c(\sigma) = \frac{1}{2}Q\left(\frac{A_1 - A_2\rho}{\sigma}\right) + \frac{1}{2}Q\left(\frac{A_1 + A_2\rho}{\sigma}\right). \tag{5.48}$$

The behaviors of (5.47) and (5.48) as a function of the interferer's amplitude A_2 are sharply different (Figure 5.7). Whereas (5.47) is independent of A_2, (5.48) takes values in the range

$$Q\left(\frac{A_1}{\sigma}\right) \le P_1^c(\sigma) \le \frac{1}{2}$$

with the lower and upper bounds being arbitrarily tight for $A_2 \to 0$ and $A_2 \to \infty$, respectively. Therefore, if the interfering amplitude is small enough, the single-user matched filter detector is preferable to the decorrelator. The reason for this phenomenon is that even though the components in the respective decision statistics due to the desired user are identical in both cases, the component due to the noise has variance σ^2 for the single-user matched filter detector versus variance $\sigma^2/(1 - \rho^2)$ for the decorrelating

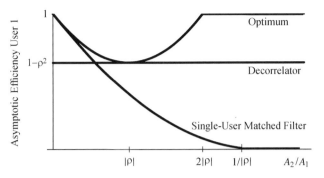

FIGURE 5.8.
Asymptotic
multiuser
efficiencies for
two synchronous
users.

detector. Thus, the price paid for the complete elimination of multiaccess interference is *noise enhancement*.

Figure 5.8 (plotted with $\rho = 0.6$) compares the asymptotic multiuser efficiencies of the optimum, single-user matched filter and decorrelating detectors as a function of the relative amplitude of the interferer. Note that the decorrelator efficiency is

1. independent of A_2/A_1;
2. lower than the single-user matched filter detector asymptotic multiuser efficiency when the interferer is weak:

$$\frac{A_2}{A_1} \leq \frac{1 - \sqrt{1 - \rho^2}}{|\rho|} = \frac{|\rho|}{2} + O(|\rho|^3);$$

3. equal to the worst-case optimum efficiency at the point $A_2 = |\rho| A_1$ and it is strictly suboptimal at all other ratios A_2/A_1.

Since the decorrelating detector bit-error-rate is independent of the amplitude of the interferers, the power-tradeoff region (permissible region of signal-to-noise ratios so that the bit-error-rate of all users does not exceed P) is always a quadrant as shown in Figures 5.9 and 5.10:

$$\frac{A_k^2}{\sigma^2} \geq (Q^{-1}(\text{P}))^2 R_{kk}^+. \tag{5.49}$$

It is interesting to compare the power-tradeoff region (5.49) with that of the single-user matched filter (Figure 5.9) and the optimum detector (Figure 5.10). Note that in the cases depicted in Figure 5.10 the two-user optimum detector offers marginal gains with respect to the decorrelating detector when both amplitudes are equal.

Returning to the K-user case, since (5.42) is the bit-error-rate of a single-user channel with signal-to-noise ratio equal to

$$\frac{A_k^2}{\sigma^2 R_{kk}^+}$$

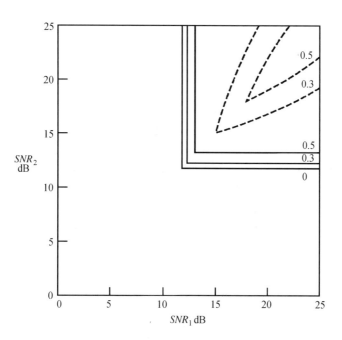

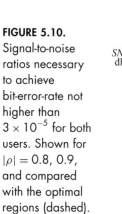

FIGURE 5.9. Signal-to-noise ratios necessary to achieve bit-error-rate not higher than 3×10^{-5} for both users. Shown for $|\rho| = 0, 0.3, 0.5$, and compared with the single-user matched filter detector regions (dashed).

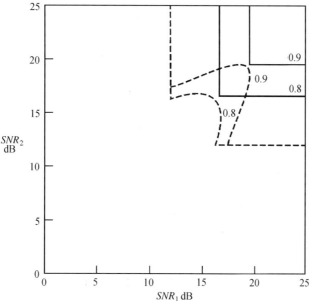

FIGURE 5.10. Signal-to-noise ratios necessary to achieve bit-error-rate not higher than 3×10^{-5} for both users. Shown for $|\rho| = 0.8, 0.9$, and compared with the optimal regions (dashed).

the multiuser efficiency is equal to

$$\eta_k^d = \frac{1}{R_{kk}^+}, \tag{5.50}$$

which does not depend on either the noise level or the interfering amplitudes, and thus, it is equal to the asymptotic multiuser efficiency and to the near–far

253

resistance, that is,

$$\bar{\eta}_k^d = \frac{1}{R_{kk}^+}.$$

We have just obtained the remarkable property that the decorrelating detector achieves the maximum near–far resistance (cf. (4.97)). Accordingly, knowledge of the received amplitudes is not required to combat the near–far problem optimally, and the same degree of robustness against imbalances in the received amplitudes as the exponentially complex optimum detector can be achieved with a computational complexity per demodulated bit similar to that of the single-user matched filter detector.

Why does the decorrelating detector achieve maximum near–far resistance? In order to visualize this property, define the set of signals

$$\Xi_k = \left\{ s_k + \sum_{i \neq k} a_i s_i ; a_i \in R \right\},$$

that is, Ξ_k is the hyperplane of all multiuser signals over all energies and data, except that $A_k = 1$ and $b_k = 1$ (Figure 5.11). Reflecting on (4.93) we see that the optimum near–far resistance $\bar{\eta}_k$ is equal to the square of the distance from the origin to Ξ_k. How do we relate this distance to the decorrelating detector? Because of the orthogonality property in (5.10), the decorrelating waveform $\tilde{s}_k$ is (a scaled version of) the unit-norm signal d_k, which is orthogonal to Ξ_k (Figure 5.11). The output of the (unit-energy) decorrelating filter d_k due to the desired user is equal to

$$\pm A_k \langle d_k, s_k \rangle = \pm A_k \sqrt{\bar{\eta}_k}$$

where we have used the fact that both s_k and d_k have unit norm. Thus, their inner product is simply the distance from Ξ_k to the origin. The output due to

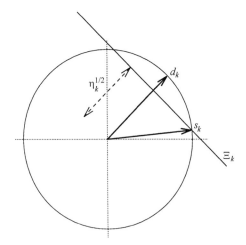

FIGURE 5.11.
Optimality of decorrelating detector near–far resistance.

the noise is zero-mean Gaussian with variance σ^2. Therefore, the multiuser efficiency and near–far resistance of the decorrelating detector are equal to $\bar{\eta}_k$.

The worst asymptotic effective energy (3.115) of the decorrelator readily follows from (5.50):

$$\omega^d(A_1, \ldots, A_K) = \min_{k=1,\ldots,K} \frac{A_k^2}{R_{kk}^+}. \tag{5.51}$$

As we discussed in Section 4.4, it is interesting to analyze performance for direct-sequence systems with random signatures. It was shown in (4.111) that if $\beta < 1$, then the optimum near–far resistance converges in mean-square sense to $1 - \beta$, where

$$\beta = \lim_{K \to \infty} \frac{K}{N}.$$

Since the optimum near–far resistance is equal to the asymptotic efficiency of the decorrelator, the same result holds for the decorrelator asymptotic efficiency regardless of the received powers assuming, as usual, that K denotes the number of users with nonzero power. Furthermore, as a consequence of (4.111) it can be shown (Problem 5.27) that if $\beta < 1$, then the bit-error-rate of the decorrelator converges in mean-square sense:

$$\lim_{K \to \infty} Q\left(\frac{A_k}{\sigma}\sqrt{\bar{\eta}_k}\right) = Q\left(\frac{A_k}{\sigma}\sqrt{1-\beta}\right). \tag{5.52}$$

As we have seen, the decorrelator tunes out the multiaccess interference at the expense of a smaller projection with the signal of interest, or in other words, at the expense of an enhancement of the noise variance at the output of the filter. In Chapter 6 we discuss how to design linear transformations with improved performance by incorporating knowledge of received energies.

Any linear detector independent of the received energies other than the decorrelator has zero near–far resistance. This is because if the output of the linear transformation is contaminated by the signal of an interfering user, and the linear transformation is not allowed to depend on the received powers, then for sufficiently large amplitude of that interferer, the error probability goes to $1/2$ as $\sigma \to 0$.

Regarding the behavior of the decorrelating detector when the background Gaussian noise is dominant, we can assert that unless the kth user is orthogonal to all the interferers, the single-user matched filter has lower bit-error-rate than the decorrelating detector for sufficiently low signal-to-noise ratio. Indeed it can be shown (Problem 5.108) that

$$\lim_{\sigma \to \infty} \frac{\sigma}{A_k} \left(P_k^d(\sigma) - P_k^c(\sigma)\right) = \frac{1 - \sqrt{\bar{\eta}_k}}{\sqrt{2\pi}}. \tag{5.53}$$

255

Furthermore, there are crosscorrelation matrices for which the single-user matched filter gives better bit-error-rate than the decorrelating detector for *all* signal-to-noise ratios (Problem 5.30). In problems of practical interest, the crosscorrelations and desired bit-error-rates are such that the region where the decorrelating detector is bested by the single-user matched filter is rarely encountered.

To conclude this section, we analyze the bit-error-rate when the cross-correlation matrix is singular. If s_k does not belong to the space spanned by the other signature waveforms, then the bit-error-rate in (5.42) is still valid provided that $\mathbf{R}$ is understood to be the crosscorrelation matrix of s_k and a linearly independent set of interfering signature waveforms. If s_k belongs to the space spanned by the other signature waveforms then the generalized decorrelating detector found in Proposition 5.1 is no longer able to tune out all the interfering waveforms and its bit-error-rate can be written as a sum of Q-functions in a way akin to the expression in (3.90).

5.6 PERFORMANCE ANALYSIS: ASYNCHRONOUS CASE

The decision statistic of the kth user asynchronous decorrelating detector consists of two components: $A_k b_k$ and a zero-mean Gaussian random variable. In order to analyze the bit-error-rate of the receiver, all we need to do is evaluate the variance of the noise component. Recall that the noise vector sequence $\{\mathbf{n}[i]\}$ at the output of the matched filter bank has an autocorrelation matrix sequence whose z-transform is (cf. (2.107) and (2.112))

$$\sum_{l=-\infty}^{\infty} E[\mathbf{n}[i]\mathbf{n}^T[i+l]]z^l = \sigma^2 \mathbf{S}(z).$$

The output of the linear transformation with transfer function $\mathbf{S}^{-1}(z)$ due to the sequence $\{\mathbf{n}[i]\}$ will be denoted by $\{\tilde{\mathbf{n}}[i]\}$. Its autocorrelation matrix sequence is denoted by

$$\mathbf{D}[l] = E[\tilde{\mathbf{n}}[i]\tilde{\mathbf{n}}^T[i+l]],$$

whose transform is easily shown to be

$$\sum_{l=-\infty}^{\infty} \mathbf{D}[l]z^l = \sigma^2 \mathbf{S}^{-1}(z)\mathbf{S}(z)[\mathbf{S}^{-1}(z^{-1})]^T \qquad (5.54)$$

$$= \sigma^2 [\mathbf{S}^{-1}(z^{-1})]^T, \qquad (5.55)$$

from where we can recover the matrix $\mathbf{D}[0]$ as

$$
\begin{aligned}
\mathbf{D}[0] &= \frac{\sigma^2}{2\pi} \int_{-\pi}^{\pi} [\mathbf{S}^{-1}(e^{j\omega})]^T \, d\omega \\
&= \frac{\sigma^2}{2\pi} \int_{-\pi}^{\pi} \left([\mathbf{R}^T[1]e^{j\omega} + \mathbf{R}[0] + \mathbf{R}[1]e^{-j\omega}]^{-1}\right)^T \, d\omega.
\end{aligned}
$$

But we are only interested in the kth diagonal element of $\mathbf{D}[0]$:

$$
\begin{aligned}
\mathrm{var}(\tilde{n}_k[i]) &= D_{kk}[0] \\
&= \frac{\sigma^2}{2\pi} \int_{-\pi}^{\pi} [\mathbf{R}^T[1]e^{j\omega} + \mathbf{R}[0] + \mathbf{R}[1]e^{-j\omega}]^+_{kk} \, d\omega \\
&\overset{\text{def}}{=} \frac{\sigma^2}{\bar{\eta}_k^d}.
\end{aligned}
\tag{5.56}
$$

It follows from (5.56) that the bit-error-rate of the asynchronous decorrelating detector is equal to

$$
P_k^d(\sigma) = Q\left(\frac{A_k\sqrt{\bar{\eta}_k^d}}{\sigma}\right).
\tag{5.57}
$$

Therefore, the decorrelating (asymptotic) multiuser efficiency and near–far resistance are equal to

$$
\bar{\eta}_k^d = \left(\frac{1}{2\pi} \int_{-\pi}^{\pi} [\mathbf{R}^T[1]e^{j\omega} + \mathbf{R}[0] + \mathbf{R}[1]e^{-j\omega}]^+_{kk} \, d\omega\right)^{-1}.
\tag{5.58}
$$

Since the argument we used to justify the optimality of the near–far resistance of the decorrelating detector carries over verbatim to the asynchronous case, we can conclude that the expression in (5.58) is the optimum asynchronous near–far resistance. As a point of reference, recall that in the two-user case the optimum near–far resistance is equal to (4.132)

$$
\bar{\eta}_k = \sqrt{[1 - (\rho_{12} + \rho_{21})^2][1 - (\rho_{12} - \rho_{21})^2]}.
\tag{5.59}
$$

Figure 5.12 (Lupas and Verdú [255]) shows the bit-error-rates of the decorrelating and single-user matched filter detectors in a $K = 6$ asynchronous channel where the signature waveforms are maximal-length shift-register sequences of length 31. The interferers are assumed to have the same

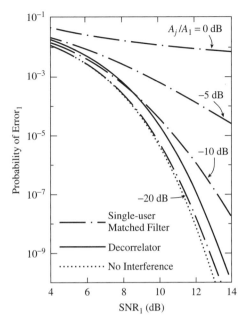

FIGURE 5.12.
Bit-error-rate of decorrelating detector and single-user matched filter detector. Five equal-energy interferers.

received amplitudes. The curves corresponding to the single-user matched filter detector are parametrized by the value of A_k/A_1. We see that with this set of signature waveforms the single-user matched filter detector is unable to achieve an acceptable bit-error-rate level whereas the decorrelating detector performance is about 1 dB below the optimum attainable in the absence of multiuser interference.

The performance analysis of the one-shot asynchronous decorrelating detector can be carried out using the results obtained in Section 5.5 together with the equivalent "synchronous" crosscorrelation matrix. In the two-user case, we obtain from (5.38):

$$\bar{\eta}_1^{1s} = \frac{1}{R_{11}^+} = 1 - \frac{\rho_{12}^2}{1 - \vartheta_2} - \frac{\rho_{21}^2}{\vartheta_2}, \qquad (5.60)$$

which is always smaller than the optimum near–far resistance (4.132) regardless of the value of $\rho_{21}^2 \le \vartheta_2 \le 1 - \rho_{12}^2$.

5.7 COHERENT DECORRELATOR IN THE PRESENCE OF FADING

In this section we examine the average performance of the coherent decorrelator in the presence of Rayleigh fading.

5.7.1 FREQUENCY-FLAT FADING

The observed signal is

$$y(t) = \sum_{k=1}^{K} |A_k| e^{j\theta_k} b_k s_k(t) + \sigma n(t). \tag{5.61}$$

The decorrelating receiver consists of a bank of matched filters that produce the observables

$$y_k = \langle y, s_k \rangle = \int_0^T y(t) s_k^*(t) \, dt$$

followed by multiplication of the vector of observables by the inverse of the Hermitian crosscorrelation matrix (cf. (2.84)). Finally, each component of $\mathbf{R}^{-1}\mathbf{y}$ is compared to the incoming phase:

$$\hat{b}_k = \text{sgn}(\Re\{(\mathbf{R}^{-1}\mathbf{y})_k e^{-j\theta_k}\}).$$

It is interesting to notice that this coherent receiver requires knowledge of the phases of only those users it needs to demodulate. Thus a receiver that does not need to demodulate user i may assume that the signature waveform of user i is $s_i(t) e^{j\phi_i}$ for an arbitrary (but constant over $[0, T]$) ϕ_i. Let us verify that the decision $\hat{b}_k$ is not affected by the choice of $\{\phi_i, i \neq k\}$. Let $\mathbf{P} = \text{diag}\{e^{j\phi_1}, \dots, e^{j\phi_K}\}$. Denote the output of the bank of filters matched to $s_i(t) e^{j\phi_i}, i = 1, \dots, K$ by

$$\bar{\mathbf{y}} = \mathbf{P}^* \mathbf{y}$$

and the crosscorrelation matrix of $s_i(t) e^{j\phi_i}, i = 1, \dots, K$ by (cf. (2.83))

$$\bar{\mathbf{R}} = \mathbf{P}^* \mathbf{R} \mathbf{P}.$$

Thus, the kth component of

$$\bar{\mathbf{R}}^{-1} \bar{\mathbf{y}} = \mathbf{P}^* \mathbf{R}^{-1} \mathbf{P} \mathbf{P}^* \mathbf{y}$$

$$= \mathbf{P}^* \mathbf{R}^{-1} \mathbf{y} \tag{5.62}$$

is independent of $\{\phi_i, i \neq k\}$.

It follows from (2.84) that the decision variable of the kth user is equal to

$$\Re\{(\mathbf{R}^{-1}\mathbf{y})_k e^{-j\theta_k}\} = |A_k| b_k + \tilde{n}_k,$$

where $\tilde{n}_k$ has variance equal to $\sigma^2 (\mathbf{R}^{-1})_{kk}$. Therefore, conditioned on $|A_k|$, the probability of error is equal to

$$Q\left(\frac{|A_k|}{\sigma \sqrt{R_{kk}^+}}\right). \tag{5.63}$$

If the fading coefficient of the kth user is Rayleigh-distributed with $E[|A_k|^2]$ $= 2A_k^2$ (i.e., the RMS value of each component is equal to A_k), then the average of (5.63) is equal to

$$P_k^{Fd}(\sigma) = \int_0^\infty r e^{-r^2/2} Q\left(\frac{A_k r}{\sigma\sqrt{R_{kk}^+}}\right) dr$$

$$= \frac{1}{2}\left(1 - \frac{1}{\sqrt{1 + \frac{\sigma^2 R_{kk}^+}{A_k^2}}}\right). \tag{5.64}$$

According to the definition of asymptotic multiuser efficiency in (3.132), the asymptotic efficiency of the decorrelating detector when the desired user is Rayleigh faded is equal to

$$\eta_k^{Fd} = \frac{1}{R_{kk}^+}. \tag{5.65}$$

Since (5.65) is independent of the strength of the interferers, the near–far resistance is given by the same expression. But the optimum near–far resistance for Rayleigh fading is equal to 1 (Section 4.6), so the optimum near–far resistance property of the decorrelator for deterministic amplitudes does not carry over in the presence of Rayleigh fading.

5.7.2 HOMOGENEOUS FADING

We analyze now the coherent decorrelator when each spread-spectrum signature waveform undergoes the homogeneous fading mechanism in (2.66) under the assumption that the fading parameters therein are independent, zero-mean, complex Gaussian.

Assuming that the decorrelator can perfectly acquire the fading coefficients (assumed to be zero-mean complex Gaussian), the receiver is exactly as in Section 5.7.1. To proceed with the analysis we will make the key assumption that $N \geq K$. In such case the unnormalized crosscorrelation matrix is invertible with probability 1 and, according to (5.45), the error probability is given by

$$P_k^{Sd}(\sigma) = E\left[Q\left(\frac{1}{\sigma\sqrt{H_{kk}^+}}\right)\right], \tag{5.66}$$

where the expectation is with respect to the signature waveforms and received amplitudes. Since P_k^{Sd} is surely independent of the energy of the interferers, we may as well view the matrix $\mathbf{H}$ in (5.66) as the unnormalized crosscorrelation matrix of a system with independent and identically distributed Rayleigh amplitudes and normalized signature waveforms uniformly distributed on the unit complex sphere. It turns out that the distribution of the reciprocal of H_{kk}^+ is particularly simple: standard chi-square with $2(N - K + 1)$ degrees of freedom. To see this, we recall from (4.100) that we can write

$$\frac{1}{A_k^2 H_{kk}^+} = \mathbf{s}_k^* [\mathbf{I} - \mathbf{S}_k (\mathbf{S}_k^* \mathbf{S}_k)^{-1} \mathbf{S}_k^*] \mathbf{s}_k, \tag{5.67}$$

where the KN components of $\mathbf{S}$ are independent complex Gaussian with zero mean and unit variance. Since $\mathbf{S}_k$ has full rank the matrix in the right side of (5.67) is the projection matrix (Problem 5.36)

$$[\mathbf{I} - \mathbf{S}_k (\mathbf{S}_k^* \mathbf{S}_k)^{-1} \mathbf{S}_k^*] = \sum_{i=1}^{N-K+1} \mathbf{v}_i \mathbf{v}_i^*, \tag{5.68}$$

where $(\mathbf{v}_1, \ldots, \mathbf{v}_{N-K+1})$ are orthonormal vectors orthogonal to the columns of $\mathbf{S}_k$. Uniting (5.67) and (5.68), we obtain

$$\frac{1}{A_k^2 H_{kk}^+} = \sum_{i=1}^{N-K+1} |\mathbf{s}_k^* \mathbf{v}_i|^2. \tag{5.69}$$

The orthonormal vectors $\mathbf{v}_i$ depend on the vectors $\mathbf{s}_j$, $j \neq k$, which are independent of $\mathbf{s}_k$. Therefore, the inner products $\{\mathbf{s}_k^* \mathbf{v}_i\}$ are independent with identical distribution. Conditioned on $\mathbf{v}_i$, $\{\mathbf{s}_k^* \mathbf{v}_i\}$ is zero-mean complex Gaussian with unit variance. The same distribution results upon averaging with respect to $\mathbf{v}_i$ because it is independent of $\mathbf{s}_k$.

Finally, we can apply formula (3.65) with $L = N - K + 1$ and $\gamma = A_k / \sigma$. Therefore, performance depends on N and K only through $N - K$. This means that a K-user system with N chips per bit has the same performance as a single-user system with spreading gain equal to $N - K + 1$. Note that this conclusion does not hold in the absence of homogeneous fading when the signature waveforms are randomly chosen from the set of ± 1-valued sequences.

5.7.3 DIVERSITY RECEPTION

We consider here the D-channel diversity model (2.74) whose equivalent[2] complex-valued discrete-time model is (2.95)

$$\mathbf{y}_d = \mathbf{R}(d) \mathbf{A}(d) \mathbf{b} + \mathbf{n}_d, \tag{5.70}$$

---

[2] See Problem 4.11.

where

$$\mathbf{A}(d) = \mathrm{diag}\{\mathsf{A}_{d1}, \ldots, \mathsf{A}_{dK}\},$$

and $\mathbf{n}_d$ is complex Gaussian with real invertible covariance matrix $2\sigma^2\mathbf{R}(d)$. Furthermore, $\mathbf{n}_\ell$ and $\mathbf{n}_d$ are independent for $\ell \neq d$. For each of the D channels, we can apply a decorrelator, the output of which is denoted by

$$\tilde{\mathbf{y}}_d = \mathbf{R}^{-1}(d)\mathbf{y}_d$$

$$= \mathbf{A}(d)\mathbf{b} + \tilde{\mathbf{n}}_d,$$

where the covariance matrix of $\tilde{\mathbf{n}}_d$ is $2\sigma^2\mathbf{R}^{-1}(d)$. Since b_k only affects $\{\tilde{\mathsf{y}}_{dk}, d = 1, \ldots, D\}$, it is reasonable to base a decision on b_k only on those observables for the sake of reduced complexity. However, this incurs loss of optimality because the noise components are correlated. Once we decide to obtain $\hat{b}_k$ on the basis of

$$\tilde{\mathsf{y}}_{1k} = \mathsf{A}_{1k}b_k + \tilde{\mathsf{n}}_{1k},$$

$$\vdots \tag{5.71}$$

$$\tilde{\mathsf{y}}_{Dk} = \mathsf{A}_{Dk}b_k + \tilde{\mathsf{n}}_{Dk},$$

we face a single-user channel. We can minimize the error probability by combining the observables in (5.71) according to the rule (Problem 3.9):

$$\hat{b}_k = \mathrm{sgn}\left(\Re\left\{\sum_{d=1}^{D} \frac{\mathsf{A}_{dk}}{R_{kk}^+(d)}\tilde{\mathsf{y}}_{dk}^*\right\}\right). \tag{5.72}$$

We observe that the receiver in (5.72) is a combination of the decorrelator and the maximal-ratio combining rule (3.138) properly modified to take into account the different noise variances. Coherence is only required with respect to each of the D signals due to the desired user. Using (5.71) we can write the observable in (5.72) as

$$\Re\left\{\sum_{d=1}^{D} \frac{\mathsf{A}_{dk}}{R_{kk}^+(d)}\tilde{\mathsf{y}}_{dk}^*\right\} = b_k \sum_{d=1}^{D} \frac{|\mathsf{A}_{dk}|^2}{R_{kk}^+(d)} + n,$$

where the variance of the zero-mean Gaussian random variable n is

$$E[n^2] = \sigma^2 \sum_{d=1}^{D} \frac{|\mathsf{A}_{dk}|^2}{R_{kk}^+(d)}.$$

Thus, conditioned on $\mathsf{A}_{1k}, \ldots, \mathsf{A}_{Dk}$, the error probability is

$$Q\left(\frac{1}{\sigma}\sqrt{\sum_{d=1}^{D} \frac{|\mathsf{A}_{dk}|^2}{R_{kk}^+(d)}}\right).$$

In order to average with respect to independent Rayleigh-distributed $|A_{dk}|$ we can use (3.64) with $L = D$ and

$$\lambda_i = \frac{E[|A_{ik}|^2]}{\sigma^2 R_{kk}^+(i)},$$

provided those quantities are all distinct.

5.8   DIFFERENTIALLY-COHERENT DECORRELATION

In Section 3.7 we studied single-user differentially-coherent demodulation, whereby the transmitted bits are obtained from the original data stream via

$$e_k[i] = e_k[i - 1]b_k[i],$$

and the decisions are given by

$$\hat{b}_k[i] = \text{sgn}(\Re\{y_k^*[i - 1]y_k[i]\}), \tag{5.73}$$

where

$$y_k[i] = \int_{iT}^{iT+T} y(t)s_k^*(t - iT)\, dt, \tag{5.74}$$

and the incoming waveform is

$$y(t) = \sum_{k=1}^{K} |A_k| e^{j\theta_k} (e_k[i - 1]s_k(t - iT + T) + e_k[i]s_k(t - iT)) + \sigma n(t), \tag{5.75}$$

on the interval $t \in [iT - T, iT + T]$. The demodulator in (5.73) does not require knowledge of the amplitudes or phases of any of the users.

To robustify (5.73) with respect to the presence of multiaccess interference, we will replace the matched filter outputs therein by decorrelator outputs. Retaining the notation of (5.8) in the complex-valued channel,

$$\tilde{s}_k(t) = \sum_{j=1}^{K} R_{jk}^+ s_j(t), \tag{5.76}$$

we will denote the decorrelator outputs by

$$\tilde{y}_k[i] = \int_{iT}^{iT+T} y(t)\tilde{s}_k^*(t - iT)\, dt. \tag{5.77}$$

And (5.73) is replaced by

$$\hat{b}_k[i] = \text{sgn}(\Re\{\tilde{y}_k^*[i - 1]\tilde{y}_k[i]\}). \tag{5.78}$$

263

No knowledge of the amplitudes or phases of any of the users is required to produce (5.78). Indeed, according to (5.62), $\tilde{y}_k[i]$ is independent of the phase assumed by the receiver for users other than k. Furthermore, the phase dependence of $\tilde{y}_k[i]$ on the phase of the kth user assumed by the receiver is canceled with that of $\tilde{y}_k^*[i-1]$ if the incoming phase remains constant over consecutive intervals, which is a basic underlying assumption of differentially-coherent demodulation.

To analyze the error probability of (5.78), we make explicit the dependence of the decorrelator outputs on the data:

$$\tilde{y}_k[i] = A_k e^{j\theta_k} \langle s_k, \tilde{s}_k \rangle e_k[i] + n_k[i] \tag{5.79}$$

with $n_k[i]$ a complex Gaussian random variable with independent real and imaginary parts, each with variance

$$\sigma^2 \|\tilde{s}_k\|^2 = \sigma^2 \left\| \sum_{j=1}^{K} R_{jk}^+ s_j \right\|^2$$

$$= \sigma^2 \sum_{l=1}^{K} \sum_{j=1}^{K} R_{kl}^+ \rho_{lj} R_{jk}^+$$

$$= \sigma^2 R_{kk}^+. \tag{5.80}$$

The complex-valued version of (5.9) holds:

$$\langle s_k, \tilde{s}_k \rangle = \sum_{j=1}^{K} R_{kj}^+ \langle s_k, s_j \rangle$$

$$= \sum_{j=1}^{K} R_{kj}^+ R_{jk}$$

$$= 1. \tag{5.81}$$

By symmetry, the probability of error of the differentially-coherent decorrelator is

$$P_k^{dcd}(\sigma) = P[\hat{b}_k[i] = -1 \mid e_k[i] = e_k[i-1] = 1]$$

$$= P[\Re\{XY^*\} < 0], \tag{5.82}$$

where X and Y are complex Gaussian with independent real and imaginary parts, each with variance $\sigma^2 R_{kk}^+$ and means

$$m_X = m_Y = A_k e^{j\theta_k}.$$

Recalling Proposition 3.6, we obtain the closed-form expression for the bit-error-rate of the differentially-coherent decorrelator conditioned on $|A_k|$:

$$P_k^{dcd}(\sigma) = \frac{1}{2}e^{-|A_k|^2/(2\sigma^2 R_{kk}^+)}. \tag{5.83}$$

Accordingly, the (asymptotic) multiuser efficiency and the near–far resistance of the kth user are equal to

$$\bar{\eta}_k^{dcd} = \frac{1}{R_{kk}^+}, \tag{5.84}$$

if $|A_k|$ is deterministic. So, in this case, the differentially-coherent demodulator retains the optimum near–far resistance property.

In the presence of fading, we need to average (5.83) with respect to the complex coefficients $A_1, \ldots, A_K$; however, R_{kk}^+ is not influenced by those coefficients; therefore it is just a matter of computing the average

$$\frac{1}{2}E\left[e^{-A_k^2 R^2/(2\sigma^2 R_{kk}^+)}\right]$$

with respect to R. In the case of Rayleigh fading, we obtain

$$\frac{1}{2}E\left[e^{-A_k^2 R^2/(2\sigma^2 R_{kk}^+)}\right] = \frac{1}{2}\frac{1}{1 + \frac{A_k^2}{\sigma^2 R_{kk}^+}}. \tag{5.85}$$

Comparing to the single-user result (3.157) we see that the efficiency remains equal to (5.84) in the presence of Rayleigh fading.

5.9 DECORRELATION FOR NONLINEAR MODULATION

We have seen that the decorrelator converts the received signal into a single-user signal by tuning out the multiuser interference with a linear transformation. As we illustrate in this section, this principle does not hinge either on the linearity of the modulation or on receiver coherence.

Consider the m-ary K-user channel (3.170)

$$y(t) = \sum_{k=1}^{K} |A_k| e^{j\theta_k} s_k(t; b_k) + \sigma n(t). \tag{5.86}$$

As we saw in Section 2.5, we can view this channel as one with mK users and linear modulation:

$$\sum_{k=1}^{K} |A_k| e^{j\theta_k} s_k(t; b_k) = \sum_{n=1}^{mK} \bar{A}_n \bar{b}_n \bar{s}_n(t), \tag{5.87}$$

with

$$\bar{s}_{j+(k-1)m}(t) = s_k(t; j), \tag{5.88}$$

$$\bar{b}_{j+(k-1)m} = 1\{b_k = j\}, \tag{5.89}$$

$$\bar{A}_{j+(k-1)m} = A_k. \tag{5.90}$$

Sufficient statistics are obtained by projecting along each of the mK signatures appearing in (5.87):

$$\bar{y}_n = \int_0^T y(t)\bar{s}_n^*(t)\,dt. \tag{5.91}$$

As in (2.84), the dependence of these observables on the data can be displayed as an mK-dimensional model

$$\bar{y} = \bar{R}\bar{A}\bar{b} + \bar{n}, \tag{5.92}$$

where $\bar{R}$ is the covariance matrix of the signature waveforms $\bar{s}_1, \ldots, \bar{s}_{mK}$, $\bar{y}$ is defined in (5.91), $\bar{b}$ is defined in (5.89),

$$\bar{A} = \text{diag}\{\bar{A}_1, \ldots, \bar{A}_{mK}\},$$

and $\bar{n}$ is a Gaussian complex vector with covariance matrix equal to $\bar{R}$. Note that $\bar{b}$ has only K nonzero components at positions $b_1, b_2 + m, \ldots, b_K + (K-1)m$.

We cannot apply verbatim the multiuser detectors we have derived for the basic linear antipodal channels to the alternative linear model in (5.92) because of the interdependence in the binary components of $\bar{b}$. If $\bar{R}$ is invertible, then it is natural to consider the observables

$$\bar{z} = \bar{R}^{-1}\bar{y}.$$

Even though the data of the kth user affects only components $mk - m + 1, \ldots, mk$, it is suboptimal to discard the other components because their noise samples are correlated with those in the $mk - m + 1, \ldots, mk$ components. Nevertheless, for the sake of complexity reduction it makes sense to demodulate the kth user information based only on:

$$\begin{bmatrix} \bar{z}_{m(k-1)+1} \\ \vdots \\ \bar{z}_{mk} \end{bmatrix} = A_k \begin{bmatrix} 1\{b_k = 1\} \\ \vdots \\ 1\{b_k = m\} \end{bmatrix} + \sigma \begin{bmatrix} n_{m(k-1)+1} \\ \vdots \\ n_{mk} \end{bmatrix} \tag{5.93}$$

where the covariance matrix $Q(k)$ of the vector $[n_{m(k-1)+1} \cdots n_{mK}]$ is the kth diagonal $m \times m$ block of $\bar{R}$. The optimum noncoherent detector for b_k

based on the single-user channel (5.93) is (Problem 3.51):

$$\hat{b}_k = \arg\max_i \left\{ \exp\left(-\frac{Q_{ii}^+(k)}{\sigma^2}\right) I_0\left(2\frac{\left|\sum_{j=1}^m Q_{ij}^+(k)\bar{z}_{m(k-1)+j}\right|}{\sigma^2}\right) \right\},$$

(5.94)

which does not require knowledge of the received energies.

Analyzing the probability of error of (5.94) for arbitrary m-ary modulation poses significant challenges. When additional structure is placed on the nonlinear modulation, it is possible to proceed further. For example, consider the special case of Walsh orthogonal modulation. We saw in Section 2.5 that in this type of m-ary nonlinear modulation ($m = 2^J$), user k is assigned a unit-energy signature waveform $s_k(t)$ and symbol $b_k \in \{1, \ldots, m\}$ is transmitted by sending the signal

$$\sum_{i=1}^m h[b_k, i]s_k(t - iT),$$

where the $m \times m$ matrix

$$\mathcal{H} = \begin{bmatrix} h[1, 1] & \cdots & h[1, m] \\ \ldots & \ddots & \ldots \\ h[m, 1] & \cdots & h[m, m] \end{bmatrix}$$

with $\{-1, +1\}$ entries is constructed in Problem 1.4 and has orthogonal rows.

Assuming a synchronous system, the receiver obtains

$$y(t) = \sum_{k=1}^K A_k \sum_{n=1}^m h[b_k, n]s_k(t - nT) + \sigma n(t),$$

(5.95)

where the real and imaginary components of $n(t)$ are independent and have unit spectral density. Suppose that we construct the $K \times m$ matrix of matched filter outputs $\mathbf{Y}$ whose (k, i) entry is

$$y_{ki} = \int_{iT}^{iT+T} y(t)s_k^*(t - iT)\, dt,$$

and denote the corresponding noise matrix by $\mathbf{N}$ whose (k, i) entry is

$$n_{ki} = \int_{iT}^{iT+T} n(t)s_k^*(t - iT)\, dt.$$

Furthermore, denote the $K \times m$ matrix of channel data

$$\mathbf{B} = \begin{bmatrix} h[b_1, 1] & \cdots & h[b_1, m] \\ \ldots & \ddots & \ldots \\ h[b_K, 1] & \cdots & h[b_K, m] \end{bmatrix}.$$

267

It is easy to verify that

$$\mathbf{Y} = \mathbf{RAB} + \sigma\mathbf{N}.\tag{5.96}$$

Now, let us assume that the crosscorrelation matrix is invertible and consider the $K \times m$ matrix

$$\mathbf{Z} \stackrel{\text{def}}{=} \mathbf{R}^{-1}\mathbf{Y}\mathcal{H}^T\tag{5.97}$$

$$= m\mathbf{AX} + \sigma\mathbf{R}^{-1}\mathbf{N}\mathcal{H}^T,\tag{5.98}$$

where we have used the orthogonality of the rows of $\mathcal{H}$ and we have defined the $K \times m$ matrix

$$\mathbf{X} = \begin{bmatrix} 1\{b_1 = 1\} & \cdots & 1\{b_1 = m\} \\ \vdots & \ddots & \vdots \\ 1\{b_K = 1\} & \cdots & 1\{b_K = m\} \end{bmatrix}.$$

According to (5.98), it makes sense to decide

$$\hat{b}_k = \arg\max_{j=1,\dots,m} |Z_{kj}|.\tag{5.99}$$

Once more, the decorrelator has succeeded in converting the problem to a single-user one. Indeed, the kth row is independent of the data sent by the interferers or their strengths. As before, we cannot expect (5.99) to be optimal because it neglects useful information present in $\mathbf{Y}$. In order to analyze the probability of error of the decision rule in (5.99) it is crucial to realize that the noise variables affecting the vector $(Z_{k1}, \dots, Z_{km})$ therein are independent. Indeed, for any $(i, j) \in \{1, \dots, m\}^2$:

$$E[(\mathbf{R}^{-1}\mathbf{N}\mathcal{H}^T)_{ki}(\mathbf{R}^{-1}\mathbf{N}\mathcal{H}^T)_{kj}]$$

$$= \sum_{\ell=1}^{K}\sum_{q=1}^{m}\sum_{r=1}^{K}\sum_{p=1}^{m} R_{k\ell}^+ R_{kr}^+ E[n_{\ell q}n_{rp}] h[i,q]h[j,q]$$

$$= \sum_{\ell=1}^{K}\sum_{r=1}^{K} R_{\ell r} R_{k\ell}^+ R_{kr}^+ \sum_{q=1}^{m} h[i,q]h[j,q]$$

$$= m R_{kk}^+ \delta_{ij},\tag{5.100}$$

where we have used the fact that if $q \neq p$, then $n_{\ell q}$ and n_{rp} are independent because they are projections of nonoverlapping segments of white noise. The orthogonality of the rows of the Walsh matrix has been used in (5.100). We conclude that we have an m-ary orthogonal single-user noncoherent problem that fits the framework of Section 3.8 with $m|A_k|$ in lieu of $|A|$ and

$mR_{kk}^+\sigma^2$ in lieu of σ^2. We just need to use (3.167) to arrive at the following expression for the probability of symbol error:

$$\mathsf{P}^{ncd} = \sum_{n=1}^{m-1}(-1)^{n+1}\binom{m-1}{n}\frac{1}{n+1}\exp\left(-\frac{nm|\mathsf{A}_k|^2}{2(n+1)\sigma^2R_{kk}^+}\right), \quad (5.101)$$

which is consistent with (3.167) in the single-user case since, in the present setting, signals have m times the energy of those in Section 3.8. Thus, in the case of orthogonal modulation the multiuser efficiency of the noncoherent decorrelator is equal to the familiar expression

$$\eta_k^{ncd} = \frac{1}{R_{kk}^+}.$$

5.10 BIBLIOGRAPHICAL NOTES

The problem of eliminating multiuser interference with a linear receiver was posed as early as 1967 (Shnidman [409]). The derivation of the asymptotic efficiency of the decorrelating detector for synchronous channels and the proof of its optimum near–far resistant property are due to Verdú [489] in the case of nonsingular correlation matrices. Its justification as a generalized maximum-likelihood detector when the energies are not known appeared in Lupas and Verdú [254]. The results in Verdú [489] were generalized to the case of linearly dependent signature waveforms in Lupas and Verdú [254]. Reference Schneider [392] claimed erroneously that the decorrelating detector minimizes bit-error-rate for a synchronous channel with equal-amplitude users. An independent optimality claim for the synchronous decorrelator was made in Kohno et al. [215].

The asynchronous decorrelating detector was obtained in Lupas and Verdú [253, 255]. A decorrelating detector for the asynchronous channel involving the inversion of the $K(2M+1)$-dimensional crosscorrelation matrix of fictitious users was proposed in Kohno et al. [214]. Further justification for the decorrelating detector was given in Lupas and Verdú [255], showing that the decorrelating detector emerged as the solution to a minimax problem. Reference Lupas and Verdú [255] noticed that the asynchronous decorrelator can be viewed as a two-step process: first eliminate multiuser interference by introducing single-user intersymbol interference; this is then removed by a single-user zero-forcing equalizer. The zero-forcing equalizer can be replaced by other single-user equalization methods, such

as maximum-likelihood sequence detection (Liang et al. [240])
(cf. Problem 5.4).

The forerunner of the decorrelating detector in the single-user
intersymbol interference channel is the *zero-forcing equalizer*
(e.g., Proakis [345]). The optimality of the near–far resistance
of the decorrelating detector has no counterpart in the single-
user setting. Decorrelation for multiuser channels subject to intersymbol interference has been considered in, among others,
Klein and Baier [209], Klein et al. [210], and Wang and Poor
[523].

The one-shot decorrelating detector for asynchronous problems was proposed in Verdú [499]. Other truncated-window
decorrelators are put forward in Wijayasuriya et al. [532], Bravo
[35], Zheng and Barton [562], Juntti and Aazhang [188], and
Wijayasuriya et al. [536]. At the expense of some performance
degradation, quasi-synchronous users (Problem 2.40) can be
decorrelated without knowledge of exact chip timing as shown
in van Heeswyk et al. [460] and IItis and Mailaender [173].

The approximate decorrelator of Section 5.4 was proposed
in Kohno et al. [214] and Verdú [498] and analyzed in Mandayam and Verdú [265], Patel and Holtzman [324], Divsalar and
Simon [66], Wu and Duel-Hallen [547], I et al. [168], and Seskar
and Mandayam [402] in both coherent and noncoherent versions. Higher-order approximations (cf. Problem 5.11 (d)) are
considered in Ghazi-Moghadam et al. [112], Agashe and
Woerner [13], and Moshavi [288]. The computational complexi-
ties of various algorithms for decorrelation that avoid the
computation of matrix inverses are studied in Juntti [187]. Implementation of a Gram–Schmidt-based computation of the decorrelator (Problem 5.1) is considered in Myers and Magaña [296].
Finite-precision arithmetic implementation of the decorrelator
is studied in Paris [317]. The impact of timing jitter on the decorrelator and other linear multiuser detectors is analyzed in Zheng
and Barton [562], Parkvall et al. [320], Buehrer et al. [36], and
Riba et al. [359].

The analysis of the coherent decorrelator in the presence of
Raleigh-faded amplitudes is due to Zvonar and Brady [570]. Its
performance when coupled with trellis-coded modulation and
diversity is studied in Caire et al. [41]. Generally in conjunction with rake correlators, the decorrelator has been adapted
to single-user multipath channels in Mydlow et al. [295] and
to multiuser multipath channels in Wijayasuria and McGeeham
[533], Wijayasuriya et al. [534], Chen and Roy [51], Huang and
Schwartz [166], Zvonar and Brady [571], Kawahara and Matsumoto [205], Papproth and Kaleh [315], Kandala et al. [198],

Huang [165], Zvonar [568], Zvonar and Brady [573], Wijaya-suriya et al. [536], and Wang and Poor [521]. The result in Section 5.7.2 on the error probability with homogeneous fading is due to Winters et al. [538] and the proof we have given follows that of Caire et al. [41].

Decorrelation with multidimensional array models was studied in Miller [276], Miller and Schwartz [277], Hosur et al. [161], Huang [165], Muñoz-Medina and Fernández-Rubio [293], Zvonar and Stojanovic [574], Stojanovic and Zvonar [427], and Kim et al. [208].

Differentially-coherent decorrelation is studied in Varanasi and Aazhang [475] for synchronous channels (including the case of linearly dependent signatures) and Varanasi [466], Zvonar and Brady [572], and in Liu and Siveski [245] for asynchronous channels.

The use of the decorrelator for nonlinearly modulated signals was originally suggested in a coherent setting by Schneider [393]. The analysis of the decorrelator for noncoherent demodulation of Walsh-modulated signals is due to Hegarty and Vojcic [141]. Nonorthogonal nonlinear modulation is considered in Russ and Varanasi [370], and Varanasi and Russ [477].

Adaptive versions of the decorrelator are investigated in Chen and Roy [51], Mitra and Poor [279], Mitra and Poor [281], Lim and Rasmussen [241], and Ulukus and Yates [457]. A corpus of signal processing results under the heading of "adaptive signal separation" (reviewed in Cardoso [44]) investigates the computation of the inverse of a linear transformation based on noiseless observations and (partial) knowledge of the distribution of the inputs, which need not be discrete-valued as in our setting.

The counterpart of decorrelation in antenna arrays subject to undesired sources is called *null steering*, for which a wealth of fast adaptive methods are known (Rader [349]).

The decorrelating detector has been investigated in Felhauer et al. [90] for the estimation of unknown received powers. Using the decorrelator for detecting whether a user is active has been studied in Mitra and Poor [280] and Mitra and Poor [282]. The performance degradation suffered by the decorrelator due to unaccounted-for multiaccess interference is studied in Kajiwara and Nakagawa [194] and Esteves and Scholtz [79]. The degradation in coherent decorrelation due to imperfect phase estimates is treated in Zvonar and Brady [569], Rusch and Poor [369], and Orten and Ottosson [311]. Recursive updating algorithms for the computation of decorrelating transformations are given in Wijayasuriya et al. [535], and Juntti [186, 187].

Decorrelating-based timing acquisition schemes have been proposed in Iltis [169].

The generalization of the decorrelation approach mentioned at the end of Section 5.1 whereby a front-end linear transformation projects onto the orthogonal space of a subset of unwanted users has been explored in Scharf and Friedlander [385], Varanasi [467], and Schlegel et al. [388]. A hybrid decorrelating-optimum approach to obtain suboptimum multiuser detectors is proposed in Varanasi [467] and Varanasi [468] for coherent multiuser demodulation without and with fading, respectively: a demodulator for a subset of users can be implemented by first decorrelating with respect to the users not in the subset and then performing optimum detection for the users in the subset (cf. Problems 5.45 and 5.46). The computational complexity is then exponential in the size of the subset. The performance and implementation of a similar receiver in conjunction with error-correcting codes is analyzed in Schlegel and Xiang [390], Schlegel et al. [388], Schlegel and Mathews [387], and Alexander et al. [17].

The capacity achievable with the decorrelator in a frequency-selective Rayleigh-fading channel has been considered in Goeckel and Stark [121]. A constructive scheme that uses error control coding and decorrelator outputs is proposed in Alexander et al. [16].

A version of the decorrelator that is robustified against non-Gaussian noise is proposed and analyzed in Wang and Poor [520].

The excision of narrowband interference using the decorrelating detector is explored in Rusch and Poor [367].

5.11 PROBLEMS

PROBLEM 5.1. Suppose that $(z_1, \dots, z_K)$ are obtained from $(s_1, \dots, s_K)$ by means of the Gram–Schmidt procedure:

$$z_1 = s_1,$$

$$\vdots$$

$$z_k = s_k - \sum_{j=1}^{k-1} \frac{\langle s_k, z_j \rangle}{\|z_j\|^2} z_j.$$

(a) Show that z_K is a decorrelating transformation for the Kth user.
(b) Show that $\bar{\eta}_K = \|z_K\|^2$.

PROBLEM 5.2. Find a necessary and sufficient condition on $\{s_1, \ldots, s_K\}$ for $\|\tilde{s}_k\| = 1$.

PROBLEM 5.3. Given an invertible crosscorrelation matrix $\mathbf{R}$ and a K-vector $\mathbf{e}$ whose components are drawn from $\{-1, 1\}$, we construct the matrix $\bar{\mathbf{R}}$ with entries:

$$\bar{\rho}_{jk} = \rho_{jk} e_j e_k.$$

Show that the diagonal elements of $\bar{\mathbf{R}}^{-1}$ do not depend on $\mathbf{e}$.

PROBLEM 5.4. We saw in Section 5.2 that the asynchronous decorrelator detector consists of a linear transformation adj $\mathbf{S}(z)$ that eliminates multiuser interference followed by a bank of identical single-user zero-forzing equalizers $[\det \mathbf{S}(z)]^{-1}$. Show that if the single-user zero-forcing equalizer is replaced by a maximum-likelihood sequence detector, the multiuser efficiency remains the same as that of the decorrelator.

PROBLEM 5.5. Suppose that the unit-energy signature waveforms $\{s_1, \ldots, s_K\}$ are linearly independent. Show that the unit-energy signal h that maximizes $\langle h, s_1 \rangle$ subject to $\langle h, s_k \rangle = 0, k = 2, \ldots, K$ is the normalized decorrelating transformation $\tilde{s}_1 / \|\tilde{s}_1\|$ (5.8).

PROBLEM 5.6. Consider a maximal-length shift-register sequence of period N and suppose that K users are assigned different periods of that sequence as their spreading code. Show that if the users are synchronous, the probability of error of the kth user's decorrelator is equal to

$$P_k^d(\sigma) = Q\left(\frac{A_k}{\sigma} \sqrt{\frac{(N+1)(N-K+1)}{N(N-K+2)}} \right). \tag{5.102}$$

PROBLEM 5.7. Show that if $\rho_{kj} = \alpha^{|k-j|}$ with $0 \leq \alpha < 1$, the probability of error of the decorrelator for user 1 is equal to

$$P_1^d(\sigma) = Q\left(\frac{A_1}{\sigma} \sqrt{1 - \alpha^2} \right). \tag{5.103}$$

PROBLEM 5.8.

(a) Show that the near–far resistance of a linear detector $\mathbf{v}$ for the kth user is

$$\bar{\eta}_k^v = \min_{\substack{\mathbf{c} \in R^K \\ c_k = 1}} \frac{\max^2\{0, \mathbf{c}^T \mathbf{R} \mathbf{v}\}}{\mathbf{v}^T \mathbf{R} \mathbf{v}}. \tag{5.104}$$

273

(b) Verify that if $\mathbf{v}^*$ equals the kth column of $\mathbf{R}^{-1}$ and

$$\mathbf{c}^* = \frac{\mathbf{v}^*}{v_k^*},$$

then $(\mathbf{v}^*, \mathbf{c}^*)$ is a saddle point of the function

$$f(\mathbf{v}, \mathbf{c}) = \frac{\max^2\{0, \mathbf{c}^T \mathbf{R} \mathbf{v}\}}{\mathbf{v}^T \mathbf{R} \mathbf{v}}$$

when the feasible set for $\mathbf{c}$ is all K-vectors with $c_k = 1$, and the feasible set for $\mathbf{v}$ is all K-vectors.

PROBLEM 5.9. Consider two CDMA K-user synchronous systems where the kth user transmits

(A) $$A_k \sum_i s_k(t - iT) b_k[i],$$

(B) $$\alpha A_k \sum_i \tilde{s}_k(t - iT) b_k[i],$$

where $\tilde{s}_k$ is defined in (5.8). Suppose that α is chosen so that the sum of the powers transmitted by the users are equal in both systems:

$$\alpha^2 = \frac{\sum_{k=1}^K A_k^2}{\sum_{k=1}^K A_k^2 \|\tilde{s}_k\|^2}.$$

Find a set of parameters for which a decorrelator for user 1 in system A obtains lower probability of error than an optimum detector for user 1 in system B. (cf. Vojcic and Jang [513])

PROBLEM 5.10. Consider a K-user synchronous CDMA channel with crosscorrelation matrix $\mathbf{R}$:

$$y(t) = \sum_{i=-M}^M \sum_{k=1}^K A_k b_k[i] s_k(t - iT) + \sigma n(t).$$

Nothing is known about the received amplitudes except that they are constant over time. Consider a joint maximum-likelihood detector for

$$(A_1, \ldots, A_K, b_1[-M], \ldots, b_1[M], \ldots, b_K[-M], \ldots, b_K[M]).$$

Describe the decisions $(b_1[0], \ldots, b_K[0])$ in the asymptotic regime $M \to \infty$.

PROBLEM 5.11. Suppose that the normalized crosscorrelation matrix $\mathbf{R}$ is invertible. Let $\mathbf{M} = \mathbf{R} - \mathbf{I}$. Thus $\mathbf{M}$ is a symmetric matrix with null diagonal elements. The *spectral radius* of a matrix is the largest magnitude of its eigenvalues.

(a) Show that the eigenvalues of $\mathbf{M}$ are strictly greater than -1.
(b) Give an example of a crosscorrelation matrix $\mathbf{R}$ whose spectral radius is larger than 2.
(c) Show that if the spectral radius of $\mathbf{R}$ is strictly less than 2, then

$$\lim_{n \to \infty} \mathbf{M}^n = \mathbf{0}.$$

(d) Under the assumption of (c) verify that

$$\mathbf{R}^{-1} = \mathbf{I} - \mathbf{M} + \mathbf{M}^2 - \mathbf{M}^3 + \mathbf{M}^4 - \mathbf{M}^5 + \cdots.$$

PROBLEM 5.12.

(a) Show that if $\lambda_{\max}$ denotes the largest eigenvalue of $\mathbf{R}$, and

$$0 < \alpha < \frac{1}{\lambda_{\max}},$$

then

$$\mathbf{R}^{-1} = \alpha \sum_{j=0}^{\infty} [\mathbf{I} - \alpha \mathbf{R}]^j. \tag{5.105}$$

(b) Suppose that $\mathbf{R}^{-1}$ is approximated by neglecting all but the first two terms in (5.105). Choose α that minimizes the trace of the error matrix.
(c) Modify the approximate decorrelator of Section 5.4 based on the result in (b).

PROBLEM 5.13. A *matrix norm* is a norm in the linear space of square matrices of given dimension $R^{n \times n}$ ($\|\mathbf{C}\| > 0$ if $\mathbf{C} \neq \mathbf{0}$; $\|\beta \mathbf{C}\| = |\beta| \|\mathbf{C}\|$; $\|\mathbf{C} + \mathbf{D}\| \leq \|\mathbf{C}\| + \|\mathbf{D}\|$) that satisfies

$$\|\mathbf{CD}\| \leq \|\mathbf{C}\| \|\mathbf{D}\|.$$

Examples of matrix norms are:

$$\|\mathbf{C}\|_p = \left(\sum_{i=1}^{n} \sum_{j=1}^{n} |C_{ij}|^p \right)^p, \quad 1 \leq p \leq 2,$$

$$\|\mathbf{C}\|_\infty = \max_{1 \leq i \leq n} \sum_{j=1}^{n} |C_{ij}|.$$

Show that the spectral radius (cf. Problem 5.11) of a matrix is upper bounded by any of its norms (e.g., Lancaster and Tismenetsky [277]).

PROBLEM 5.14. Show that if λ is an eigenvalue of $\mathbf{R}$, then it satisfies

$$2 - \|\mathbf{R}\|_\infty \leq \lambda.$$

[*Hint:* Use Geršgorin's theorem (Lancaster and Tismenetsky [277]).]

PROBLEM 5.15. (Bulumulla and Venkatesh [38]) Suppose that the K-user synchronous decorrelator operates on uniformly quantized matched-filter outputs:

$$\mathbf{R}^{-1}[q(y_1)\cdots q(y_K)]^T,$$

where the quantizer has M levels (M is assumed to be odd for convenience):

$$q(x) \stackrel{\text{def}}{=} \begin{cases} n\Delta, & \text{if } 0 < \left(n - \frac{1}{2}\right)\Delta < x \leq \left(n + \frac{1}{2}\right)\Delta, \\ 0, & \text{if } -\frac{\Delta}{2} \leq x \leq \frac{\Delta}{2}, \\ -n\Delta, & \text{if } \left(-n - \frac{1}{2}\right)\Delta \leq x < \left(-n + \frac{1}{2}\right)\Delta < 0, \end{cases}$$

and the dynamic range of the quantizer is adjusted for each realization of the matched-filter output vector so that

$$\Delta = \frac{2\|\mathbf{y}\|_\infty}{M}.$$

Show that error-free demodulation in the absence of background noise can be achieved if the number of quantization levels exceeds

$$M > \frac{\kappa(\mathbf{R})}{\min_k A_k},$$

where the *condition number* is defined as

$$\kappa(\mathbf{R}) \stackrel{\text{def}}{=} \|\mathbf{R}\|_\infty \|\mathbf{R}^{-1}\|_\infty.$$

PROBLEM 5.16. Let $\{\mathbf{e}_1, \ldots, \mathbf{e}_L\}$ and $\{\lambda_1, \ldots, \lambda_L\}$ denote the orthonormal eigenvectors and corresponding eigenvalues of the matrix

$$E[\mathbf{r}\mathbf{r}^T] = \left[\sigma^2\mathbf{I} + \sum_{k=1}^{K} A_k^2 \mathbf{s}_k \mathbf{s}_k^T\right].$$

Suppose that the eigenvalues are numbered in decreasing order. Define the $L \times K$ matrix

$$\mathbf{P} = [\mathbf{e}_1 \cdots \mathbf{e}_K],$$

which satisfies $\mathbf{P}^T\mathbf{P} = \mathbf{I}$. According to Problem 2.47,

$$\lambda_i = \begin{cases} \gamma_i^2 + \sigma^2, & \text{if } i = 1, \ldots, K, \\ \sigma^2, & \text{if } i = K+1, \ldots, L, \end{cases}$$

and

$$SA^2S^T = P \operatorname{diag}\{\gamma_1^2, \ldots, \gamma_K^2\}P^T.$$

Henceforth, we will assume that the set of signature vectors is linearly independent. Therefore $\gamma_k^2 > 0$ for $k = 1, \ldots, K$.

(a) Verify that

$$S^T P \operatorname{diag}^{-1}\{\gamma_1^2, \ldots, \gamma_K^2\}P^T S = A^{-2}.$$

(b) Conclude that the decorrelator for the kth user admits the expression (Wang and Poor [522]]):

$$d_k = P \operatorname{diag}^{-1}\{\lambda_1 - \sigma^2, \ldots, \lambda_K - \sigma^2\}P^T s_k.$$

PROBLEM 5.17. Show that in the two-user asynchronous channel, the necessary and sufficient condition for the existence of a stable realization of the decorrelating detector is

$$|\rho_{12}| + |\rho_{21}| < 1.$$

PROBLEM 5.18. (Lupas and Verdú [255]) Assume that the offsets $\{\tau_1, \ldots, \tau_K\}$ are continuous and independent random variables and $A_k > 0$ for all k. Show that there is a sequence of data such that the energy of the signal,

$$\sum_{k=1}^{K} \sum_{i=-M}^{M} A_k b_k[i] s_k(t - iT - \tau_k),$$

is zero with zero probability.

PROBLEM 5.19. (Lupas [252]) Show that condition (5.32),

$$\det[R^T[1]e^{j\omega} + R[0] + R[1]e^{-j\omega}] \neq 0 \quad \text{for all } \omega \in [0, 2\pi],$$

is equivalent to

$$(x^* R[0]x)^2 > (x^*(R[1] + R^T[1])x)^2 + (x^*(R[1] - R^T[1])x)^2.$$

for all complex-valued K-vectors x.

PROBLEM 5.20. Assume that the K users in an asynchronous CDMA channel are numbered chronologically. Let us consider an alternative approach to asynchronous one-shot decorrelation where the following observables are available:

$$y_{kl} = \int_{\tau_l}^{\tau_{l+1}} y(t)(s_k(t - \tau_k) + s_k(t - \tau_k + T))\,dt,$$

$k = 1, \ldots, K$ and $l = 1, \ldots, K$ with $\tau_1 = 0$ and $\tau_{K+1} = T$ by convention. Assume that the signature waveforms are linearly

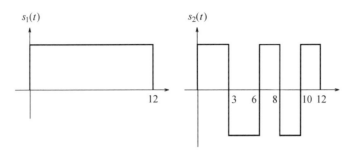

FIGURE 5.13. Signature waveforms for Problem 5.20.

independent on each subinterval $[\tau_l, \tau_{l+1})$. Define the subinterval crosscorrelations

$$\gamma_{kjl} = \int_{\tau_l}^{\tau_{l+1}} (s_k(t - \tau_k) + s_k(t - \tau_k + T))$$

$$(s_j(t - \tau_j) + s_j(t - \tau_j + T)) \, dt.$$

(a) Express the partial crosscorrelations ρ_{kj} in terms of γ_{kjl}.
(b) How can a_{kjl} be chosen so that

$$r_{kl} = \sum_{j=1}^{K} a_{kjl} y_{jl}$$

contains no interference from users other than k?
(c) Describe the decision rule for user 1 that minimizes the error probability among those rules that use $\{r_{1l}, l = 1, \ldots, K\}$ as observables.
(d) Particularize your answer in (c) to a two-user channel with $\tau_2 = T/2$ and whose signature waveforms are shown in Figure 5.13.
(e) Find the near–far resistance for the channel and receiver of (d).
(f) Find the near–far resistance of the one-shot decorrelator of Section 5.3 in the special case of (d).
(g) Among all one-shot receivers that use the observables y_{kl} and whose error probability does not depend on the interfering amplitudes, which receiver minimizes the error probability?

PROBLEM 5.21. Show that for an asynchronous channel, the one-shot decorrelator achieves the best near–far resistance of any one-shot multiuser detector.

PROBLEM 5.22. Consider a two-user asynchronous channel where the signature waveforms are given in Figure 5.14. Plot the efficiencies of the one-shot decorrelator and of the asynchronous (infinite-window) decorrelator as a function of the offset.

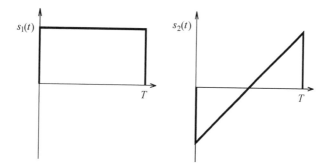

FIGURE 5.14.
Signature
waveforms for
Problem 5.22.

PROBLEM 5.23. Show that if $\mathbf{F}$ is the lower triangular factor of the nonsingular crosscorrelation matrix $\mathbf{R} = \mathbf{F}^T\mathbf{F}$, then

$$1 \leq F_{kk}^2 R_{kk}^+.$$

PROBLEM 5.24. (Bravo [35]) Find a recursive equation for the K-vectors $\mathbf{a}_j$ that satisfy

$$\begin{bmatrix} \mathbf{R}[0] & \mathbf{R}^T[1] & \mathbf{0} & \cdots & \mathbf{0} & \mathbf{0} \\ \mathbf{R}[1] & \mathbf{R}[0] & \mathbf{R}^T[1] & \cdots & \mathbf{0} & \mathbf{0} \\ \mathbf{0} & \mathbf{R}[1] & \mathbf{R}[0] & \cdots & \mathbf{0} & \mathbf{0} \\ \vdots & \vdots & \vdots & \cdots & \vdots & \vdots \\ \mathbf{0} & \mathbf{0} & \mathbf{0} & \cdots & \mathbf{R}[1] & \mathbf{R}[0] \end{bmatrix} \begin{bmatrix} \mathbf{a}_0 \\ \mathbf{a}_1 \\ \mathbf{a}_2 \\ \vdots \\ \mathbf{a}_J \end{bmatrix} = \mathbf{e}[i,k],$$

where the components of the $J \times K$ vector $\mathbf{e}[i,k]$ are all 0 except

$$e_{k+iK}[i,k] = 1$$

with $i = 0, \ldots, J$.

PROBLEM 5.25. (Juntti [186])

(a) Suppose that the decorrelating matrix $\mathbf{R}_K^{-1}$ for signature waveforms $s_1, \ldots, s_K$ has been computed, and an additional user with signature s_{K+1} is powered on. Find an expression for $\mathbf{R}_{K+1}^{-1}$ in terms of $\mathbf{R}_K^{-1}$ and $\rho_{K+1} = (\rho_{1\,K+1}, \ldots, \rho_{K\,K+1})$. [*Hint:* Use Proposition 4.3.]

(b) Suppose that the decorrelating matrix $\mathbf{R}_K^{-1}$ for signature waveforms $s_1, \ldots, s_K$ has been computed, and user K is powered off. Find an expression for $\mathbf{R}_{K-1}^{-1}$ in terms of $\mathbf{R}_K^{-1}$.

(c) (Juntti [187]) In the setting of (a) show how to compute the Cholesky factor (Proposition 2.2) of $\mathbf{R}_{K+1}^{-1}$ in terms of the Cholesky factor of $\mathbf{R}_K^{-1}$ and ρ_{K+1}.

PROBLEM 5.26. Consider a K-user synchronous channel where the signature waveforms are such that $\rho_{jk} = \rho$ for all j and k.

(a) Find an explicit expression for the decorrelating detector.

(b) Find P_k^d as a function of A_k/σ, ρ, and K.

(c) Find the minimum value of α such that

$$\lim_{K \to \infty} P_k^d(\sigma) = Q\left(\frac{A_k}{\sigma}\right), \qquad (5.106)$$

if

$$\rho = \frac{1}{K^\alpha}.$$

(d) Assume that $A_2 = \cdots = A_K$. Find the maximum value of A_2/A_1 for which

$$\eta_1^d < \eta_1^c.$$

(e) Assume that $A_1 = \cdots = A_K = A$. Find the signal-to-interference ratio at the output of the approximate decorrelator of Section 5.4.

(f) Solve (c) for the approximate decorrelator in lieu of the decorrelator.

PROBLEM 5.27. Consider a K-user synchronous CDMA channel with direct-sequence spread-spectrum waveforms with spreading factor equal to N chips per bit. Assume that

$$\beta = \lim_{K \to \infty} \frac{K}{N} < 1,$$

and that the binary signature sequences are randomly and independently chosen. Show that the error probability of the decorrelating detector converges in mean-square to

$$\lim_{K \to \infty} Q\left(\frac{A_k}{\sigma}\sqrt{\bar{\eta}_k}\right) = Q\left(\frac{A_k}{\sigma}\sqrt{1 - \beta}\right). \qquad (5.107)$$

Hint: Use (3.52), (4.111), and

$$\left(\sqrt{a} - \sqrt{1 - \beta}\right)^2 \leq \frac{(a - (1 - \beta))^2}{1 - \beta}.$$

PROBLEM 5.28. Consider a K-user synchronous CDMA channel with direct-sequence spread-spectrum waveforms with spreading factor equal to N chips per bit. In this problem we focus on the signal-to-interference ratio (SIR) of the output of a linear transformation where the second moments of desired component and interference (background noise plus multiaccess interference) component are averaged with respect to transmitted data and random selection of signature waveforms. We will consider an asymptotic regime of large number of users and large bandwidth:

$$\beta \stackrel{\text{def}}{=} \lim_{K\to\infty} \frac{K}{N} < 1$$

and

$$\mu \stackrel{\text{def}}{=} \lim_{K\to\infty} \frac{1}{K} \sum_{j=2}^{K} \frac{A_j^2}{\sigma^2}.$$

Show the following asymptotic behaviors of the normalized SIR of user 1:

$$\lim_{K\to\infty} \frac{\sigma^2 \text{SIR}}{A_1^2}.$$

(a) Decorrelator:

$$1 - \beta.$$

(b) Approximate decorrelator:

$$\frac{(1-\beta)^2}{1 - \beta + \beta^2 + \mu\beta^2 + \mu\beta^3}.$$

(c) Single-user matched filter:

$$\frac{1}{1 + \mu\beta}.$$

(d) Consider a linear receiver, which is a linear combination of the single-user matched filter and the approximate decorrelator:

$$s_1(t) - \alpha \sum_{k=2}^{K} \rho_{1k} s_k(t).$$

Give an equation that describes the value of α (as a function of β and μ) that maximizes the asymptotic SIR. Plot the optimum α as a function of μ for $\beta = 1/3$.

[*Hint:* See Problem 4.51.]

PROBLEM 5.29. Show that in any K-user synchronous or asynchronous channel

$$\lim_{\sigma\to\infty} \frac{\sigma}{A_k}\left(P_k^d(\sigma) - P_k^c(\sigma)\right) = \frac{1 - \sqrt{\bar{\eta}_k}}{\sqrt{2\pi}}. \tag{5.108}$$

PROBLEM 5.30. Consider a three-user synchronous channel with

$$A_1 = A_2 = A_3,$$

$$\rho_{12} = -\rho_{13},$$

and

$$\frac{2 - 2\rho_{23}^2}{3 + \rho_{23} - 2\rho_{23}^2} < |\rho_{12}| < \frac{1}{2}.$$

Show that the error probability of the decorrelating detector for user 1 is strictly higher than the error probability of the single-user matched-filter detector for all signal-to-noise ratios.

PROBLEM 5.31. Consider a two-user synchronous channel. Find values of A_1/σ, A_2/σ, and ρ such that the decorrelating detector has lower probability of error than the single-user matched filter, but the signal-to-interference ratio at the output of the single-user matched filter is higher than that of the decorrelator.

PROBLEM 5.32. Consider four unit-energy signals $\{s_1, s_2, s_3, s_4\}$ such that

$$\langle s_1, s_2 \rangle = \rho,$$

$$\langle s_1, s_4 \rangle = 0,$$

$$\langle s_3, s_4 \rangle = 0,$$

$$s_3 + s_4 = \sqrt{2}s_2.$$

(a) What is the decorrelating receiver for user 1 if user 4 is known not to be active?
(b) Show that the solution obtained in (a) maximizes $\langle s_1, h \rangle$ among all h that have fixed energy and are orthogonal to s_2, s_3, and s_4.
(c) Find the optimum near–far resistance for user 1 and the probability of error achieved by the receiver obtained in (a) when all four users are active.
(d) Find the optimum near–far resistance for user 2 with and without user 4.

PROBLEM 5.33. Find the probability of error for a Moore–Penrose generalized decorrelator for user k as a function of the matrix S and the received amplitudes. Particularize your result to the three-user case where

$$s_3 = \alpha(s_1 + s_2).$$

PROBLEM 5.34. Show that if R is invertible, then $S^+S = I$.

PROBLEM 5.35. Show that the L-vector c is a decorrelating transformation for the kth user if and only if

$$SDS^T c = \alpha s_k, \quad \alpha \neq 0,$$

for all $K \times K$ nonsingular diagonal matrices D.

282

PROBLEM 5.36. Let $\mathbf{Q}$ be an $n \times m$ matrix and let $\mathbf{Q}^+$ be the Moore–Penrose generalized inverse of $\mathbf{Q}$. Denote the $n \times n$ matrix

$$\mathbf{P} = \mathbf{I} - \mathbf{Q}\mathbf{Q}^+,$$

and let

$$r = n - \text{rank}\,\mathbf{Q}.$$

Prove the following properties of $\mathbf{P}$:

(a) The matrix $\mathbf{P}$ is idempotent (i.e., $\mathbf{P}^2 = \mathbf{P}$).
(b) $\mathbf{P}$ is the projector onto the space orthogonal to that spanned by the columns of $\mathbf{Q}$. In other words, if $\mathbf{x}$ is a linear combination of the columns of $\mathbf{Q}$, then

$$\mathbf{P}\mathbf{x} = 0,$$

and if $\mathbf{y}$ is orthogonal to every column of $\mathbf{Q}$, then

$$\mathbf{P}\mathbf{y} = \mathbf{y}.$$

(c) If $r = 0$, then $\mathbf{P} = \mathbf{0}$.
(d) The eigenvalues of $\mathbf{P}$ cannot be different from 0 or 1.
(e)

$$\mathbf{P} = \sum_{i=1}^{r} \mathbf{v}_i \mathbf{v}_i^T,$$

where $\{\mathbf{v}_1, \ldots, \mathbf{v}_r\}$ are orthonormal and orthogonal to every column of $\mathbf{Q}$.
(f) $\mathbf{P}^+ = \mathbf{P}$.

PROBLEM 5.37. Show the following properties of the Moore–Penrose generalized inverse:

(a) $(\mathbf{S}^T)^+ = (\mathbf{S}^+)^T$.
(b) $(\mathbf{S}^T\mathbf{S})^+ = \mathbf{S}^+(\mathbf{S}^+)^T$.
(c) $(\mathbf{S}^T\mathbf{S})^+\mathbf{S}^T = \mathbf{S}^+$.
(d) If $\mathbf{B}$ is invertible, then $(\mathbf{B}\mathbf{S})^+\,\mathbf{B} = \mathbf{S}^+$.

PROBLEM 5.38. Show that the Moore–Penrose generalized inverse of a vector $\mathbf{x}$ is $\mathbf{x}^T/\|\mathbf{x}\|^2$. Thus, if $\mathbf{x}$ is a column vector, then $\mathbf{x}^+\mathbf{x} = 1$.

PROBLEM 5.39. Show that if the kth column of $\mathbf{S}$ is not spanned by the other columns of $\mathbf{S}$, then the kth row of $\mathbf{S}^+$ is not spanned by the other rows of $\mathbf{S}^+$.

PROBLEM 5.40. Assume that $\mathbf{s}_k$ is not spanned by the other signature vectors.

(a) Show that

$$(\mathbf{S}^+\mathbf{s}_j)_k = \delta_{jk}.$$

(b) Conclude that the kth row of $\mathbf{S}^+$ is a decorrelating transformation for the kth user.

(c) Show that as long as $A_k \neq 0$, the solution to the equation

$$\mathbf{S}\mathbf{A}^2\mathbf{S}^T\mathbf{x} = \mathbf{s}_k$$

is the kth row of $\mathbf{S}^+$ divided by A_k^2.

PROBLEM 5.41. Show that for a synchronous CDMA system that employs a Welch-bound-equality set (Problem 2.41), the decorrelator for user k is the single-user matched filter for user k.

PROBLEM 5.42. Suppose that all K synchronous users are assigned the same signature waveform. Show that the Moore–Penrose generalized inverse of the crosscorrelation matrix is

$$\frac{1}{K^2}\begin{bmatrix} 1 & \cdots & 1 \\ \vdots & \ddots & \vdots \\ 1 & \cdots & 1 \end{bmatrix}.$$

PROBLEM 5.43. Consider a generalization of (5.16)

$$\mathbf{r} = \mathbf{S}\mathbf{A}\mathbf{b} + \sigma\mathbf{m}, \tag{5.109}$$

where Σ, the covariance matrix of $\mathbf{m}$, is nonsingular. Show that

$$\mathbf{x}(\mathbf{r}) = \mathbf{S}^+\mathbf{r}$$

is the solution to

$$\min_{\mathbf{x}\in R^K}[\mathbf{S}\mathbf{x} - \mathbf{r}]^T\Sigma^{-1}[\mathbf{S}\mathbf{x} - \mathbf{r}]. \tag{5.110}$$

[*Hint:* Use Problem 5.37(d).]

PROBLEM 5.44. Verify Equation (5.54).

PROBLEM 5.45. A class of the detectors that are hybrids of the optimum multiuser detector and the decorrelating detector can be designed by generalizing the following construction. Consider a three-user synchronous channel:

$$y(t) = A_1b_1s_1(t) + A_2b_2s_2(t) + A_3b_3s_3(t) + \sigma n(t)$$

and a receiver for users 1 and 2 whose observables are the decorrelator outputs:

$$\langle y, s_1 - \rho_{13}s_3 \rangle,$$

$$\langle y, s_2 - \rho_{23}s_3 \rangle.$$

(a) Find the jointly optimum detector for users 1 and 2 using the foregoing observables.
(b) Find the asymptotic efficiency of user 1 as a function of cross-correlations and A_2/A_1 obtained with the receiver in (a).
(c) Find the near–far resistance for user 1.

PROBLEM 5.46. (Schlegel et al. [388]) Suppose that in addition to the K users of interest, there are J users who need not be demodulated and for which no information on their received amplitudes is known. The orthogonal-space observables in (2.97) become

$$\mathbf{r} = \mathbf{SAb} + \mathbf{U\bar{A}e} + \sigma\mathbf{m},$$

where $\mathbf{U}$ is the $L \times J$ matrix of signature vectors of unwanted users, $\bar{\mathbf{A}}$ is the $J \times J$ diagonal matrix of unknown amplitudes, and $\mathbf{e}$ is the L-vector of undesired users' bits. Thus, the unknown quantities in this model are $\mathbf{b}$, $\bar{\mathbf{A}}$, and $\mathbf{e}$. Show that the maximum-likelihood decision for $\mathbf{b}$ minimizes

$$\|[\mathbf{I} - \mathbf{UU}^+](\mathbf{r} - \mathbf{SAb})\|,$$

where $\mathbf{U}^+$ is the Moore–Penrose generalized inverse of $\mathbf{U}$.

PROBLEM 5.47. (Zheng and Barton [563]) Consider a QPSK asynchronous CDMA channel where every other symbol period is empty, that is,

$$y(t) = \sum_{k=1}^{K} \sum_{i=-M}^{M} A_k \, [b_k[2i] \cos(2\pi f_c t + \phi_k)$$

$$+ b_k[2i+1] \sin(2\pi f_c t + \phi_k) \,] \, s_k(t - 2iT - \tau_k) + \sigma n(t).$$

$$(5.111)$$

(a) Show that there is no loss of performance by restricting attention to one-shot detectors.
(b) In the two-user case, find the average (over phases and delays) error probability of a decorrelating detector for user 1 as a function of the partial crosscorrelations $\rho_{12}(\tau)$ and $\rho_{21}(\tau)$. Neglect integrals of nonbaseband signals.

PROBLEM 5.48. The diversity model in (2.74) is

$$y_1(t) = \sum_{k=1}^{K} A_{1k}b_k s_{1k}(t) + \sigma n_1(t),$$

$$\vdots$$

$$y_D(t) = \sum_{k=1}^{K} A_{Dk}b_k s_{Dk}(t) + \sigma n_D(t),$$

where $n_1(t), \ldots, n_D(t)$ are independent white Gaussian processes. Consider the decorrelating-diversity receiver (cf. (3.141))

$$\hat{b}_k = \text{sgn}\left(\Re\left\{\sum_{d=1}^{D} A_{dk}^*(\mathbf{R}^{-1}(d)\mathbf{y}_d)_k\right\}\right), \quad (5.112)$$

where the entries of the nonsingular matrices $\mathbf{R}(d)$ are

$$\rho_{kj}(d) = \int_0^T s_{dk}^*(t)s_{dj}(t)\,dt$$

and

$$\mathbf{y}_d = [y_{d1} \cdots y_{dK}],$$

$$y_{dk} = \int_0^T y_d(t)s_{dk}^*(t)\,dt.$$

Show that the probability of error of the kth user conditioned on $(|A_{1k}| \cdots |A_{Dk}|)$ is

$$P_k^{Dd}(\sigma) = Q\left(\frac{\sum_{d=1}^{D} |A_{dk}|^2}{\sigma\sqrt{\sum_{d=1}^{D} |A_{dk}|^2 R_{kk}^+(d)}}\right). \quad (5.113)$$

Conclude that in the special case in which

$$\bar{\eta}_k = \frac{1}{R_{kk}^+(d)}$$

does not depend on d and $(|A_{1k}|, \ldots, |A_{Dk}|)$ are independent Rayleigh random variables with

$$E[|A_{dk}|^2] = 2A_k^2,$$

the average of (5.113) becomes

$$P_k^{Dd}(\sigma) = \frac{1}{2} - \frac{1}{2}\frac{1}{\sqrt{1 + \sigma^2/(A_k^2\bar{\eta}_k)}}$$

$$\times \left(1 + \sum_{n=1}^{D-1} \frac{1 \cdot 3 \cdot 5 \cdots (2n-1)}{n!\,2^n\,((A_k^2\bar{\eta}_k)/\sigma^2 + 1)^n}\right). \quad (5.114)$$

PROBLEM 5.49. Show that the near–far resistance of a K-user Code-Time-Division Multiple Access system (cf. Problem 2.59) with assigned waveform $s(t)$ and aggregate bit-rate equal to R is equal to the harmonic mean of its folded spectrum:

$$\bar{\eta}_k = \left[R \int_0^1 \left[\sum_{n=-\infty}^{\infty} |S(R\lambda - Rn)|^2 \right]^{-1} d\lambda \right]^{-1}$$

where

$$S(f) = \int_{-\infty}^{\infty} s(t)e^{-j2\pi ft} dt.$$

CHAPTER SIX

NONDECORRELATING LINEAR MULTIUSER DETECTION

Linear multiuser detectors can be implemented in a decentralized fashion where only the user or users of interest need be demodulated. When the received amplitudes are completely unknown the decorrelating detector is a sensible choice as we saw in Chapter 5. In this chapter, we examine the level of improvement in performance attainable by incorporating information about the received signal-to-noise ratios in the linear transformation. Since the single-user matched filter is better than the decorrelator for sufficiently low signal-to-noise ratios, it is evident that such a performance improvement is feasible.

We will take two different approaches. In the first approach, we choose the linear transformation to maximize asymptotic multiuser efficiency for every vector of received amplitudes. The second, and more fruitful, approach is to choose the linear transformation that minimizes the mean-square error between its outputs and the data.

This chapter includes the study of adaptive implementations of linear receivers requiring no prior knowledge of received signal-to-noise ratios or interfering signature waveforms.

6.1 OPTIMUM LINEAR MULTIUSER DETECTION

In general, it is possible to achieve a certain tradeoff of interference rejection and attenuation of desired signal component in order to maximize asymptotic multiuser efficiency within the constraint of a linear multiuser

288

detector. Denote the kth user linear transformation by $\mathbf{v}_k$, and let

$$\hat{b}_k = \text{sgn}(\mathbf{v}_k^T \mathbf{y}), \qquad (6.1)$$

where $\mathbf{y}$ is the vector of normalized matched filter outputs (3.70). Then

$$\mathbf{v}_k^T \mathbf{y} = \sum_{j=1}^{K} A_j b_j \mathbf{v}_k^T \mathbf{r}_j + \mathbf{v}_k^T \mathbf{n}, \qquad (6.2)$$

where $\mathbf{r}_j$ is the jth column of the normalized crosscorrelation matrix $\mathbf{R}$. The probability of error achieved by $\mathbf{v}_k$ can be expressed as

$$P_k^{\mathbf{v}_k} = E\left[Q\left(\frac{A_k \mathbf{v}_k^T \mathbf{r}_k + \sum_{j \neq k} A_j b_j \mathbf{v}_k^T \mathbf{r}_j}{\sigma \sqrt{\mathbf{v}_k^T \mathbf{R} \mathbf{v}_k}} \right) \right], \qquad (6.3)$$

where the expectation is with respect to b_j, $j \neq k$. The asymptotic efficiency of the kth user is given by the square of the smallest argument of the Q-function normalized by A_k^2/σ^2:

$$\eta_k(\mathbf{v}_k) = \frac{1}{\mathbf{v}_k^T \mathbf{R} \mathbf{v}_k} \max^2 \left\{ 0, \mathbf{v}_k^T \mathbf{r}_k - \sum_{j \neq k} \frac{A_j}{A_k} |\mathbf{v}_k^T \mathbf{r}_j| \right\}. \qquad (6.4)$$

To illustrate the optimization of (6.4) with respect to $\mathbf{v}_k$, we will examine the two-user case in detail. Without loss of generality, if we let $\mathbf{v}_1 = [1 \ x]^T$, then we obtain

$$\eta_1(\mathbf{v}_1) = \max^2 \left\{ 0, f\left(x, \rho, \frac{A_2}{A_1} \right) \right\} \qquad (6.5)$$

with

$$f\left(x, \rho, \frac{A_2}{A_1} \right) = \frac{1 + x\rho - \frac{A_2}{A_1}|x + \rho|}{\sqrt{1 + 2\rho x + x^2}}. \qquad (6.6)$$

The argument that maximizes (6.6) is

$$x^* = \begin{cases} -\frac{A_2}{A_1}\text{sgn}\rho, & \text{if } A_2/A_1 < |\rho|, \\ -\rho, & \text{otherwise} \end{cases} \qquad (6.7)$$

(see Problem 6.2). Note that when the relative energy of the interferer is strong enough, $A_2 \geq A_1|\rho|$, then the decorrelating detector maximizes asymptotic efficiency among all linear transformations (which are allowed knowledge of the values of A_2 and A_1). Otherwise, the received signal is correlated with

$$s_1(t) - \frac{A_2}{A_1}\text{sgn}(\rho)s_2(t), \qquad (6.8)$$

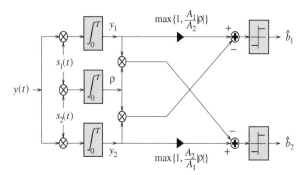

FIGURE 6.1.
Linear detector that maximizes asymptotic efficiency for the two-user synchronous channel.

or, equivalently, with

$$\frac{A_1}{A_2}|\rho|s_1(t) - \rho s_2(t)$$

(Figure 6.1). Comparing Figure 6.1 to Figure 5.3 we can view the maximum asymptotic efficiency linear detector as a compromise solution between the decorrelating detector and the single-user matched filter, which approaches the latter as the relative power of the interferer decreases.

Substituting the solution (6.7) into (6.5), we obtain the maximum asymptotic efficiency achievable by a linear transformation (Figure 6.2):

$$\eta_1(\mathbf{v}_1^*) = \begin{cases} \eta_1 = 1 + \frac{A_2^2}{A_1^2} - 2|\rho|\frac{A_2}{A_1}, & \text{if } A_2/A_1 < |\rho|, \\ \eta_1^d = 1 - \rho^2, & \text{otherwise.} \end{cases} \quad (6.9)$$

We can conclude from (6.9) that if the background noise is small relative to the strength of the transmitted signals, and if $A_2/A_1 \geq |\rho|$, then there is no point (as far as near–far resistance is concerned) in utilizing the actual values of the received energies; if $A_2/A_1 < |\rho|$, then in view of (4.65), there is no point in using the more complex optimum multiuser detector, as the modified correlator in (6.8) will achieve the same performance (in the high signal-to-noise ratio region).

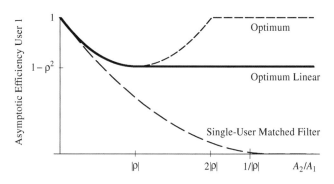

FIGURE 6.2.
Asymptotic multiuser efficiencies for two synchronous users.

Due to the presence of the absolute value function in (6.4), solving for the optimum linear detector in the K-user case entails the solution of a nonlinear optimization problem that does not admit a closed-form solution. An algorithm is given in Lupas and Verdú [254] along with a sufficient condition for the best linear detector to achieve optimum asymptotic kth user efficiency:

$$A_k > \max_{j=1,\dots,K} \sum_{i \neq k} A_i \left| \frac{\rho_{ij}}{\rho_{kj}} \right| \tag{6.10}$$

(which is also necessary in the two-user case), as well as a sufficient condition for the decorrelating detector to be the best kth user linear detector, namely

$$\frac{A_j}{A_k} \geq \frac{\left| R_{jk}^+ \right|}{R_{kk}^+} \quad \text{for all } j \neq k. \tag{6.11}$$

The computational complexity of solving for the maximal-asymptotic-efficiency linear transformation is prohibitive for large number of users, particularly in the asynchronous case.

The solution we have reviewed in this section is geared toward low background noise situations and does not incorporate knowledge of the background noise level. The behavior of the maximal-asymptotic-efficiency linear multiuser detector in the low signal-to-noise ratio region suffers from shortcomings similar to those of the decorrelating detector, (e.g., for sufficiently high σ, the single-user matched filter has lower bit-error-rate) (Problem 6.1). If the signal-to-noise ratios are available, then one can pose the minimization (still over linear transformations) of bit-error-rate. Such an optimization problem is even more complex than the one considered here (Problem 6.16). A more fruitful route to incorporate knowledge of received signal-to-noise ratios is discussed in the next section.

6.2 MINIMUM MEAN-SQUARE ERROR (MMSE) LINEAR MULTIUSER DETECTION

A common approach in estimation theory to the problem of estimating a random variable W on the basis of observations Z is to choose the function $\hat{W}(Z)$ that minimizes the mean-square error (MSE):

$$E[(W - \hat{W}(Z))^2].$$

Under very general conditions, it can be shown that the solution is the conditional-mean estimator:

$$\hat{W}(Z) = E[W|Z].$$

In most problems, it is challenging to derive the conditional-mean estimator from the joint distribution of W and Z. For that reason, it is common to minimize MSE within the restricted set of linear (or affine) transformations of Z. The linear minimum MSE (MMSE) estimator is, in general, easy to compute and it depends on the joint distribution of W and Z only through their variances and covariance.

We can turn the problem of linear multiuser detection into a problem of linear estimation, by requiring that the MSE between the k user bit b_k and the output of the kth linear transformation $v_k^T y$ be minimized. Although this approach does not lead to the minimization of the bit-error-rate

$$P[b_k \neq \mathrm{sgn}(v_k^T y)],$$

it is a sensible optimality criterion, particularly when the multiuser receiver, rather than demodulating the data, supplies soft decisions to an error-control decoder (cf. Section 2.4)).

The MMSE linear detector for the kth user chooses the waveform c_k of duration T that achieves

$$\min_{c_k} E[(b_k - \langle c_k, y \rangle)^2] \tag{6.12}$$

and outputs the decision

$$\hat{b}_k = \mathrm{sgn}(\langle c_k, y \rangle).$$

The MMSE linear transformation maximizes the signal-to-interference ratio at the output of the linear transformation (Problem 6.5):

$$\frac{1}{\min_{c_k} E[(b_k - \langle c_k, y \rangle)^2]} = 1 + \max_{c_k} \frac{E[(\langle c_k, A_k b_k s_k \rangle)^2]}{E[(\langle c_k, y - A_k b_k s_k \rangle)^2]}. \tag{6.13}$$

We can always express the linear transformation as

$$c_k = c_k^s + c_k^o,$$

where c_k^s is spanned by the signature waveforms $s_1, \ldots, s_K$ and c_k^o is orthogonal to c_k^s; then

$$E[(b_k - \langle c_k, y \rangle)^2] = E[(b_k - \langle c_k^s, y \rangle)^2] + \sigma^2 \| c_k^o \|^2. \tag{6.14}$$

Consequently, it is best to restrict attention to c_k spanned by $s_1, \ldots, s_K$. This implies that the linear MMSE detector outputs a weighted combination of the matched filter outputs, and enables us to turn the problem in (6.12) into a finite-dimensional optimization one, namely choosing the K-vector $\mathbf{m}_k$ that minimizes

$$E\left[\left(b_k - \mathbf{m}_k^T \mathbf{y}\right)^2\right]. \tag{6.15}$$

We now have K uncoupled optimization problems (one for each user), which can be solved simultaneously by choosing the $K \times K$ matrix $\mathbf{M}$ (whose k column is equal to $\mathbf{m}_k$) that achieves

$$\min_{\mathbf{M} \in R^{K \times K}} E[\|\mathbf{b} - \mathbf{M}\mathbf{y}\|^2], \tag{6.16}$$

where

$$\mathbf{y} = \mathbf{R}\mathbf{A}\mathbf{b} + \mathbf{n}, \tag{6.17}$$

and the expectation is with respect to the vector of transmitted bits $\mathbf{b}$ and the noise vector $\mathbf{n}$, which has zero mean and covariance matrix equal to $\sigma^2 \mathbf{R}$. Since

$$\|\mathbf{x}\|^2 = \text{trace}\{\mathbf{x}\mathbf{x}^T\},$$

the first step to solve (6.16) will be to obtain an expression for the covariance matrix of the error vector:

$$\begin{aligned}
\text{cov}\{\mathbf{b} - \mathbf{M}\mathbf{y}\} &= E[(\mathbf{b} - \mathbf{M}\mathbf{y})(\mathbf{b} - \mathbf{M}\mathbf{y})^T] \\
&= E[\mathbf{b}\mathbf{b}^T] - E[\mathbf{b}\mathbf{y}^T]\mathbf{M}^T \\
&\quad - \mathbf{M}E[\mathbf{y}\mathbf{b}^T] + \mathbf{M}E[\mathbf{y}\mathbf{y}^T]\mathbf{M}^T.
\end{aligned} \tag{6.18}$$

Using (6.17) and the fact that noise and data are uncorrelated we get

$$E[\mathbf{b}\mathbf{b}^T] = \mathbf{I}, \tag{6.19}$$

$$E[\mathbf{b}\mathbf{y}^T] = E[\mathbf{b}\mathbf{b}^T\mathbf{A}\mathbf{R}] = \mathbf{A}\mathbf{R}, \tag{6.20}$$

$$E[\mathbf{y}\mathbf{b}^T] = E[\mathbf{R}\mathbf{A}\mathbf{b}\mathbf{b}^T] = \mathbf{R}\mathbf{A}, \tag{6.21}$$

$$\begin{aligned}
E[\mathbf{y}\mathbf{y}^T] &= E[\mathbf{R}\mathbf{A}\mathbf{b}\mathbf{b}^T\mathbf{A}\mathbf{R}] + E[\mathbf{n}\mathbf{n}^T] \\
&= \mathbf{R}\mathbf{A}^2\mathbf{R} + \sigma^2\mathbf{R}.
\end{aligned} \tag{6.22}$$

Substituting these expressions in (6.18) we express the covariance matrix of the error vector as

$$\text{cov}\{\mathbf{b} - \mathbf{M}\mathbf{y}\} = \mathbf{I} + \mathbf{M}(\mathbf{R}\mathbf{A}^2\mathbf{R} + \sigma^2\mathbf{R})\mathbf{M}^T - \mathbf{A}\mathbf{R}\mathbf{M}^T - \mathbf{M}\mathbf{R}\mathbf{A} \tag{6.23}$$

$$= [\mathbf{I} + \sigma^{-2}\mathbf{A}\mathbf{R}\mathbf{A}]^{-1} + (\mathbf{M} - \bar{\mathbf{M}})(\mathbf{R}\mathbf{A}^2\mathbf{R} + \sigma^2\mathbf{R})(\mathbf{M} - \bar{\mathbf{M}})^T, \tag{6.24}$$

where

$$\bar{\mathbf{M}} \stackrel{\text{def}}{=} \mathbf{A}^{-1}[\mathbf{R} + \sigma^2 \mathbf{A}^{-2}]^{-1}, \tag{6.25}$$

and we have assumed that $\mathbf{A}$ is nonsingular (i.e., only those users who are active are taken into account). The identity in (6.24) is readily checked by means of

$$\bar{\mathbf{M}}(\mathbf{R}\mathbf{A}^2\mathbf{R} + \sigma^2\mathbf{R}) = \mathbf{A}\mathbf{R}$$

and

$$(\mathbf{I} - \mathbf{A}\mathbf{R}\bar{\mathbf{M}}^T)(\mathbf{I} + \sigma^{-2}\mathbf{A}\mathbf{R}\mathbf{A}) = \mathbf{I},$$

which in turn follow from (6.25). Thanks to (6.24), we are now ready to solve

$$\min_{\mathbf{M} \in R^{K \times K}} E[\|\mathbf{b} - \mathbf{M}\mathbf{y}\|^2] = \min_{\mathbf{M} \in R^{K \times K}} \text{trace} \{\text{cov}\{\mathbf{b} - \mathbf{M}\mathbf{y}\}\}. \tag{6.26}$$

The matrix $\mathbf{R}\mathbf{A}^2\mathbf{R} + \sigma^2\mathbf{R}$ is nonnegative definite. Therefore the trace of the second term in the right side of (6.24) is always nonnegative. We conclude that the matrix $\bar{\mathbf{M}}$ defined in (6.25) achieves the minimum sum of mean-square errors (6.26):

$$\min_{\mathbf{M} \in R^{K \times K}} E[\|\mathbf{b} - \mathbf{M}\mathbf{y}\|^2] = \text{trace} \{[\mathbf{I} + \sigma^{-2}\mathbf{A}\mathbf{R}\mathbf{A}]^{-1}\}. \tag{6.27}$$

According to (6.25) the MMSE linear detector outputs the following decisions:

$$\hat{b}_k = \text{sgn}\left(\frac{1}{A_k}([\mathbf{R} + \sigma^2\mathbf{A}^{-2}]^{-1}\mathbf{y})_k\right)$$

$$= \text{sgn}(([\mathbf{R} + \sigma^2\mathbf{A}^{-2}]^{-1}\mathbf{y})_k). \tag{6.28}$$

Therefore, the MMSE linear detector (Figure 6.3) replaces the transformation $\mathbf{R}^{-1}$ of the decorrelating detector by

$$[\mathbf{R} + \sigma^2\mathbf{A}^{-2}]^{-1},$$

where

$$\sigma^2\mathbf{A}^{-2} = \text{diag}\left\{\frac{\sigma^2}{A_1^2}, \ldots, \frac{\sigma^2}{A_K^2}\right\}.$$

Note that we have not required $\mathbf{R}$ to be nonsingular. Note also that the dependence of the MMSE detector on the received amplitudes is only through the signal-to-noise ratios A_k/σ. The derivation of the MMSE linear multiuser

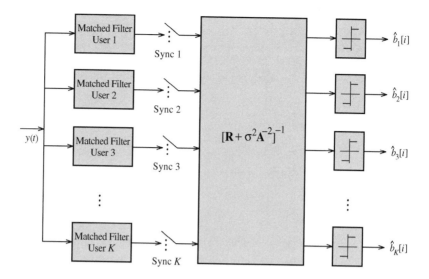

FIGURE 6.3.
MMSE linear detector for the synchronous channel.

detector assumed a great deal less than the basic CDMA model. We did not need to assume that the background noise is Gaussian. Furthermore, in order to reach the solution (6.25) we did not use the fact that the transmitted bits are binary valued; we only required that (a) they be uncorrelated from user to user and (b) $E[b_k^2] = 1$. The latter restriction is easily dropped by absorbing a factor $E[b_k^2]$ into A_k^2.

Another formulation of the MMSE problem, which will be useful in Chapter 7, substitutes (6.16) by

$$\min_{\mathbf{M} \in R^{K \times K}} E[\|\mathbf{A}\mathbf{b} - \mathbf{M}\mathbf{y}\|^2]. \tag{6.29}$$

This is equivalent to trying to replicate $A_k b_k$ (instead of b_k) at the output of the linear transformation. But A_k is assumed known, so it should not be surprising that the solution to the optimization problem in (6.29) is given by

$$\mathbf{M}^* = [\mathbf{R} + \sigma^2 \mathbf{A}^{-2}]^{-1}, \tag{6.30}$$

which leads to the same detector as before (6.28), and is obtained following the approach used to solve (6.16).

In the two-user case we have (Figure 6.4)

$$[\mathbf{R} + \sigma^2 \mathbf{A}^{-2}]^{-1} = \left[\left(1 + \frac{\sigma^2}{A_1^2}\right)\left(1 + \frac{\sigma^2}{A_2^2}\right) - \rho^2\right]^{-1} \begin{bmatrix} 1 + \frac{\sigma^2}{A_2^2} & -\rho \\ -\rho & 1 + \frac{\sigma^2}{A_1^2} \end{bmatrix}. \tag{6.31}$$

As we know, the single-user matched filter receiver is optimized to fight the background white noise exclusively, whereas the decorrelating detector

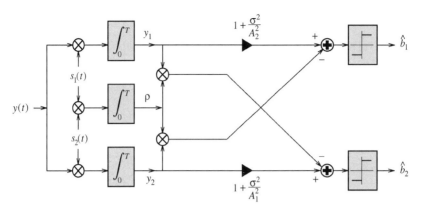

FIGURE 6.4.
MMSE linear receiver for two synchronous users.

eliminates the multiuser interference disregarding the background noise. In contrast, the MMSE linear detector can be seen as a compromise solution that takes into account the relative importance of each interfering user and the background noise. In fact both the conventional receiver and the decorrelating receiver are limiting cases of the MMSE linear detector. If we hold A_1 fixed and let $A_2, \ldots, A_K \to 0$, then the first row of $[\mathbf{R} + \sigma^2 \mathbf{A}^{-2}]^{-1}$ tends to

$$\left[\frac{A_1^2}{A_1^2 + \sigma^2}, 0, \ldots, 0 \right],$$

which corresponds to the matched filter for user 1. As σ grows, $[\mathbf{R} + \sigma^2 \mathbf{A}^{-2}]^{-1}$ becomes a strongly diagonal matrix, and the MMSE detector approaches the conventional detector as $\sigma \to \infty$.

If we hold all the amplitudes fixed and let $\sigma \to 0$, then

$$[\mathbf{R} + \sigma^2 \mathbf{A}^{-2}]^{-1} \to \mathbf{R}^{-1}. \tag{6.32}$$

Therefore, as the signal-to-noise ratios go to infinity, the MMSE linear detector converges to the decorrelating detector. This fact implies that the MMSE linear detector has the same asymptotic efficiency and near–far resistance as the decorrelating detector. In particular, the MMSE linear detector also achieves optimum near–far resistance.

In the asynchronous case, the MMSE linear detector is a K-input, K-output, linear, time-invariant filter with transfer function

$$[\mathbf{R}^T[1]z + \mathbf{R}[0] + \sigma^2 \mathbf{A}^{-2} + \mathbf{R}[1]z^{-1}]^{-1}. \tag{6.33}$$

The simplest way to verify this is to notice that parallel to the asynchronous decorrelating detector, (6.33) is the limiting form of the inverse of the equivalent crosscorrelation matrix that we would obtain in a case with finite frame length:

$$
\begin{bmatrix}
\mathbf{R}[0] + \sigma^2 \mathbf{A}^{-2} & \mathbf{R}^T[1] & 0 & \cdots & 0 \\
\mathbf{R}[1] & \mathbf{R}[0] + \sigma^2 \mathbf{A}^{-2} & \mathbf{R}^T[1] & \cdots & \vdots \\
0 & \mathbf{R}[1] & \vdots & \vdots & 0 \\
\vdots & \vdots & \cdots & \mathbf{R}[0] + \sigma^2 \mathbf{A}^{-2} & \mathbf{R}^T[1] \\
0 & 0 & \cdots & \mathbf{R}[1] & \mathbf{R}[0] + \sigma^2 \mathbf{A}^{-2}
\end{bmatrix}
$$

(6.34)

In contrast to the decorrelating detector, the inverse in (6.33) always exists because the matrix therein is the sum of a nonnegative definite matrix and the diagonal positive definite matrix $\sigma^2 \mathbf{A}^{-2}$. We interpreted the asynchronous decorrelating detector as a cascade of a linear structure that combines matched filter outputs so as to eliminate multiuser interference followed by a zero-forcing equalizer that eliminates single-user intersymbol interference. The situation is somewhat different for the MMSE detector. In this case, the combination of matched filter outputs does not quite cancel all the multiuser interference, and the single-user filter does not act as a zero-forcing equalizer (nor as an MMSE equalizer for the single-user sequence).

The truncated-window strategy followed in Section 5.3 can also be applied to the MMSE linear multiuser detector with straightforward modifications. This observation turns out to be important in the adaptive implementations of the MMSE multiuser detector to be discussed in Sections 6.4 and 6.6.

Returning now to the synchronous case, an alternative representation for the MMSE linear detector (which will be useful in later Sections) can be obtained using the discrete-time model introduced in (2.96) in which the observables are projections along a basis of orthonormal signals:

$$
\mathbf{r} = \sum_{k=1}^{K} A_k b_k \mathbf{s}_k + \sigma \mathbf{m},
$$

(6.35)

where $\mathbf{m}$ is an L-dimensional Gaussian vector with independent unit-variance components. Considering user 1 as the user of interest, the vector $\mathbf{v}$ that minimizes

$$
E[(b_1 - \mathbf{v}^T \mathbf{r})^2]
$$

must force the gradient to the zero-vector, that is,

$$
E[b_1 \mathbf{r} - \mathbf{r}\mathbf{r}^T \mathbf{v}] = 0.
$$

(6.36)

To compute the left side of (6.36) we note that

$$
E[b_1 \mathbf{r}] = A_1 \mathbf{s}_1,
$$

and

$$E[\mathbf{rr}^T] = \sigma^2 \mathbf{I} + \sum_{k=1}^{K} A_k^2 \mathbf{s}_k \mathbf{s}_k^T. \tag{6.37}$$

Therefore, the solution to (6.36) is given by the vector

$$\mathbf{v}^* = E[\mathbf{rr}^T]^{-1} E[b_1 \mathbf{r}] \tag{6.38}$$

$$= A_1 \left[\sigma^2 \mathbf{I} + \sum_{k=1}^{K} A_k^2 \mathbf{s}_k \mathbf{s}_k^T \right]^{-1} \mathbf{s}_1. \tag{6.39}$$

This is the MMSE linear transformation in the space of (2.97), which we refer to as the *Wiener* characterization of the linear MMSE detector. The minimum mean-square error achievable with a linear transformation is

$$E[(b_1 - \mathbf{r}^T \mathbf{v}^*)^2] = 1 - (E[b_1 \mathbf{r}])^T (E[\mathbf{rr}^T])^{-1} E[b_1 \mathbf{r}]$$

$$= 1 - A_1^2 \mathbf{s}_1^T E[\mathbf{rr}^T]^{-1} \mathbf{s}_1$$

$$= 1 - A_1^2 \mathbf{s}_1^T \left[\sigma^2 \mathbf{I} + \sum_{k=1}^{K} A_k^2 \mathbf{s}_k \mathbf{s}_k^T \right]^{-1} \mathbf{s}_1. \tag{6.40}$$

Let us denote the covariance matrix of the interference by

$$\Sigma \overset{\text{def}}{=} \sigma^2 \mathbf{I} + \sum_{k=2}^{K} A_k^2 \mathbf{s}_k \mathbf{s}_k^T, \tag{6.41}$$

where notice that we are excluding user 1 from the sum. Using Σ^{-1}, we can obtain (Problem 6.11) alternative expressions for the MMSE transformation (6.39) and the minimum mean-square error (6.40):

$$\mathbf{v}^* = \frac{A_1}{1 + A_1^2 \mathbf{s}_1^T \Sigma^{-1} \mathbf{s}_1} \Sigma^{-1} \mathbf{s}_1, \tag{6.42}$$

and

$$E[(b_1 - \mathbf{r}^T \mathbf{v}^*)^2] = \frac{1}{1 + A_1^2 \mathbf{s}_1^T \Sigma^{-1} \mathbf{s}_1}. \tag{6.43}$$

Furthermore, recall that $\mathbf{v}^*$ achieves the maximum output signal-to-interference ratio of any linear transformation, which thanks to (6.42) and (6.43) can be expressed as

$$SIR_1 = \frac{E\left[\left(A_1 b_1 \mathbf{s}_1^T \mathbf{v}^* \right)^2 \right]}{E\left[\left((\mathbf{r}^T - A_1 b_1 \mathbf{s}_1^T) \mathbf{v}^* \right)^2 \right]}$$

$$= \frac{1}{E[(b_1 - \mathbf{r}^T \mathbf{v}^*)^2]} - 1$$

$$= A_1^2 \mathbf{s}_1^T \Sigma^{-1} \mathbf{s}_1. \tag{6.44}$$

6.3 PERFORMANCE OF MMSE LINEAR MULTIUSER DETECTION

Since the MMSE detector converges to the decorrelating detector as $\sigma \to 0$ its asymptotic multiuser efficiency and near–far resistance are identical to those of the decorrelator:

$$\bar{\eta}_k = \frac{1}{R_{kk}^+}$$

in the synchronous case, and

$$\bar{\eta}_k = \left(\frac{1}{2\pi} \int_{-\pi}^{\pi} [\mathbf{R}^T[1]e^{j\omega} + \mathbf{R}[0] + \mathbf{R}[1]e^{-j\omega}]_{kk}^+ \, d\omega \right)^{-1} \qquad (6.45)$$

in the asynchronous case.

In high signal-to-noise ratio channels with linearly independent signature waveforms, it would not seem judicious to incur the extra complexity of having to incorporate the received signal-to-noise ratios, only to attain a very small performance improvement over the decorrelating detector. As we discuss in Section 6.4, the chief advantage of the MMSE detector is the ease with which it can be implemented adaptively.

Because the linear MMSE detector does not tune out the interfering users, the analysis of its bit-error-rate is not as straightforward as that of the decorrelator. As in the case of the single-user matched filter (Section 3.4.1), the decision statistic (conditioned on the bit of interest) is not Gaussian but the sum of a Gaussian random variable (due to the background noise) and a binomial random variable (due to the multiaccess interference). In the synchronous case, the first component of the output of the linear MMSE transformation in (6.30) can be written as

$$(\mathbf{M}^*\mathbf{y})_1 = ([\mathbf{R} + \sigma^2\mathbf{A}^{-2}]^{-1}\mathbf{y})_1 = B_1\left(b_1 + \sum_{k=2}^{K} \beta_k b_k \right) + \sigma\tilde{n}_1 \qquad (6.46)$$

with

$$\beta_k = \frac{B_k}{B_1},$$

$$B_k = A_k(\mathbf{M}^*\mathbf{R})_{1k},$$

$$\tilde{n}_1 \sim \mathcal{N}(0, (\mathbf{M}^*\mathbf{R}\mathbf{M}^*)_{11}).$$

The *leakage coefficient* β_k quantifies the contribution of the kth interferer to the decision statistic, relative to the contribution of the desired user. The

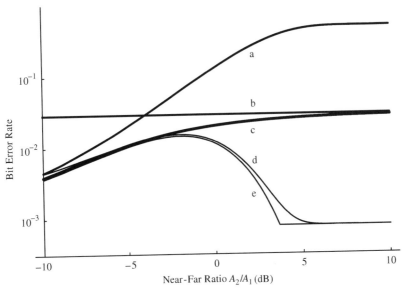

FIGURE 6.5.
Bit-error-rate with
two users and
crosscorrelation
$\rho = 0.8$:
(a) single-user
matched filter;
(b) decorrelator;
(c) MMSE;
(d) minimum
(upper bound);
(e) minimum
(lower bound).

probability of error readily follows:

$$P_1^m(\sigma) = 2^{1-K} \sum_{b_2,\dots,b_K \in \{-1,1\}^{K-1}} Q\left(\frac{A_1}{\sigma} \frac{(\mathbf{M}^*\mathbf{R})_{11}}{\sqrt{(\mathbf{M}^*\mathbf{R}\mathbf{M}^*)_{11}}} \left(1 + \sum_{k=2}^{K} \beta_k b_k\right)\right).$$

$$(6.47)$$

Figure 6.5 evaluates (6.47) in the special case of two users with crosscorrelation equal to 0.8. The signal-to-noise ratio of the desired user is equal to 10 dB. The probability of error is shown as a function of the near–far ratio A_2/A_1 and it is compared to the probability of error of the single-user matched filter, the decorrelating detector, and the optimum multiuser detector. Note that for sufficiently low interferer power, the MMSE probability of error is better than that of the decorrelating detector, and even better than the upper bound to the minimum probability of error found in (4.61). For relatively high-power interferers, the MMSE detector, while far from achieving the single-user bit-error-rate (which is essentially achievable by the optimum detector), performs quite similarly to the decorrelator and much better than the single-user matched filter.

To evaluate (6.47) we face the same difficulty we encountered with (3.90): an exponential (in K) number of terms. This computational burden is exacerbated by the need to compute the leakage coefficients. However, in contrast to the behavior of the single-user matched filter, the error probability P_1^m can be generally accurately approximated by replacing the multiaccess interference by a Gaussian random variable with identical variance: $Q(\sqrt{\text{SIR}_1})$, where SIR_1 is given in (6.44). Equivalently, we can use (3.66),

$$E[Q(\mu + \lambda X)] = Q\left(\frac{\mu}{\sqrt{1 + \lambda^2}}\right), \qquad (6.48)$$

where X is unit normal, and

$$\mu = \frac{A_1}{\sigma} \frac{(\mathbf{M}^*\mathbf{R})_{11}}{\sqrt{(\mathbf{M}^*\mathbf{R}\mathbf{M}^*)_{11}}}$$

and

$$\lambda^2 = \mu^2 \sum_{k=2}^{K} \beta_k^2.$$

The accuracy of this approximation has been supported by several analytical results in Poor and Verdú [338]. On an intuitive basis, the small deviation from normality of the decision statistic can be seen by analyzing the asymptotic cases $\sigma \to 0$ and $\sigma \to \infty$ first. As $\sigma \to 0$ the MMSE detector approaches the decorrelator, and, thus, the leakage coefficients vanish; as $\sigma \to \infty$, the background-noise contribution at the output of the linear transformation dominates the multiaccess interference. In either case, the decision statistic is asymptotically Gaussian. For finite nonzero σ, the maximization of signal-to-interference ratio dictates negligible leakage coefficients unless the background noise is relatively powerful. Representative numerical results are shown in Figure 6.6, which plots the bit-error-rates

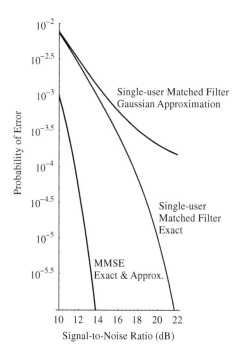

FIGURE 6.6.
Bit-error-rate with eight equal-power users and identical crosscorrelations $\rho_{kl} = 0.1$.

achieved by the single-user detector and by the linear MMSE detector in
the case of eight synchronous users with identical crosscorrelations (equal
to 0.1) and perfect power control (Problem 6.32). The Gaussian approximations are shown in both cases; whereas for the matched filter, the approximation is only accurate for very low signal-to-noise ratios, for the MMSE
detector both curves are indistinguishable. Note also the large gap in performance between the MMSE detector and the single-user matched filter
(ranging from about 4 dB to 8 dB for the signal-to-noise ratios shown in
Figure 6.6).

Based on numerical and analytical evidence (cf. Problem 6.31), Poor
and Verdú [338] conjectured that the bit-error-rate of the MMSE detector
is better than that of the decorrelator for all levels of background Gaussian
noise, number of users, and crosscorrelation matrices.

We now turn our attention to the analysis of minimum mean-square error
when using random binary-valued signature sequences. According to the
solution found in (6.27), if all the users have equal power, then the minimum
mean-square error for the kth user is the kth diagonal element of the matrix

$$\left[\mathbf{I} + \frac{A^2}{\sigma^2} \mathbf{R} \right]^{-1}.$$

Averaging the minimum mean-square error with respect to the signature
waveforms we obtain

$$E\left[\min_{c_k} E[(b_k - \langle c_k, y \rangle)^2] \right] = E\left[\left[\mathbf{I} + \frac{A^2}{\sigma^2} \mathbf{R} \right]^{-1}_{kk} \right]$$

$$= \frac{1}{K} E\left[\text{trace} \left[\mathbf{I} + \frac{A^2}{\sigma^2} \mathbf{R} \right]^{-1} \right]$$

$$= \frac{1}{K} E\left[\sum_{j=1}^{K} \lambda_j \left(\left[\mathbf{I} + \frac{A^2}{\sigma^2} \mathbf{R} \right]^{-1} \right) \right]$$

$$= E\left[\frac{1}{1 + \frac{A^2}{\sigma^2} \lambda_k(\mathbf{R})} \right]. \tag{6.49}$$

If the ratio of the number of users to the spreading gain is, or converges to,
a constant:

$$\lim_{K \to \infty} \frac{K}{N} = \beta \in (0, +\infty),$$

then we can compute the limit as $K \to \infty$ of (6.49) using the result in

302

Proposition 2.1:

$$\lim_{K \to \infty} E\left[\frac{1}{1 + \frac{A^2}{\sigma^2}\lambda_k(\mathbf{R})}\right] = \int_0^\infty \frac{1}{1 + \frac{A^2}{\sigma^2}x} f_\beta(x)\, dx \qquad (6.50)$$

$$= 1 - \frac{1}{4\beta} \frac{\sigma^2}{A^2} \mathcal{F}\left(\frac{A^2}{\sigma^2}, \beta\right) \qquad (6.51)$$

with

$$\mathcal{F}(x, z) \stackrel{\text{def}}{=} \left(\sqrt{x(1 + \sqrt{z})^2 + 1} - \sqrt{x(1 - \sqrt{z})^2 + 1}\right)^2. \qquad (6.52)$$

The density in (6.50) is defined in (2.28). The explicit expression (6.51) for the definite integral can be verified in Gradshteyn and Ryzhik [125]. Note that the result in (6.51) holds regardless of whether K is greater or smaller than N. It can be checked that as the individual signal-to-noise ratio and the number of users grow without bound the averaged mean-square error behaves as

$$\lim_{K \to \infty} E\left[\min_{c_k} E[(b_k - \langle c_k, y\rangle)^2]\right] = \begin{cases} \frac{1}{1-\beta} \frac{\sigma^2}{A^2} + o\left(\frac{\sigma^2}{A^2}\right), & \text{if } \beta < 1; \\ \frac{\sigma}{A} + o\left(\frac{\sigma^2}{A^2}\right), & \text{if } \beta = 1; \\ 1 - \beta^{-1} + O\left(\frac{\sigma^2}{A^2}\right), & \text{if } \beta > 1, \end{cases} \qquad (6.53)$$

where the outer expectation is with respect to the signature waveforms and the inner expectation is with respect to noise and data. In parallel with the mean-square convergence result we showed for the optimum near–far resistance in (4.111), we will show that the minimum mean-square error with random binary sequences converges to the right side of (6.51):

$$\lim_{K \to \infty} \min_{c_k} E[(b_k - \langle c_k, y\rangle)^2] = 1 - \frac{1}{4\beta} \frac{\sigma^2}{A^2} \mathcal{F}\left(\frac{A^2}{\sigma^2}, \beta\right),$$

where the expectation is with respect to the noise and transmitted data but not with respect to the choice of signature sequences. Since the minimum mean-square error is upper bounded by 1, it is enough (cf. (4.113)) to show the asymptotic identity:

$$\lim_{K \to \infty} E\left[\frac{1}{\min_{c_k} E[(b_k - \langle c_k, y\rangle)^2]}\right]$$

$$= \lim_{K \to \infty} \frac{1}{E\left[\min_{c_k} E[(b_k - \langle c_k, y\rangle)^2]\right]}. \qquad (6.54)$$

For that purpose, we use (6.41) and (6.44) to write the average signal-to-interference ratio as

$$E\left[\frac{1}{\min_{c_1} E[(b_1 - \langle c_1, y\rangle)^2]}\right] - 1$$

$$= A^2 E\left[s_1^T \Sigma^{-1} s_1\right]$$

$$= \frac{A^2}{N} \sum_{n=1}^{N} E\left[(\Sigma^{-1})_{nn}\right] \qquad (6.55)$$

$$= \frac{A^2}{N} E[\text{trace}(\Sigma^{-1})]$$

$$= \frac{A^2}{N} E\left[\sum_{n=1}^{N} \lambda_n(\Sigma^{-1})\right]$$

$$= \frac{A^2}{N} E\left[\sum_{n=1}^{N} \frac{1}{\lambda_n(\Sigma)}\right]$$

$$= E\left[\frac{A^2}{\lambda_1(\Sigma)}\right]$$

$$= E\left[\frac{A^2}{\sigma^2 + A^2 \beta \lambda_1\left(\frac{1}{\beta} \sum_{k=2}^{K} s_k s_k^T\right)}\right], \qquad (6.56)$$

where (6.55) follows by taking the expectation with respect to s_1 because its components are uncorrelated and have zero mean and variance equal to $1/N$. To take the limit as $K \to \infty$ of (6.56) we make use of the result in Problem 2.48:

$$\lim_{K \to \infty} E\left[\frac{A^2}{\sigma^2 + A^2 \beta \lambda_1\left(\frac{1}{\beta} \sum_{k=2}^{K} s_k s_k^T\right)}\right] = \int_0^{+\infty} \frac{A^2}{\sigma^2 + A^2 \beta x} f_{1/\beta}(x) \, dx$$

$$= \frac{A^2}{\sigma^2} - \frac{1}{4} \mathcal{F}\left(\frac{A^2}{\sigma^2}, \beta\right). \qquad (6.57)$$

Having obtained the average signal-to-interference ratio in (6.57) we just need to check that

$$\left[1 - \frac{1}{4\beta} \frac{\sigma^2}{A^2} \mathcal{F}\left(\frac{A^2}{\sigma^2}, \beta\right)\right]^{-1} = 1 + \frac{A^2}{\sigma^2} - \frac{1}{4} \mathcal{F}\left(\frac{A^2}{\sigma^2}, \beta\right). \qquad (6.58)$$

But this is equivalent to checking that the parabola

$$z^2 - 4\left(1 + \frac{A^2}{\sigma^2}(1 + \beta)\right) z + 16\beta \frac{A^4}{\sigma^4}$$

has a root at

$$z = \mathcal{F}\left(\frac{A^2}{\sigma^2}, \beta\right).$$

Thus, the proof of (6.54) is now complete. Since the signal-to-interference ratio achieved by the MMSE detector is upper bounded (by A^2/σ^2), the foregoing reasoning shows that the signal-to-interference ratio converges in mean-square sense:

$$\lim_{K \to \infty} \max_{\mathbf{c}} \frac{E\left[\left(Ab_k \mathbf{s}_k^T \mathbf{c}\right)^2 \right]}{E\left[\left((\mathbf{r}^T - Ab_k \mathbf{s}_k^T)\mathbf{c}\right)^2 \right]} = \frac{A^2}{\sigma^2} - \frac{1}{4}\mathcal{F}\left(\frac{A^2}{\sigma^2}, \beta \right) \quad (6.59)$$

$$\stackrel{\text{def}}{=} \frac{A^2_{\text{MMSE}}}{\sigma^2}. \quad (6.60)$$

Using (6.60) we can solve (Problem 6.19) for the β required to achieve any $A_{\text{MMSE}} < A$:

$$\beta = \left(1 + \frac{\sigma^2}{A^2_{\text{MMSE}}} \right)\left(1 - \frac{A^2_{\text{MMSE}}}{A^2} \right). \quad (6.61)$$

From (6.61) it can be verified that the asymptotic output signal-to-noise ratio $A^2_{\text{MMSE}}/\sigma^2$ of the MMSE linear filter satisfies the equation

$$\frac{A^2_{\text{MMSE}}}{\sigma^2} = \frac{A^2}{\sigma^2 + \beta \frac{A^2}{1 + \frac{A^2_{\text{MMSE}}}{\sigma^2}}}. \quad (6.62)$$

It is interesting to compare (6.62) with the corresponding expression for the single-user matched filter (3.184):

$$\frac{A^2}{\sigma^2 + \beta A^2}.$$

In the case of the single-user matched filter, every interferer contributes an interfering power equal to its power divided by the spreading gain N, whereas in MMSE processing that interfering power is further reduced by a factor equal to

$$\frac{1}{1 + \frac{A^2_{\text{MMSE}}}{\sigma^2}}.$$

Figure 6.7 shows the probability of error of the MMSE detector and the single-user matched filter with $K = 100$ and eighteen different realizations of randomly chosen direct-sequence signature waveforms with $N = 1,000$. As we mentioned in Section 3.4.1, the bit-error-rate of the single-user matched filter converges for any signal-to-noise ratio as $K \to \infty$ while K/N is kept constant. However, this convergence may be very slow, as illustrated in Figure 6.7, where despite the large number of users in this channel, the performance variability of the single-user matched filter is very

305

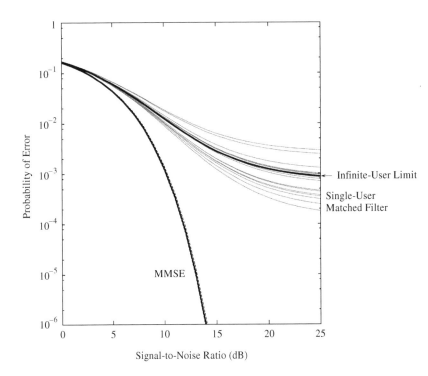

FIGURE 6.7.
Bit-error-rate with 100 equal-power users and random direct sequence signatures with $N = 1,000$.

noticeable. The infinite-user limit shown in Figure 6.7 is the right side of Equation (3.94). The bit-error-rates of the single-user matched filter and MMSE receiver in Figure 6.7 are evaluated via (3.69). For the parameters considered in this figure, the performance of the MMSE receiver is insensitive with respect to the actual realization of random signatures and is indistinguishable from the infinite-user limit:

$$Q\left(\sqrt{\frac{A^2}{\sigma^2} - \frac{1}{4}\mathcal{F}\left(\frac{A^2}{\sigma^2}, \beta \right)} \right).$$

6.4 ADAPTIVE MMSE LINEAR MULTIUSER DETECTION

The structure of linear multiuser detectors (such as the decorrelating detector and the MMSE detector) is attractive because they can be implemented in decentralized fashion with what amounts to a modified matched filter. However, this modified matched filter depends on the signal crosscorrelations (and on the received signal-to-noise ratios in the case of the MMSE detector), and the computation of its impulse response involves matrix inversion. Although there are ways to speed up such a computation

(for example in the decorrelating detector, Gram–Schmidt orthogonalization can be used), it is of interest to examine ways in which it could be
avoided altogether. It is particularly important to eliminate the need to compute the linear detector impulse response in asynchronous channels where
crosscorrelations are time varying and in channels with time-varying received powers (in the case of the MMSE linear multiuser detector). It would
be very desirable to obtain a linear multiuser detector that would not only
eliminate the need for on-line computation of its impulse response, but it
would also eliminate the need to know the crosscorrelations, or in general,
the signature waveforms of the interfering users. We will show how that
goal can be accomplished with an adaptive implementation of the MMSE
linear detector, which "learns" the desired filter impulse response from the
received waveform, *provided that the data of the desired user is known to
the receiver*. This last requirement may seem like too much to ask since
after all, the data are what we are after. In practice it requires the transmission of a *training sequence* (i.e., a string of data known to the receiver)
prior to the transmission of actual data. The receiver uses an adaptive law
to adjust its linear transformation while the training sequence is being sent.
If the crosscorrelations and amplitudes vary over time, training sequences
can be sent periodically to readjust the receiver. In practice, it is common
to perform fine adjustment of the linear transformation (once the adaptive
algorithm has converged and the transmission of the training sequence is
over) by letting the adaptive algorithm run with the decisions made by the
detector in lieu of the true transmitted data. This type of operation is usually
referred to as *decision-directed*.

To justify the adaptive law for the MMSE linear detector, we will first
review the elements of gradient descent stochastic optimization. Suppose
that we wish to find the (multidimensional) parameter θ^* that minimizes
the following function:

$$\Psi(\theta) = E[g(X, \theta)] \tag{6.63}$$

where X is a random variable. If the function Ψ is *convex*, then starting at
any initial condition θ_0, it would be possible to converge to the minimum of
Ψ by following the direction of steepest descent (i.e., the opposite direction
to the gradient $\nabla\Psi$):

$$\theta_{j+1} = \theta_j - \mu\nabla\Psi(\theta_j). \tag{6.64}$$

If the step size is arbitrarily small, then eventually θ_j will be as close as
desired to θ^* (in practice, the step size can be progressively shrunk as the
algorithm converges). Other than using the convexity of Ψ we have not yet

invoked the structure in (6.63), according to which we need to know the probability distribution of X in order to compute the gradient. But what if the distribution of X is not known? Then, we cannot compute the expectation in (6.63) and neither the cost function Ψ nor its gradient are known. So it is not possible to implement (6.64) if the distribution of X is unknown. Instead, let us assume that the optimization algorithm is allowed to observe an independent sequence $\{X_1, X_2, \ldots\}$ where each of the random variables has the same distribution as X. From this sequence we could estimate the distribution of X and compute an approximation to $\nabla\Psi$. However, a more direct approach can be devised. At each step of the algorithm, we could replace the unknown

$$\nabla\Psi(\theta_j) = E\nabla g(X, \theta_j)$$

by its "noisy" version, $\nabla g(X_{j+1}, \theta_j)$:

$$\theta_{j+1} = \theta_j - \mu\nabla g(X_{j+1}, \theta_j). \tag{6.65}$$

The rationale for (6.65) is that even though the negative gradient in (6.65) does not point in the true direction of steepest descent, its average does indeed point in that direction at each step in the algorithm because

$$E[\nabla g(X_j, \theta_j)] = E[\nabla g(X, \theta_j)] = \nabla\Psi(\theta_j).$$

According to the *law of large numbers*, if the step size is infinitesimal then deviations from the average will tend to cancel, and the trajectory of the algorithm will track very closely the lines of steepest descent of Ψ. This algorithm is known as stochastic gradient descent or *stochastic approximation*.[1] In the special case of a quadratic-error cost function g, stochastic approximation is also known as the *LMS* (least-mean-square) algorithm. We motivated (6.65) as a method of computing the minimum of (6.63) when the distribution of the random variable X therein is unknown but a sequence of independent identically distributed realizations of X is available. In practice, it may be computationally advantageous to use the simple stochastic approximation algorithm in (6.65) to compute the minimum of Ψ even when the distribution of X is known. Moreover, if the stationary sequence of realizations of X is dependent, we can confidently use the stochastic approximation algorithm as long as the sequence is ergodic, and the time average of the "noisy" gradients converges to its expectation.

---

[1] The term *stochastic approximation* is usually reserved for versions of stochastic gradient descent where the step size is progressively decreased.

In practice, the penalty for making μ exceedingly small is very slow convergence to the vicinity of the solution θ^*; larger values of μ accelerate convergence but introduce larger steady-state fluctuations around θ^*. To overcome these shortcomings, the step size can be shrunk as the algorithm progresses.

Let us now apply the foregoing general framework to the derivation of an adaptive MMSE linear detector. We will do so in the synchronous setting and then we will briefly address the asynchronous case. As we saw in Section 6.2, the MMSE linear detector for user 1 correlates the received waveform with the signal c_1 that minimizes

$$E[(b_1 - \langle c_1, y \rangle)^2].\tag{6.66}$$

The minimization of (6.66) fits the above stochastic approximation framework with

$$g(X, c_1) = (b_1 - \langle c_1, y \rangle)^2,\tag{6.67}$$

where X represents the received waveform y and the bit b_1. It is immediate to check that (6.66) is a strictly convex function in c_1. The independent identically distributed observations used in the stochastic approximation algorithm are

$$X_j = (b_1[j], y[j]),\tag{6.68}$$

where

$$y[j](t) = \sum_{i=1}^{K} A_i b_i[j] s_i(t - jT) + n(t), \quad t \in [jT, jT + T].\tag{6.69}$$

Hence all we need to do to specify the gradient descent algorithm (6.65) is to obtain the gradient of $(b_1 - \langle c_1, y \rangle)^2$ with respect to c_1, which is equal to

$$2(\langle c_1, y \rangle - b_1)y.\tag{6.70}$$

We conclude that the adaptive algorithm is

$$c_1[j] = c_1[j - 1] - \mu\left(\langle c_1[j - 1], y[j] \rangle - b_1[j]\right)y[j].\tag{6.71}$$

Therefore, the update of the impulse response is equal to the received waveform scaled by the error between the known data and the filter output.

The solution to which the adaptation law in (6.71) converges (with adequately decreasing step size) is the MMSE linear detector for user 1. Consequently, it depends on all the received signal-to-noise ratios and signature waveforms. However, the algorithm requires no a priori knowledge of the signature waveforms (not even that of the desired user) nor of the received signal-to-noise ratios. Besides the need for the training sequence of the desired user, the only underlying requirement of the algorithm is that it has acquired bit-epoch synchronism with the desired user. Why does it converge to the desired user's MMSE detector, rather than to another user's detector (e.g., the strongest one)? Because the algorithm is governed by the data (training sequence) of the desired user. Another noteworthy aspect of the derivation of the adaptive MMSE detector is that it did not have to take into account the structure of the interference in (6.69), namely white Gaussian noise plus linearly modulated signature waveforms. The adaptive linear transformation self-tunes to exploit such a structure.

As we have seen, other than to initialize the algorithm, the adaptive MMSE linear detector does not use knowledge of the signature waveform of the desired user. A different adaptive strategy is discussed in Section 6.6 that makes effective use of the signature waveform of the desired user.

Even though the signals $(c_1[j], y[j])$ in (6.71) are continuous-time signals of duration T, practical implementation of the adaptive law will ordinarily be carried out with finite-dimensional vectors. If the signature waveforms of all the users are known, then the dimensionality of the adaptive vector need not be larger than K. Although this knowledge may be realistic in some situations, it is important to capitalize on a major advantage of the adaptive linear MMSE detector: no signature waveforms need be known beforehand. Fortunately, any finite-dimensional basis known to span all received signature waveforms will serve the purpose of implementing a linear finite-dimensional adaptive algorithm. For example, in a synchronous CDMA channel with direct-sequence spread-spectrum waveforms with identical chip waveforms, no optimality is lost (cf. Section 2.9) by working with the discrete-time outputs of chip matched filters, in which case, the algorithm in (6.71) would operate with vectors whose dimension is equal to the number of chips per bit. For approximately band-limited chip waveforms, sampling at the Nyquist rate provides sufficient statistics for (synchronous and) asynchronous channels.[2] In either case, the sequence of received vectors obeys the model (cf. (2.97))

$$\mathbf{r}[n] = \mathbf{SAb}[n] + \sigma\mathbf{m}[n]. \tag{6.72}$$

[2] Nyquist sampling sidesteps the need for the receiver to acquire synchronism with asynchronous interferers.

Now, the finite-dimensional version of the adaptive algorithm in (6.71) is

$$\mathbf{v}_1[n] = \mathbf{v}_1[n-1] - \mu \left(\mathbf{v}_1^T[n-1]\mathbf{r}[n] - b_1[n] \right) \mathbf{r}[n].$$
(6.73)

We can prove that the algorithm in (6.73) is globally convergent[3] when μ decreases suitably. This convergent behavior is a consequence of the following general result, which puts the foregoing intuitive ideas on stochastic approximation on a sound footing.

Consider the convex penalty function

$$\Psi(\mathbf{v}) = E[(u - \mathbf{v}^T \mathbf{w})^2],$$
(6.74)

where u is a random scalar and $\mathbf{w}$ is a random L-dimensional vector. The minimization of (6.74) results in the linear MMSE estimate of u based on the observation of $\mathbf{w}$. As in (6.39), the argument that minimizes (6.74) is

$$\mathbf{v}^* = E[\mathbf{w}\mathbf{w}^T]^{-1} E[u\mathbf{w}].$$
(6.75)

If realizations of $(u, \mathbf{w})$ are available, then an alternative to computing the right-hand side of (6.75) is to use stochastic approximation. In this case, convergence of the algorithm is fully justified by the following result:

Proposition 6.1 *(Gyorfi [130]) Consider the recursion*

$$\mathbf{v}[n] = \mathbf{v}[n-1] - \frac{1}{n}(\mathbf{v}^T[n-1]\mathbf{w}[n] - u[n])\mathbf{w}[n],$$
(6.76)

where $\mathbf{v}[n]$, $\mathbf{w}[n]$ *are L-dimensional vectors and* $u[n]$ *are scalars that satisfy the following conditions:*

(a) $\{\mathbf{w}[n], u[n]\}$ *is a stationary ergodic sequence.*
(b) $E[\mathbf{w}[1]\mathbf{w}[1]^T]$ *is invertible.[4]*
(c) $E[\|\mathbf{w}[1]\|^4] < \infty.$
(d) $E[u^2[1]] < \infty.$

Then, for any $\mathbf{v}[0]$,

$$\lim_{n\to\infty} \mathbf{v}[n] = E[\mathbf{w}[1]\mathbf{w}[1]^T]^{-1} E[u[1]\mathbf{w}[1]] \quad \textit{almost surely.}$$
(6.77)

---

[3] An adaptive algorithm is globally convergent if it converges to the desired solution regardless of the initialization.
[4] Because of the stationarity assumed in (a), conditions (b), (c), and (d) apply automatically to $\mathbf{w}[n]$ and $u[n]$.

Now (6.73) can be seen as a special case of (6.76) with

$$u[n] = b_1[n], \tag{6.78}$$

$$\mathbf{w}[n] = \mathbf{r}[n]. \tag{6.79}$$

As long as the training sequence is independent identically distributed, the process $\{\mathbf{w}[n], u[n]\}$ is independent identically distributed in the synchronous case and the sufficient condition (a) in Proposition 6.1 is satisfied. Condition (b) is satisfied as can be seen from (6.39). Condition (c) is satisfied because all moments of the norm of a Gaussian random vector are finite. Condition (d) is obviously satisfied because $u^2[n] = 1$. Assuming that all K training sequences are mutually independent (cf. Problem 6.35) the right side of (6.77) becomes the Wiener characterization of the MMSE linear multiuser detector upon substitution of (6.78) and (6.79). Thus, Proposition 6.1 shows that the adaptive MMSE algorithm (with $\mu_n = \frac{1}{n}$) converges regardless of the initialization to the desired solution.

In practical implementations in nonstationary environments, a nonzero lower bound must be imposed on μ_n so as to be able to track channel variations. We turn to a brief examination of the factors that affect the speed of convergence to steady state of the adaptive law in (6.73) in the case of a fixed step size μ. As we have seen in (6.39), the MMSE filter corresponds to

$$\mathbf{v}_1^* = E[\mathbf{r}\mathbf{r}^T]^{-1} E[b_1 \mathbf{r}]. \tag{6.80}$$

The expected error between the adaptive filter and the desired solution follows the evolution:

$$E[\mathbf{v}_1[n] - \mathbf{v}_1^*] = E[\mathbf{v}_1[n-1] - \mathbf{v}_1^*] - \mu\, E[\mathbf{r}[n]\mathbf{r}^T[n]\mathbf{v}_1[n-1]]$$
$$+ \mu E[\mathbf{r}\mathbf{r}^T]\mathbf{v}_1^*$$
$$= [\mathbf{I} - \mu E[\mathbf{r}\mathbf{r}^T]]E[\mathbf{v}_1[n-1] - \mathbf{v}_1^*], \tag{6.81}$$

where (6.81) follows because $\mathbf{v}_1[n-1]$ depends on observations

$$\mathbf{r}[0], \dots, \mathbf{r}[n-1]$$

but is uncorrelated with $\mathbf{r}[n]$. The error vector is seen to obey a simple linear equation (6.81), which is easily solvable using matrix exponentials. Rather than doing that, we can get a feel for the speed of decay of the average error by choosing a different basis for the linear transformation. Let us orthogonalize the symmetric matrix in the right side of (6.81):

$$\mathbf{Q}_\mu \stackrel{\text{def}}{=} [\mathbf{I} - \mu E[\mathbf{r}\mathbf{r}^T]] = \mathbf{G}^T \mathbf{\Lambda} \mathbf{G},$$

where $\mathbf{G}$ is an orthonormal matrix, $\mathbf{G}\mathbf{G}^T = \mathbf{I}$, and $\mathbf{\Lambda}$ is a diagonal matrix containing the eigenvalues of $\mathbf{Q}_\mu$. Then (6.81) is equivalent to

$$\mathbf{G}E[\mathbf{v}_1[n] - \mathbf{v}_1^*] = \mathbf{\Lambda}\mathbf{G}E[\mathbf{v}_1[n-1] - \mathbf{v}_1^*].$$

This means that each component of the expected error vector rotated by $\mathbf{G}$ evolves autonomously with a decay factor equal to its corresponding eigenvalue:

$$a_{1,j}[n] = (1 - \mu\sigma^2 - \mu\lambda_j)^n a_{1,j}[0], \tag{6.82}$$

where

$$\mathbf{a}_1[n] = (a_{1,1}[n], \ldots, a_{1,L}[n])$$

$$= \mathbf{G}E[\mathbf{v}_1[n-1] - \mathbf{v}_1^*],$$

and λ_j denotes the jth eigenvalue of

$$\sum_{k=1}^{K} A_k^2 \mathbf{s}_k \mathbf{s}_k^T.$$

The geometric decay rate of (6.82) is

$$|1 - \mu\sigma^2 - \mu\lambda_j|,$$

which is stable (i.e., less than 1) if and only if

$$0 < \mu < \frac{2}{\sigma^2 + \lambda_j}.$$

It follows that stability of (6.73) dictates a step size smaller than

$$\mu_{\max} = \frac{2}{\sigma^2 + \lambda_{\max}},$$

where $\lambda_{\max}$ is the maximum eigenvalue of $\sum_{k=1}^{K} A_k^2 \mathbf{s}_k \mathbf{s}_k^T$. Within that constraint, the speed of convergence is maximized by choosing the step size that maximizes the minimum of the decay rates (Problem 6.36). Regardless of the choice of μ the speed of convergence is the same for all users, although the steady-state mean-square error need not be the same for different users.

In adaptive filtering there are well-known ways to speed up convergence based on recursive least-squares (Problem 6.46) and other techniques.

The approach we have followed so far in this section has to be modified only slightly to deal with the asynchronous channel. If we take the one-shot approach of Section 5.3, the foregoing algorithm holds verbatim. The adaptive algorithm (6.66) will converge to the one-shot version of the MMSE linear detector. The adaptive receiver needs to know the data of the desired user and needs to synchronize to its bit epochs (not to those of other interferers). The near–far resistance of the resulting one-shot MMSE linear detector is the same as the one-shot decorrelating detector, which is not as good as that of the asynchronous decorrelating detector. In order to improve the performance of the MMSE linear detector, we need to lengthen the observation interval to a truncated window that spans more than one bit period. This implies that the inner product in the penalty function (6.66) must now be taken over the whole truncated window. This has no effect on the convexity of the function, and the desired MMSE linear detector is the unique local minimum of this function. The derivation of the adaptive gradient descent algorithm is the same as before with the only change being that the linear transformation being adapted and the observed vector now span the whole window of observation. Note that in the asynchronous case $\{\mathbf{w}[n], u[n]\}$ is not an independent sequence even if the training sequence is independent identically distributed because of the overlapping of interfering waveforms in consecutive windows. However, $\{\mathbf{w}[n], u[n]\}$ is stationary and Gaussian, and hence ergodic, which is enough to invoke Proposition 6.1. Nonetheless, the evolution of the expected error in (6.81) does not hold in the asynchronous case because of the dependence of the observations in adjacent blocks.

Naturally, convergence to the MMSE solution hinges on the received amplitudes and crosscorrelations being constant. When these parameters are slowly varying (relative to the speed of convergence of the algorithm) then it is still possible to track those variations with the adaptive algorithm driven either by a training sequence or in decision-directed mode. However, if there is a sudden change in the CDMA channel such as a strong interferer suddenly being powered on, the decision-directed algorithm will start using unreliable decisions in lieu of the true data and may not converge. It may become necessary that the receiver be able to detect those sudden channel changes and request the desired user to interrupt its transmission in order to send a training sequence. Thus, reliance on training sequences is more cumbersome and disruptive in CDMA channels than in single-user channels. It is therefore of considerable interest to study adaptive multiuser detectors that operate in *blind* mode (i.e., without the need to know the data). Before we do this in Section 6.6, we need to introduce a general representation of linear multiuser detectors.

6.5   CANONICAL REPRESENTATION OF LINEAR MULTIUSER DETECTORS

In this section we show a property of the MMSE linear multiuser detector that turns out to be pivotal in the derivation of a blind adaptive

314

algorithm in Section 6.6. To show this property we will introduce a canonical representation for any linear multiuser detector.

A linear multiuser detector for user 1 is characterized by a waveform c_1, such that

$$\hat{b}_1 = \operatorname{sgn}(\langle y, c_1 \rangle). \tag{6.83}$$

Although the representation we derive in this section holds for both synchronous and asynchronous channels, it is helpful to focus on the synchronous case, in which all the signals have duration T. Let us introduce the following *canonical representation* for c_1:

$$c_1 = s_1 + x_1, \tag{6.84}$$

where x_1 is such that

$$\langle s_1, x_1 \rangle = 0. \tag{6.85}$$

The representation in (6.84) and (6.85) is canonical in that every linear multiuser detector for user 1 can be expressed in that form. This is because the set of signals c_1 that can be written as in (6.84) and (6.85) are those that satisfy

$$\langle s_1, c_1 \rangle = \|s_1\|^2 = 1, \tag{6.86}$$

and the decision in (6.83) is invariant to positive scaling; thus, the only linear transformations ruled out by (6.86) are the signals c_1 orthogonal to s_1, which may as well be ruled out since they result in error probability equal to $1/2$.

Thanks to the canonical linear transformation, every linear transformation for multiuser detection is characterized by its corresponding orthogonal signal x_1. Given a desired (up to a scale factor) linear transformation d_1, the orthogonal component to s_1 is given by

$$x_1 = \frac{1}{\langle s_1, d_1 \rangle} d_1 - s_1. \tag{6.87}$$

The linear transformation outputs

$$\langle y, s_1 + x_1 \rangle = A_1 b_1 + \sum_{k=2}^{K} A_k b_k \left(\rho_{1k} + \langle s_k, x_1 \rangle \right) + \sigma \tilde{n}_1, \tag{6.88}$$

where the variance of $\tilde{n}_1$ is $1 + \|x_1\|^2$. The energy of the component orthogonal to s_1 in an arbitrary linear transformation (relative to the energy in the direction of s_1) is called the *surplus energy*. So, for a linear transformation in canonical form $s_1 + x_1$, the surplus energy is $\|x_1\|^2$. The linear transformation with zero surplus energy is the single-user matched filter.

Using (6.88) it is readily shown (Problem 6.24) that the asymptotic multiuser efficiency and signal-to-interference ratio achieved by (6.84) are given by

$$\eta_1 = \frac{\max^2\left[0, 1 - \sum_{k=2}^{K} \frac{A_k}{A_1}|\rho_{1k} + \langle s_k, x_1 \rangle|\right]}{1 + \|x_1\|^2} \tag{6.89}$$

and

$$\text{SIR}_1 = \frac{E[(\langle A_1 b_1 s_1, s_1 + x_1 \rangle)^2]}{E\left[(\langle \sigma n + \sum_{k=2}^{K} A_k b_k s_k, s_1 + x_1 \rangle)^2\right]} \tag{6.90}$$

$$= \frac{A_1^2}{\sigma^2(1 + \|x_1\|^2) + \sum_{k=2}^{K} A_k^2 (\rho_{1k} + \langle s_k, x_1 \rangle)^2}. \tag{6.91}$$

Before we investigate the properties of the canonical representation of the linear MMSE detector let us consider the canonical representation, $s_1 + x_1^d$, of the decorrelating detector. In such case, in addition to being orthogonal to s_1, x_1^d must satisfy

$$\langle s_k, x_1^d \rangle = -\rho_{1k} \tag{6.92}$$

as a result of the decorrelating property. Substituting (6.92) into (6.89) we see that the surplus energy of the decorrelator satisfies

$$\bar{\eta}_1 = \eta_1^d = \frac{1}{1 + \|x_1^d\|^2}. \tag{6.93}$$

Thus, the closer η_1^d is to 0, the larger the surplus energy necessary to tune out the interference. This is illustrated in Figure 6.8 (cf. Figure 5.11) where

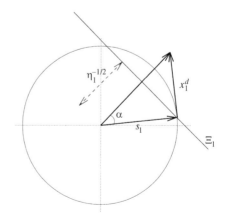

FIGURE 6.8.
Decorrelating detector in canonical form.

(using $\|s_1\| = 1$) it is apparent that

$$\cos^2 \alpha = \bar{\eta}_k,$$

$$\left(1 + \|x_k^d\|^2\right) \cos^2 \alpha = 1.$$

Regarding the canonical representation of the MMSE linear detector, it is easy to show using the Wiener form (6.39) and (6.87) that its orthogonal component is

$$\mathbf{x}_1^* = \frac{E[\mathbf{r}\mathbf{r}^T]^{-1}\mathbf{s}_1}{\mathbf{s}_1^T E[\mathbf{r}\mathbf{r}^T]^{-1}\mathbf{s}_1} - \mathbf{s}_1. \tag{6.94}$$

Although not apparent from (6.94) the orthogonal component of the MMSE linear transformation satisfies a useful property, which we discuss next. We verified in (6.13) that the MMSE linear transformation is the one that maximizes the output signal-to-interference ratio. So we need to choose x_1 that maximizes (6.91). From (6.88) we see that the choice of x_1 does not affect the desired-signal component at the output of the linear transformation. That is why the numerator in (6.91) does not depend on x_1. This is a trivial, but important observation, as it means that maximizing the ratio in (6.90) is equivalent to choosing x_1 to minimize the sum of numerator and denominator therein, that is, the sum of the variance of the desired signal component and the variance of the component due to the background noise plus multiaccess interference. That sum is nothing but the variance of the output of the linear transformation:

$$\mathsf{MOE}(x_1) = E[(\langle y, s_1 + x_1 \rangle)^2]. \tag{6.95}$$

So we can conclude that the x_1 that minimizes MMSE also minimizes MOE. Indeed, *the linear minimum output variance detector*[5] *is the linear MMSE multiuser detector*:

$$E[(A_1 b_1 - \langle y, s_1 + x_1 \rangle)^2] = A_1^2 + \mathsf{MOE}(x_1) - 2A_1^2 \langle s_1, s_1 + x_1 \rangle$$

$$= \mathsf{MOE}(x_1) - A_1^2. \tag{6.96}$$

To arrive at the key relationship (6.96) between minimum output variance and minimum mean-square error we have not used the structure of the CDMA channel model beyond the fact that b_1 is uncorrelated with the interference (multiaccess interference plus background noise). Not only does (6.96) give further motivation for the use of the linear MMSE detector but, more importantly, it has key consequences for its adaptive implementation, as discussed in the next section.

---

[5] Also called *minimum output energy* detector in the literature.

6.6 BLIND MMSE MULTIUSER DETECTION

In the single-user equalization and identification literature, adaptive algorithms that operate without knowledge of the channel input are called *blind*. In this section, we show an adaptive detector that converges to the MMSE detector without requiring training sequences. The knowledge required by the detector presented in this section is identical to the knowledge required by the single-user matched filter receiver, namely, the signature waveform and timing of the desired user.

6.6.1 GRADIENT DESCENT ALGORITHM

The approach we take to self-tune the detector is similar to that of Section 6.4, namely, stochastic gradient descent of a convex penalty function. However, in this case, the penalty function will be the output variance (6.95) instead of the mean-square error. If we were to minimize the output variance over all possible linear transformations, then $c_1 = 0$ would be the trivial (and useless) solution. Instead, we have seen in Section 6.5 that if the linear transformation is assumed to be, without loss of generality, in canonical form, and the output variance is minimized with respect to the component orthogonal to the desired user's signature waveform, then the solution is the MMSE linear transformation. As in Section 6.4 the adaptive algorithm will be derived without recourse to the structure of the multiaccess interference in the basic CDMA channel model other than using the fact that it is uncorrelated with the data of the desired user.

First let us check that the penalty function is strictly convex over the convex set of signals orthogonal to s_1. Using (6.95) we obtain

$$\mathrm{MOE}(\alpha x_1^a + (1-\alpha)x_1^b) = \alpha\,\mathrm{MOE}(x_1^a) + (1-\alpha)\mathrm{MOE}(x_1^b)$$

$$-\alpha(1-\alpha)E\big[\,(\langle y, x_1^a - x_1^b\rangle)^2\,\big] \quad (6.97)$$

$$\leq \alpha\,\mathrm{MOE}(x_1^a) + (1-\alpha)\mathrm{MOE}(x_1^b)$$

$$-\alpha(1-\alpha)\sigma^2\big\|x_1^a - x_1^b\big\|^2, \quad (6.98)$$

where to obtain the inequality we used Proposition 3.5 and the fact that the variance of the sum of uncorrelated random variables is equal to the sum of the variances. Notice from (6.98) that the higher the background noise level, the stricter the convexity of the output variance penalty function.

In order to apply the gradient descent algorithm to the penalty function (6.95), we must take into account that, at every iteration, $x_1[i]$ must be orthogonal to s_1 because we are following (on average) the steepest descent line along the subspace orthogonal to s_1. Projecting the unconstrained gradient on that subspace results in the desired steepest descent line. To check this, simply note that the unconstrained gradient can be decomposed as the sum of its projections along s_1 and its orthogonal subspace; steepest unconstrained descent requires steepest descent along each of those directions. Therefore, let us first take the unconstrained gradient of

$$\text{MOE}(x_1) = E[(\langle y, s_1 + x_1 \rangle)^2],$$

which lies in the same direction as the observed signal:

$$\nabla\text{MOE} = 2\langle y, s_1 + x_1 \rangle y. \tag{6.99}$$

The component in (6.99) orthogonal to s_1 is a scaled version of the component of y orthogonal to s_1:

$$y - \langle y, s_1 \rangle s_1.$$

Therefore, the projected gradient (orthogonal to s_1) is

$$2\langle y, s_1 + x_1 \rangle [y - \langle y, s_1 \rangle s_1]. \tag{6.100}$$

The adaptive algorithm updates proceed at the data rate. The observed waveform $y(t)$ is slotted into waveforms of duration T (cf. (6.69)):

$$\dots, \ y[i-1], \ y[i], \ y[i+1], \dots,$$

and the orthogonal component at the ith iteration, $x_1[i]$, depends on

$$\dots, \ y[i-1], \ y[i].$$

Let us denote the responses of the matched filters for s_1 and $s_1 + x_1[i-1]$ by

$$Z_{\text{MF}}[i] = \langle y[i], s_1 \rangle, \tag{6.101}$$

$$Z[i] = \langle y[i], s_1 + x_1[i-1] \rangle \tag{6.102}$$

respectively. According to (6.100) the stochastic gradient adaptation rule is

$$x_1[i] = x_1[i-1] - \mu Z[i] (y[i] - Z_{\text{MF}}[i]s_1), \tag{6.103}$$

which is depicted in Figure 6.9. In the absence of information about the interfering signature waveforms (or for ease of implementation) the natural choice for the initial condition in (6.103) is

$$x_1[i] = 0.$$

Remarkably, the adaptive algorithm in (6.103) converges to the linear MMSE detector (which achieves maximum near–far resistance) using no

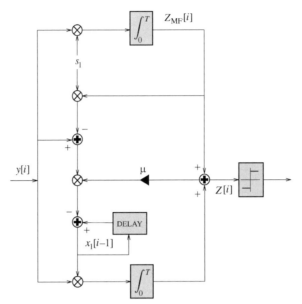

FIGURE 6.9.
Blind adaptive
multiuser detector.

more information than the single-user matched filter (which achieves zero near–far resistance), namely, the desired user's signature waveform and its timing.

In the practical implementation of (6.103), finite precision round-off error may have a cumulative effect that drives the updates in (6.103) well outside of the required orthogonal subspace. This can be remedied by occasionally replacing the update $x_1[i]$ by its orthogonal projection:[6]

$$x_1[i] - \langle x_1[i], s_1 \rangle s_1.$$

Regarding the implementation of the adaptive algorithm (6.103), we can apply verbatim the observations made in Section 6.4:

- Implementation with finite-dimensional vectors rather than continuous-time signals. For the sake of lower computational complexity and improved speed of convergence, it is desirable to use the vector space with lowest dimension that contains the desired and interfering signals. This indicates that knowledge of the signature waveforms of the interferers can be effectively used by the adaptive algorithm.
- Improved convergence speed with more complex recursions.
- Implementation in asynchronous channels.

---

[6] If the orthogonal projection is implemented at every iteration, then the algorithm is similar to that in Frost [98].

Proposition 6.1 can also be applied to the finite-dimensional implementation of (6.103) in order to prove global convergence with $\mu_i = \frac{1}{i}$ with the
following assignments:

$$u[n] = -Z_{\mathrm{MF}}[n], \tag{6.104}$$

$$\mathbf{w}[n] = \mathbf{r}[n] - Z_{\mathrm{MF}}[n]\mathbf{s}_1. \tag{6.105}$$

Proposition 6.1 then shows global convergence of the adaptive orthogonal
component to the solution given in (6.94).

6.6.2 SIGNATURE WAVEFORM MISMATCH

The blind multiuser detector trades the use of training sequences
(Section 6.4) for the knowledge of the desired signature waveform. This
brings up the issue of the robustness of the blind multiuser detector against
mismatch of the desired signature waveform. Mismatch occurs when the
receiver assumes that the desired signature waveform is a *nominal* unit-
energy waveform $\hat{s}_1$, whereas the true received signature waveform is s_1.
Whenever the channel introduces a priori unknown distortion, $\hat{s}_1$ will be
different from s_1, even if measures (such as the adaptive rake receiver,
cf. Section 2.6) are taken to estimate the true received signature waveform.

Let us consider first the benign situation where no interfering users are
present. Assuming

$$\hat{\rho}_1 \stackrel{\text{def}}{=} \langle s_1, \hat{s}_1 \rangle \geq 0,$$

the matched filter for $\hat{s}_1$ achieves error probability

$$P_1^c(\sigma) = Q\left(\frac{A_1 \hat{\rho}_1}{\sigma}\right).$$

It can be shown (Problem 6.40) that the correlator with $\hat{s}_1 + x_1$, where x_1 is
orthogonal to $\hat{s}_1$ and chosen to minimize the output variance, achieves the
following error probability:

$$P_1^o(\sigma) = Q\left(\frac{A_1}{\sigma} \frac{\hat{\rho}_1}{\sqrt{1 + (1 - \hat{\rho}_1^2)\left(\frac{A_1^4}{\sigma^4} + 2\frac{A_1^2}{\sigma^2}\right)}}\right). \tag{6.106}$$

It follows that, in the absence of multiaccess interference, the matched filter
for $\hat{s}_1$ achieves lower error probability than the minimum output variance
linear receiver regardless of the signal-to-noise ratio. Furthermore, for sufficiently *low* background noise level, the argument of the Q-function in
(6.106) is arbitrarily small. Thus, unless we have exact knowledge of the
signature waveform of the desired user, the minimum output variance solution is useless in the high signal-to-noise ratio region. Figure 6.10 exhibits

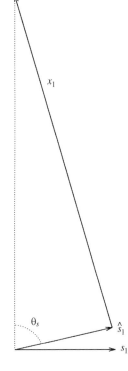

FIGURE 6.10.
Complete
desired-signal
cancellation with
mismatch.

the reason for this ill-behavior. Whenever $\hat{s}_1 \neq s_1$, there is a signal x_1 orthogonal to $\hat{s}_1$ such that the resulting signal $\hat{s}_1 + x_1$ is orthogonal to the desired signal s_1. For asymptotically low background noise, the minimum output variance solution concentrates on canceling s_1 without regard to the resulting surplus energy. We can easily find from Figure 6.10 the surplus energy required for desired signal cancellation in the absence of interferers and background noise:

$$\chi_s = \|x_1\|^2 = \|\hat{s}_1\|^2 \tan^2 \theta_s \qquad (6.107)$$

$$= \frac{\hat{\rho}_1^2}{1 - \hat{\rho}_1^2}, \qquad (6.108)$$

where we have used

$$\hat{\rho}_1 = \langle s_1, \hat{s}_1 \rangle = \sin \theta_s.$$

Therefore, small mismatch ($\hat{\rho}_1 \approx 1$) requires very large surplus energy for the cancellation of the desired signal.

Continuing our discussion of the low background noise region, let us see what happens to the minimum output variance transformation in the presence of interfering users. It is apparent that if $\hat{\rho}_1$ is comparable to $\hat{\rho}_k$, the crosscorrelation between $\hat{s}_1$ and s_k, then the minimum output variance transformation is going to devote comparable resources to suppressing the signal due to user 1 as to suppressing the signal due to user k. In such case, reliable demodulation requires narrowing down the mismatch on s_1. On the other extreme, if there is no desired-signal mismatch, then the interfering users can be tuned out with a surplus energy equal to (6.93)

$$\chi_I = \|x_1\|^2 = \frac{1}{\bar{\eta}_1} - 1. \qquad (6.109)$$

In the common regime where there is some desired-signal mismatch but $\hat{s}_1$ is much closer to s_1 than to any of the interfering waveforms, we can robustify the minimum variance receiver against desired-signal mismatch by *constraining its allowed surplus energy*. How do we choose the allowed maximum surplus energy χ_{max}? To avoid cancellation of the desired signal in the absence of interferers, we must have (6.108)

$$\chi_{max} < \chi_S = \frac{\hat{\rho}_1^2}{1 - \hat{\rho}_1^2};$$

whereas to be able to cancel the multiaccess interference we need (6.109)

$$\chi_{max} > \chi_I = \frac{1}{\bar{\eta}_1} - 1.$$

Thus, achieving both objectives simultaneously requires that the desired-signal mismatch be small enough to satisfy

$$\hat{\rho}_1^2 > 1 - \bar{\eta}_1.$$

In the two-user case, the following result holds (Problem 6.39):

Proposition 6.2 *Let s_1, s_2, and $\hat{s}_1$ be such that*

$$|\langle \hat{s}_1, s_1 \rangle| > |\langle \hat{s}_1, s_2 \rangle|.$$

Then, for every A_1, A_2, there exists a value of surplus energy χ such that the minimum output variance detector has positive asymptotic efficiency.

Let us now turn our attention to the derivation of the minimum output variance solution with constrained surplus energy. It is best to discuss the finite-dimensional version:

$$\min E[(\mathbf{r}^T \mathbf{c}_1)^2]$$

subject to

$$\|\mathbf{c}_1\|^2 \leq 1 + \chi \qquad (6.110)$$

and

$$c_1^T \hat{s}_1 = 1. \qquad (6.111)$$

Obviously, this problem is equivalent to the corresponding minimization with respect to $\hat{x}_1$ with $c_1 = \hat{s}_1 + \hat{x}_1$. We will solve this optimization problem using Lagrange multipliers with cost function

$$E[(r^T c_1)^2] + v_1(\|c_1\|^2 - 1 - \chi) - v_2(c_1^T \hat{s}_1 - 1).$$

Equating the gradient with respect to c_1 to 0, we get

$$E[rr^T]c_1^* + v_1 c_1^* - v_2 \hat{s}_1 = 0,$$

or, equivalently,

$$c_1^* = v_2 [v_1 I + E[rr^T]]^{-1} \hat{s}_1 \qquad (6.112)$$

with

$$E[rr^T] = \sigma^2 I + \sum_{k=1}^{K} A_k^2 s_k s_k^T$$

and the scalar

$$v_2^{-1} = \hat{s}_1^T [v_1 I + E[rr^T]]^{-1} \hat{s}_1$$

chosen so that (6.111) is satisfied. The resulting surplus energy χ ranges from 0 obtained with $v_1 = +\infty$ to the surplus energy of the MMSE solution (6.94) obtained with $v_1 = 0$. Comparing (6.112) to the MMSE solution in (6.39) (corresponding to unconstrained surplus energy), we notice that except for an irrelevant multiplicative factor, both solutions differ in that the *presence of a constraint on surplus energy is equivalent to an additional amount of background noise*. Note that, all other parameters being equal, an additional amount of background noise translates into an MMSE solution that is closer to the single-user matched filter and thus uses less surplus energy.

To derive the adaptive algorithm that converges to the constrained surplus energy detector we can mimic the steps that led to (6.103), except that we now take the gradient of

$$MOE(x_1) + v_1 \|x_1\|^2$$

with respect to x_1, which is

$$2\langle y, \hat{s}_1 + x_1 \rangle y + 2 v_1 x_1.$$

324

Projecting this signal to be orthogonal to $\hat{s}_1$ (cf. (6.100)) we get

$$2\langle y, \hat{s}_1 + x_1\rangle \left[y - \langle y, \hat{s}_1\rangle \hat{s}_1\right] + 2v_1 x_1.$$

Therefore, the blind adaptive rule becomes

$$x_1[i] = (1 - \mu v)x_1[i-1] - \mu Z[i]\left(y[i] - Z_{\text{MF}}[i]\hat{s}_1\right). \tag{6.113}$$

The correspondence between χ and v is complicated enough that its on-line implementation would defeat the simplicity of (6.113). A more reasonable alternative advocated in Honig et al. [156] is to use the unconstrained blind adaptive rule (6.103) starting from $x_1 = 0$, and switch to decision-directed mode before the surplus energy becomes large enough to compromise the output signal-to-noise ratio.

We have seen how to robustify the blind adaptive multiuser detector against nominal signature mismatch. However, it should be emphasized that that remedy makes no attempt to refine the estimate of the nominal desired signature waveform. The minimum output variance receiver lends itself to the task of nominal signal refinement: the closer the assumed nominal is to the true signature waveform the higher is the output variance.

6.7 BIBLIOGRAPHICAL NOTES

The results in Section 6.1 along with an algorithm to compute the K-user linear multiuser detector that maximizes asymptotic efficiency in the synchronous channel appear in Lupas-Golaszewski and Verdú [256] and Lupas and Verdú [254].

Before the inception of multiuser detection spurred by interest on multiaccess communications, there were several works on multiplexing subject to crosstalk that obtained multiple-input multiple-output MMSE linear transformations in the frequency domain (Kaye and George [206]; Nichols et al. [302]; and Salz [376]). The nonadaptive MMSE linear multiuser detector of Section 6.2 is due to Xie et al. [550] and Madhow and Honig [263] and is rederived in a different setting in Rupf et al. [365].

The analysis of the error probability of linear MMSE multiuser detection is due to Poor and Verdú [338]. Other performance results can be found in Oppermann et al. [310] and Honig and Veerakachen [159]. Performance analyses of MMSE linear multiuser detection in conjunction with error-control codes have been reported in Oppermann et al. [310], Kohno et al. [221], Burr [40], Schramm et al. [399], and Verdú and Shamai [504]. Based on the large-K analysis in Section 6.3, Verdú and Shamai

[504] gives the capacity achievable by single-user coding and MMSE demodulation in an equal-power infinite-user channel with random signature waveforms, establishing the marked suboptimality of single-user matched filters in that scenario. Formulas (6.61) and (6.62) are given in Tse and Hanly [454] along with an asymptotic analysis that allows different received powers. Fixing a target signal-to-noise ratio at the output of the MMSE demodulator, Kumar and Holtzman [225] and Ulukus and Yates [458] investigate the required received powers.

Nonadaptive implementations of the MMSE linear multiuser detector using fast recursive methods are studied in Das et al. [62].

Joint MMSE equalization and multiuser detection is analyzed for multitone CDMA in Vandendorpe and van de Wiel [464].

The conventional maximal-ratio method of diversity combining has been modified in Bernstein and Haimovich [28] using the MMSE principle to take into account the presence of multiuser interference.

Stochastic approximation was pioneered in Robbins and Monro [361]. The LMS algorithm was developed in Widrow [530]. General results on the convergence of adaptive linear filters based on stochastic approximation along with applications to data communication problems are given in Gyorfi [130]. Faster alternatives to stochastic gradient descent have received considerable attention (e.g., Falconer and Ljung [86]; Cioffi and Kailath [58]; Slock and Kailath [415]; Carayannis et al. [43]).

An early attempt at adaptive linear multiuser detection can be found in Kashihara [201]. The training-sequence-based adaptive MMSE linear multiuser detector of Section 6.4 was proposed independently in Madhow and Honig [263], Miller [273], and in Rapajic and Vucetic [352]. Its dynamics, including the choice of step size, have been studied in Miller [274]. Techniques for acceleration of convergence have been explored in Rapajic and Vucetic [353], Honig [152, 153], Oppermann and Latva-aho [309], and Wang and Poor [521]. An MMSE adaptive detector that processes the outputs of the decorrelating detector is proposed in Lee [231].

Training-sequence-based joint MMSE demodulation-timing acquisition is proposed in Madhow [259] and in Smith and Miller [416]. The robustification of the MMSE detector against timing errors is explored in Riba et al. [359].

The approach in Section 6.5 of minimizing output energy subject to an orthogonality constraint is due to Honig et al. [156]. It has been generalized in Tsatsanis [450], Schodorf and Williams [395], and Schodorf and Williams [396] to include several linear constraints. In this way, performance and dynamics can be improved by incorporating knowledge about interfering signature waveforms and possibly multipath components of the desired user.

The blind adaptive multiuser detector of Section 6.6 is due to Honig et al. [156]. A related approach that requires knowledge of all signature waveforms appeared concurrently to Honig et al. [155] in Fukawa and Suzuki [99]. The speed of convergence of the algorithm in (6.103) is examined in Roy [363]. The strategy used to robustify the blind algorithm against signal mismatch (Section 6.6.2) is reminiscent of the tap-leakage algorithm (Gitlin et al. [117]; Mayyas and Aboulnasr [268]). Robustification against finite-precision errors of the minimum output variance detector has been pursued in Schodorf and Williams [395] along with the incorporation of knowledge of a subset of interfering signature waveforms. The asynchronous version of the blind adaptive multiuser detector with insensitivity to phase offsets is reported in Gaudenzi et al. [105]. Minimum output variance algorithms subject to linear constraints are common in array processing problems in which the direction of arrival of the desired source is known (Frost [98]; Baird [20]; Johnson and Dudgeon [180]). The blind multiuser detector of Section 6.6 is extended to the setting of array observations with known direction of arrival in Subramanian and Madhow [434]. The implementation of recursive-least-squares-based MMSE blind multiuser detection with systolic arrays is considered in Wang and Poor [521].

Differentially-coherent versions of the blind multiuser detector of Section 6.6 are developed in Honig et al. [158] for frequency-flat fading and in Huang and Verdú [167] and Miller et al. [275] for multipath channels. A blind MMSE detector is obtained in Wong et al. [541] whose observables are obtained by sampling the matched filter of the desired user at a multiple of the chip rate.

Much recent work has focused on the extension of the single-user blind equalization cost-function-based methods such as the constant-modulus algorithm (Johnson [178]; Johnson et al. [179]) to multiple-input channels (e.g., Oda and Sato [305]; Batra and Barry [26]; Lee and Pickholtz [234]; Lee et al. [235]; Castedo et al. [47]; Papadias and Paulraj [313]; Tugnait [455]). Hybrid approaches that use the orthogonally anchored decomposition of Section 6.5 along with the constant-modulus algorithm are explored in Honig [152] and Kamel and Bar-Ness [196].

The blind identification methods of Soon and Tong [417] and Tsatsanis and Giannakis [451] require precoding so as to introduce cyclostationarity and a different amount of correlation in the data modulated by each user. A MUSIC-like algorithm for blind identification of signature waveforms is proposed in Liu and Xu [247].

A blind self-tuning version of the maximum-likelihood sequence detector of Verdú [490] has been proposed in Fonollosa et al. [92], Fonollosa et al. [93] and Fonollosa et al. [94] adapting the (single-user) technique of Seshadri [401] in order to recursively estimate the desired crosscorrelations. Another blind self-tuning maximum-likelihood sequence detector has been reported in Paris [316]. The combined use of second-order and higher-order statistics to perform identification of synchronous users is advocated in Slock [414]. The problem of blind decorrelation is considered in Tong [447] using linear-systems results on the existence of FIR inverses. Joint blind channel identification and data demodulation exploiting the finiteness of the source alphabet is proposed in Veen et al. [481], and Talwar and Paulraj [443].

For the purpose of reducing the number of adaptive coefficients and improved convergence, it is convenient to identify a linear subspace with the smallest possible dimension that contains all desired signature waveforms and all those interfering signatures that reach the receiver with sufficient power. Furthermore, such a subspace naturally leads to an implementation of the decorrelator (Problem 5.16) and the MMSE multiuser detector (Problem 6.12). Several methods for the identification and tracking of the principal directions of the received data have been used for multiuser detection (Haimovich and Bar-Ness [134]; Strom and Miller [429]; Honig [153]; Bar-Ness et al. [22]; Wang and Poor [522]; Lee [232]; Liu and Xu [246]; Torlak and Xu [449]; Liu and Zoltowski [248]; Tsatsanis and Giannakis [453]; Veen et al. [482]; Wang and Poor [522]; Roy [363]). Subspace tracking is studied in Strom et al. [431] and Bensley and Aazhang [27] for Music-based multiuser timing acquisition. Joint blind timing-acquisition and MMSE demodulation is explored in Madhow [260] and Madhow [261].

Surveys of adaptive multiuser detection techniques can be found in Kohno et al. [221], Verdú [500], Honig and Poor [157], and Lim and Roy [242].

Frequency-flat Rayleigh fading has been studied in Sung and Chen [437] for nonadaptive MMSE synchronous multiuser detectors and in Barbosa and Miller [25], Honig et al. [158], Zhu and Madhow [565], and Oppermann and Latvaaho [309] for adaptive MMSE detection.

The principle of MMSE multiuser detection has also been used for simultaneous suppression of narrowband interference and multiaccess interference (Rapajic and Vucetic [354]; Poor and Wang [339]; and Poor and Wang [340]).

Nondecorrelating linear receivers whose goal is the minimization of error probability are studied in Etten [80], Mandayam and Aazhang [264], Psaromiligkos and Batalama [346], and Yeh and Barry [554].

Another nondecorrelating linear detector is proposed in Monk et al. [285] and Davis et al. [64] to mitigate the effects of multiuser interference in direct-sequence spread-spectrum. This approach does not attempt to synchronize or adapt to the multiaccess interference, which is approximated as a stationary process, and the signature sequences of the interferers are assumed to be unknown and to have infinite periodicity. Thus the power spectral density of the multiaccess interference depends exclusively on the chip waveforms (and the number of users and their strength). Incorporating the nonwhiteness of the background noise results in a simple modification of the single-user matched filter where chip matched filtering is done with respect to a pseudochip waveform. A related solution was proposed in Verdú [487] and in Poor and Verdú [337] as the solution to a locally optimum detection problem. The benefits of this approach are limited since no attempt is made to exploit (or learn) the spreading codes of the interferers. A hybrid approach where some interferers are treated as stationary noise while others are synchronized to is studied in Yoon and Leib [557].

6.8 PROBLEMS

PROBLEM 6.1. Let $P_k^o(\sigma)$ be the bit-error-rate for user k of the linear K-user synchronous receiver that maximizes asymptotic efficiency.

(a) Find

$$\Delta = \lim_{\sigma \to \infty} \frac{\sigma}{A_1} \left[P_k^o(\sigma) - P_k^c(\sigma) \right].$$

(b) For which values of ρ is $\Delta = 0$ in the two-user case?

PROBLEM 6.2. Show that the argument that maximizes

$$f\left(x, \rho, \frac{A_2}{A_1}\right) = \frac{1 + x\rho - \frac{A_2}{A_1}|x + \rho|}{\sqrt{1 + 2\rho x + x^2}} \qquad (6.114)$$

is

$$x^* = \begin{cases} -\frac{A_2}{A_1}\mathrm{sgn}\rho, & \text{if } A_2/A_1 < |\rho|, \\ -\rho, & \text{otherwise.} \end{cases} \qquad (6.115)$$

PROBLEM 6.3.

(a) Find the worst asymptotic effective energy, $\omega^o(A_1, A_2)$, of the optimum linear detector in the two-user case.

(b) Verify that the equal-power power loss is the same as for the decorrelator:

$$\frac{1}{A^2}\omega^o(A_1, A_2) = 1 - \rho^2.$$

PROBLEM 6.4. (Varanasi and Aazhang [476]) Find the counterpart to the asymptotic efficiency with an arbitrary linear transformation (6.4) for differentially-coherent demodulation.

PROBLEM 6.5. Let the received process be the sum of a linearly modulated signal and a noise process:

$$y(t) = bs(t) + m(t),$$

where s is deterministic and m has zero mean and is uncorrelated with b.

(a) Show the following correspondence between the minimum mean-square error and the maximum signal-to-interference ratio achievable with a linear transformation:

$$\frac{E[b^2]}{\min_c E[(b - \langle c, y \rangle)^2]} = 1 + \max_c \frac{E[(\langle c, bs \rangle)^2]}{E[(\langle c, m \rangle)^2]}. \qquad (6.116)$$

(b) Show that the argument that achieves the minimum in the left side achieves the maximum in the right side.

(c) (Cioffi et al. [57]) Denote by $\mathcal{C}$ all the linear transformations that satisfy the "unbiasedness" condition:

$$E[\langle c, y \rangle | b] = b.$$

Show that if the minimum in the left-hand side of (6.116) is constrained to $\mathcal{C}$, then

$$\frac{E[b^2]}{\min_{c \in \mathcal{C}} E[(b - \langle c, y \rangle)^2]} = \max_c \frac{E[(\langle c, bs \rangle)^2]}{E[(\langle c, m \rangle)^2]}. \qquad (6.117)$$

PROBLEM 6.6. This problem considers the single-user channel

$$y(t) = A_1 b_1 s(t) + \sigma n(t).$$

(a) Show that the nonlinear estimator that minimizes the mean-square error $E[(b_1 - \tilde{b}_1)^2]$ is

$$\tilde{b}_1 = \tanh\left(\frac{A_1 y_1}{\sigma^2}\right).$$

[*Hint*: Use (4.18).]

(b) Find the linear MMSE estimator, that is,

$$\tilde{b}_1 = c(A_1, \sigma) y_1.$$

(c) Find the MMSE estimator that is constrained to be binary-valued.

(d) Compare the speeds at which the mean-square errors achieved by the above three estimators vanish as a function of $\sigma/A_1 \to 0$.

(e) Compare the speeds at which the mean-square errors achieved by the above three estimators go to 1 as a function of $A_1/\sigma \to 0$.

PROBLEM 6.7. Suppose that the detector knows $A_1, \ldots, A_k, \sigma$, and the timing and signature waveforms of all active users in the channel. Find the (not necessarily linear) detector whose decisions $\hat{b}_k[i] \in \{-1, +1\}$ achieve

$$\min E[(\hat{b}_k[i] - b_k[i])^2].$$

PROBLEM 6.8.

(a) Find the nonlinear estimator that minimizes the mean-square error $E[(b_k - \tilde{b}_k)^2]$ for the synchronous K-user CDMA channel.

(b) Show

$$\frac{\min E[(b_k - \tilde{b}_k)^2]}{P_k(\sigma)} \leq 4$$

with asymptotic equality for $\sigma \to 0$.

PROBLEM 6.9. A synchronous CDMA channel is used by users $2, \ldots, K$ to transmit analog information, such that $b_2[i], \ldots, b_K[i]$ are independent standard Gaussian random variables.

(a) Find the minimum probability of error detector for user 1.
[*Hint*: Use (6.35) and Proposition 3.1 with $m = 2$.]

(b) Find the minimum probability of error for user 1.

(c) Find the estimator that minimizes the mean-square error

$$E[(b_1 - \tilde{b}_1)^2].$$

PROBLEM 6.10. Consider a three-user synchronous CDMA channel with equal-power users.

(a) Find a crosscorrelation matrix $\mathbf{R}$ such that

$$\frac{|([\mathbf{R} + \sigma^2/A^2\mathbf{I}]^{-1}\mathbf{R})_{13}|}{([\mathbf{R} + \sigma^2/A^2\mathbf{I}]^{-1}\mathbf{R})_{11}} > |\rho_{13}|.$$

This shows that the MMSE transformation for user 1 may actually enhance (relative to the single-user matched filter) the interference caused by a particular user.

(b) For the crosscorrelation matrix you selected in (a) verify that

$$\frac{([\mathbf{R} + \sigma^2/A^2\mathbf{I}]^{-1}\mathbf{R})_{12}^2}{([\mathbf{R} + \sigma^2/A^2\mathbf{I}]^{-1}\mathbf{R})_{11}^2} + \frac{([\mathbf{R} + \sigma^2/A^2\mathbf{I}]^{-1}\mathbf{R})_{13}^2}{([\mathbf{R} + \sigma^2/A^2\mathbf{I}]^{-1}\mathbf{R})_{11}^2} < \rho_{12}^2 + \rho_{13}^2.$$

PROBLEM 6.11. Suppose that user 1 is the user of interest. Let

$$\Sigma = \sigma^2\mathbf{I} + \sum_{k=2}^{K} A_k^2 \mathbf{s}_k \mathbf{s}_k^T.$$

Using the expressions found in (6.39) and (6.40) and the matrix inversion lemma (4.99), show the following expressions for the MMSE linear transformation, the minimum mean-square error, and the maximum signal-to-interference ratio at the output of a linear transformation:

$$\mathbf{v}^* = \frac{A_1}{1 + A_1^2 \mathbf{s}_1^T \Sigma^{-1} \mathbf{s}_1} \Sigma^{-1} \mathbf{s}_1, \qquad (6.118)$$

$$E[(b_1 - \mathbf{r}^T \mathbf{v}^*)^2] = \frac{1}{1 + A_1^2 \mathbf{s}_1^T \Sigma^{-1} \mathbf{s}_1}, \qquad (6.119)$$

$$\mathrm{SIR}_1 = A_1^2 \mathbf{s}_1^T \Sigma^{-1} \mathbf{s}_1, \qquad (6.120)$$

respectively.

PROBLEM 6.12. Let $\{\mathbf{e}_1, \ldots, \mathbf{e}_L\}$ and $\{\lambda_1, \ldots, \lambda_L\}$ denote the orthogonal eigenvectors and eigenvalues of the matrix

$$E[\mathbf{r}\mathbf{r}^T] = \left[\sigma^2\mathbf{I} + \sum_{k=1}^{K} A_k^2 \mathbf{s}_k \mathbf{s}_k^T\right].$$

(a) Use the Wiener characterization to show that the MMSE linear transformation for the kth user can be expressed as

$$\mathbf{v}_k^* = \sum_{l=1}^{L} \alpha_{kl} \mathbf{e}_l$$

with

$$\alpha_{kl} = \frac{1}{\lambda_l} \mathbf{e}_l^T \mathbf{s}_k.$$

(b) Verify that

$$\mathbf{v}_k^* = \mathbf{E}\, \mathrm{diag}^{-1}\{\lambda_1, \ldots, \lambda_L\}\, \mathbf{E}^T \mathbf{s}_k,$$

where

$$\mathbf{E} = [\mathbf{e}_1 \cdots \mathbf{e}_L].$$

(c) Show that at most K coefficients $\alpha_{k1}, \ldots, \alpha_{kL}$ are nonzero.

(d) Suppose that the L eigenvalues and eigenvectors of the matrix

$$\frac{1}{n} \sum_{i=1}^{n} \mathbf{r}[i] \mathbf{r}^T[i]$$

are available at each n. Suggest a blind adaptive MMSE algorithm that uses knowledge of the desired signature waveform.

PROBLEM 6.13. Find the synchronous MMSE linear transformation for user 1 in the special case in which $\rho_{ij} = 0$ for $i = 2, \ldots, K$ and $j = 2, \ldots, K$.

PROBLEM 6.14. Motivated by the approximate decorrelator of Section 5.4, consider the linear transformation

$$c_k(t) = \beta_k s_k(t) - \alpha_k \sum_{j \neq k} \rho_{kj} s_j(t).$$

Choose α_k and β_k to minimize the mean-square error:

$$E[(b_k - \langle c_k, y \rangle)^2].$$

PROBLEM 6.15. (Moshavi et al. [289]) Find the coefficients α_j such that

$$[\mathbf{R} + \sigma^{-2}\mathbf{I}]^{-1} = \sum_{j=0}^{K-1} \alpha_j \mathbf{R}^j.$$

PROBLEM 6.16. Consider the two-user synchronous channel and the detector

$$\hat{b}_1 = \text{sgn}(\alpha(A_1, A_2, \sigma) y_1 - \rho y_2). \qquad (6.121)$$

(a) Find an equation that gives the value $\alpha^*(A_1, A_2, \sigma)$ that minimizes error probability.

(b) Give an explicit expression for

$$\lim_{\sigma \to 0} \alpha^*(A_1, A_2, \sigma).$$

(c) Show that as long as $A_1 > A_2 \rho$ the ratio of the bit-error-rate of the MMSE linear transformation to the minimum linear bit-error-rate can be made as large as desired with sufficiently low σ.

(d) (Mandayam and Aazhang [264]) Find a gradient descent algorithm with a penalty function whose global minimum occurs at the detector found in (a). Can you guarantee that that penalty function does not have local minima?

PROBLEM 6.17. Show that the MMSE multiuser detector for a synchronous CDMA channel employing Welch-bound-equality signature vectors (Problem 2.41) is equal to the bank of single-user matched filters if $A_1 = \cdots = A_K$.

PROBLEM 6.18. Consider a K-user synchronous CDMA direct-sequence spread-spectrum system with spreading factor N,

$$K = 2^{N-1}$$

and no pair of users (i, j) is assigned signature waveforms such that $s_i = \pm s_j$. Assume that all users have equal received signal-to-noise ratios A/σ.

(a) (Zaidel et al. [558]) Show that the MMSE linear detector is the single-user matched filter.
(b) Find the bit-error-rate as a function of A/σ and N.
(c) Find the asymptotic efficiency as a function of $N = 1, 2, \ldots$.

PROBLEM 6.19. Using the expression for $\mathcal{F}$ in (6.52) show that for all $0 < \delta < a$

$$\frac{1}{4}\mathcal{F}\left(a, \left(\frac{1}{\delta}+1\right)\left(1-\frac{\delta}{a}\right)\right) = a - \delta.$$

PROBLEM 6.20. Consider a K-user synchronous channel with linearly independent signature waveforms. Verify that if $\sigma = 0$, the Wiener characterization of the linear MMSE detector (6.39) is a decorrelating detector.

PROBLEM 6.21. We saw in Section 6.3 that for direct-sequence spread-spectrum systems with K equal-power users, N chips per bit, and randomly chosen signature waveforms, each diagonal element of the matrix

$$\left[\mathbf{I} + \frac{A^2}{\sigma^2}\mathbf{R}\right]^{-1}$$

converges in mean-square sense (as $K \to \infty$) to a constant. Show that the off-diagonal elements do not converge to a constant.

PROBLEM 6.22. Show that the sum of mean-square errors achieved by the asynchronous MMSE multiuser detector is given by

$$\frac{1}{\pi}\int_0^\pi \text{trace}\left\{(\mathbf{I} + \sigma^{-2}\mathbf{A}(\mathbf{R}^T[1]e^{j\omega} + \mathbf{R}[0] + \mathbf{R}[1]e^{-j\omega})\mathbf{A})^{-1}\right\}d\omega.$$

$$(6.122)$$

FIGURE 6.11.
Multiple-input
multiple-output
channel.

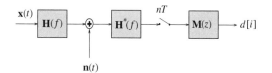

PROBLEM 6.23. (Salz [376]) Figure 6.11 depicts a dispersive channel where $\mathbf{H}(f)$ is the matrix transfer function of a K-input K-output channel,

$$\mathbf{x}(t) = \sum_{i=-\infty}^{\infty} \mathbf{b}[i]\delta(t - iT),$$

the K components of $\mathbf{n}(t)$ are independent white noise processes with spectral level σ^2, and $\mathbf{b}[i] \in \{-1, +1\}^K$ are the usual independent equally likely data bits. Show that the discrete-time filter that minimizes the mean-square error between its output and $\mathbf{b}[i]$ is

$$\mathbf{M}(z) = [\sigma^2\mathbf{I} + \mathbf{G}(z)]^{-1}, \tag{6.123}$$

where

$$\mathbf{G}(e^{j2\pi fT}) = \frac{1}{T} \sum_{i=-\infty}^{\infty} \left|\mathbf{H}\left(f - \frac{i}{T}\right)\right|^2.$$

PROBLEM 6.24. Show that a synchronous linear multiuser detector given in canonical form

$$c_1 = s_1 + x_1, \tag{6.124}$$

$$\langle s_1, x_1 \rangle = 0, \tag{6.125}$$

achieves the following asymptotic multiuser efficiency and signal-to-interference ratio, respectively:

$$\eta_1 = \frac{\max^2 \left[0, 1 - \sum_{k=2}^{K} \frac{A_k}{A_1}|\rho_{1k} + \langle s_k, x_1 \rangle|\right]}{1 + \|x_1\|^2}, \tag{6.126}$$

$$\text{SIR}_1 = \left(\frac{\sigma^2}{A_1^2}(1 + \|x_1\|^2) + \sum_{k=2}^{K} \frac{A_k^2}{A_1^2}(\rho_{1k} + \langle s_k, x_1 \rangle)^2\right)^{-1}. \tag{6.127}$$

Generalize those expressions to the asynchronous channel.

PROBLEM 6.25. Show that the surplus energy of the decorrelator is greater than or equal to the surplus energy of the linear MMSE detector.

PROBLEM 6.26.

(a) Verify that the orthogonal component of the canonical representation of the MMSE linear detector for the K-user synchronous channel is given by (6.94).

(b) Show that in the two-user case

$$x_1^m = \frac{\rho^2}{1 + \sigma^2/A_2^2 - \rho^2} s_1 - \frac{\rho}{1 + \sigma^2/A_2^2 - \rho^2} s_2.$$

PROBLEM 6.27. (Park and Doherty [319]) Find x_1 that minimizes the functional

$$J(x_1) = (\langle x_1, s_1 \rangle)^2 + E[(A_1 - \langle y, s_1 + x_1 \rangle)^2]$$

for the two-user synchronous channel.

PROBLEM 6.28. Verify the asymptotic result in (6.53).

PROBLEM 6.29. Minimize the mean output variance with respect to an L-dimensional orthogonal component $\mathbf{x}_1$ in the presence of mismatch,

$$\text{MOE}(\mathbf{x}_1) = E[(\mathbf{r}^T (\hat{\mathbf{s}}_1 + \mathbf{x}_1))^2],$$

and constrained surplus energy,

$$\|\mathbf{x}_1\|^2 = \chi.$$

PROBLEM 6.30. Particularize (6.47) to the two-user case. Show that if $|\rho| < 1$, $P_1^m(\sigma)$ is monotonically decreasing with σ.

PROBLEM 6.31. Show that the Gaussian approximation to the MMSE probability of error (6.48) is upper bounded by the error probability of the decorrelating detector.

PROBLEM 6.32. (Poor and Verdú [338]) Consider a K-user synchronous channel with crosscorrelations $\rho_{jk} = \rho$ for all $j \neq k$ and

$$A_1 = \cdots = A_K = A.$$

(a) Show that the leakage coefficients for the MMSE detector are given by

$$\beta_k = \frac{\rho}{1 + \frac{A^2}{\sigma^2}[1 + (K - 2)\rho - (K - 1)\rho^2]}.$$

(b) Show that the bit-error-rate, $P_k^m(\sigma)$, achieved by the MMSE detector is equal to

$$2^{1-K} \sum_{n=0}^{K-1} \binom{K-1}{n}$$

$$\times Q \left(\frac{A}{\sigma} \frac{1 + \frac{\sigma^2}{A^2} + (K-2)\rho - (K-1)\rho^2 + \frac{\sigma^2}{A^2}\rho(K-1-2n)}{\sqrt{\left(1 + \frac{\sigma^2}{A^2} + (K-2)\rho\right)^2 - \rho^2(K-1)\left(1 + 2\frac{\sigma^2}{A^2} + (K-2)\rho\right)}} \right).$$

(c) Assume that $\rho \geq 0$. Show that

$$\lim_{K \to \infty} P_k^m(\sigma) = Q\left(\frac{A}{\sigma}\sqrt{1-\rho}\right).$$

PROBLEM 6.33. (Poor and Verdú [338]) Consider two synchronous users with

$$|\rho| \leq \frac{1}{2}\sqrt{2 + \sqrt{3}}.$$

Show that for all $\sigma > 0$

$$P_k^m(\sigma) \leq P_k^d(\sigma).$$

PROBLEM 6.34. (Poor and Verdú [338]) Define the *non-Gaussianness* of a probability density function f_X with mean m and standard deviation v as the divergence between f_X and a Gaussian distribution with the same mean and standard deviation, that is,

$$D(f_X \| \mathcal{N}(m, v^2)) = \int_{-\infty}^{\infty} f_X(x) \log\left(\sqrt{2\pi} v f_X(x) e^{(x-m)^2/2v^2}\right) dx.$$

Consider the two-user synchronous case and let X be the contribution to the output of the MMSE linear transformation due to the background Gaussian noise and the interferer. Show that the non-Gaussianness of X is maximized when

$$\frac{\sigma^4}{A_2^4} = 1 - \rho^2.$$

PROBLEM 6.35. Consider the adaptive linear MMSE detector for user 1 in a synchronous K-user channel. Assume that the algorithm is initialized with s_1 and that all users employ the same training sequence simultaneously. Find the linear transformation to which the adaptation law converges.

PROBLEM 6.36. Consider the adaptive MMSE detector with training sequences:

$$\mathbf{v}_1[n] = \mathbf{v}_1[n-1] - \mu\left(\mathbf{v}_1^T[n-1]\mathbf{r}[n] - b_1[n]\right)\mathbf{r}[n]. \qquad (6.128)$$

Assume that $L = K = 2$. Optimize the choice of μ to maximize the speed of convergence of $E[\mathbf{v}_1[n]]$.

PROBLEM 6.37. Show that the blind detector for user 1 in an L-dimensional setting,

$$\mathbf{x}_1[i] = \mathbf{x}_1[i-1] - \mu Z[i]\,(\mathbf{r}[i] - Z_{\mathrm{MF}}[i]\mathbf{s}_1),\qquad(6.129)$$

satisfies

$$E[\mathbf{s}_1 + \mathbf{x}_1[i]] - \mathbf{c}_1^*$$
$$= \left[\mathbf{I} - \mu\left[\mathbf{I} - \mathbf{s}_1\mathbf{s}_1^T\right]E[\mathbf{r}\mathbf{r}^T]\right]\left[E[\mathbf{s}_1 + \mathbf{x}_1[i-1]] - \mathbf{c}_1^*\right],$$

where $\mathbf{c}_1^*$ is the linear MMSE solution for user 1.

PROBLEM 6.38. (Ulukus and Yates [457]) Assume that $\mathbf{s}_1$ is not spanned by the other signature vectors. Consider the following blind adaptive algorithm:

$$\mathbf{c}_1[i] = (1 + \sigma^2\mu)\mathbf{c}_1[i-1] - \mu\,(Z[i]\mathbf{r}[i] - \mathbf{s}_1).\qquad(6.130)$$

Let $\tilde{\mathbf{c}}_1$ be the solution to the equation

$$\mathbf{SA}^2\mathbf{S}^T\tilde{\mathbf{c}}_1 = \mathbf{s}_1,$$

which according to Problem 5.40 is guaranteed to exist and be a decorrelator for user 1. Show that

$$E[\mathbf{c}_1[i]] - \tilde{\mathbf{c}}_1 = (\mathbf{I} - \mu\mathbf{SA}^2\mathbf{S}^T)E[\mathbf{c}_1[i-1] - \tilde{\mathbf{c}}_1].$$

PROBLEM 6.39. Prove Proposition 6.2. [*Hint:* Positive asymptotic efficiency is obtained if and only if

$$A_1^2(\langle s_1, \hat{s}_1 + x_1^*\rangle)^2 > A_2^2(\langle s_2, \hat{s}_1 + x_1^*\rangle)^2,$$

where x_1^* is the signal orthogonal to $\hat{s}_1$ that minimizes

$$A_1^2(\langle s_1, \hat{s}_1 + x_1\rangle)^2 + A_2^2(\langle s_2, \hat{s}_1 + x_1\rangle)^2 + \nu\|x_1\|^2.$$

Work on a three-dimensional space with $\hat{s}_1 = [1\ 0\ 0]^T$.]

PROBLEM 6.40. Consider a single-user channel

$$y(t) = A_1 b_1 s(t) + \sigma n(t),$$

which is demodulated by a mismatched receiver that uses the linear transformation $\hat{s}_1 + x_1$ with x_1 orthogonal to $\hat{s}_1$ and chosen to minimize $E[(\langle y, \hat{s}_1 + x_1\rangle)^2]$. Denote

$$\hat{\rho}_1 = \langle s_1, \hat{s}_1\rangle$$

and assume

$$0 \le \hat{\rho}_1 \le 1.$$

(a) Show that the error probability is

$$Q\left(\frac{A_1}{\sigma}\frac{\langle s_1, \hat{s}_1\rangle}{\sqrt{1 + (1 - \langle s_1, \hat{s}_1\rangle^2)\left(\frac{A_1^4}{\sigma^4} + 2\frac{A_1^2}{\sigma^2}\right)}}\right).$$

(b) Find the noise level that minimizes error probability.
(c) Find the error probability if x_1 is chosen to minimize the output variance under the constraint:

$$\|x_1\|^2 = \chi.$$

(d) Find the optimum surplus energy χ as a function of $\hat{\rho}_1$ and the signal-to-noise ratio.

PROBLEM 6.41. Consider a K-user synchronous CDMA channel and let x_1 be a unit-energy signal orthogonal to s_1. Find the scalar α^* that maximizes the signal-to-interference ratio at the output of the correlator $\langle y, s_1 + \alpha x_1\rangle$ (cf. Pados and Batalama [312]).

PROBLEM 6.42. Assume that the optimum near–far resistance of user 1 satisfies

$$\bar{\eta}_1 > 0.$$

Consider a linear multiuser detector for user 1 given by the linear transformation $d_1 + x_1$, where d_1 is the decorrelating transformation for user 1, and x_1 is orthogonal to s_1 and minimizes

$$E[(\langle d_1 + x_1, y\rangle)^2].$$

Show that the resulting linear detector is the MMSE linear transformation.

PROBLEM 6.43. (Schodorf and Williams [395]) Let $\mathbf{c}_0$ be an L-dimensional vector that achieves

$$\min E[(\mathbf{c}^T \mathbf{r})^2]$$

subject to P linear constraints represented by the $P \times L$ matrix $\mathbf{D}$ and the D-vector $\mathbf{f}$:

$$\mathbf{Dc} = \mathbf{f}.$$

Assume that $P < L$.

(a) Show that

$$\mathbf{c}_0 = E[\mathbf{rr}^T]^{-1}\mathbf{D}^T(\mathbf{D}E[\mathbf{rr}^T]^{-1}\mathbf{D}^T)^{-1}\mathbf{f}.$$

(b) Design a blind algorithm (that uses independent observations of $\mathbf{r}$ but has no knowledge of its covariance matrix) to obtain $\mathbf{c}_0$. *Note*: The approach used in Section 6.6 based on the canonical representation of Section 6.5, can be generalized to the present case by letting $\mathbf{M}$ be an $(L - P) \times L$ matrix such that $\mathbf{M}\mathbf{D}^T = \mathbf{0}$ and adapting the $(L - P)$-vector $\mathbf{w}$ in

$$\mathbf{c} = \mathbf{D}^+\mathbf{f} + \mathbf{M}^T\mathbf{w}.$$

(c) Propose a choice of design constraints $(\mathbf{D}, \ \mathbf{f})$ in the following particular cases:

- The receiver knows $\mathbf{s}_1$, the signature vector of the desired user, but does not know the signatures of the interferers.
- The receiver know $\mathbf{s}_1$, the signature vector of the desired user, and $\mathbf{s}_2$, the signature vector of one of the interferers.
- The receiver does not know $\mathbf{s}_1$ but it knowns that $\mathbf{s}_1$ belongs to the space spanned by $\mathbf{g}_1, \mathbf{g}_2$.

PROBLEM 6.44. Consider a two-user synchronous channel with orthogonal signature waveforms s_1 and s_2 and a blind multiuser detector for user 1 with a mismatched nominal equal to

$$\hat{s}_1 = \frac{s_1 + s_2}{\sqrt{2}}.$$

Suppose that for every value of A_2/A_1 the receiver has the ability to choose the value of surplus energy that maximizes η_1. Derive the maximum asymptotic efficiency as a function of A_2/A_1.

PROBLEM 6.45. (*Least Squares*) In this problem we consider a method to obtain a minimum output variance detector that makes no statistical assumptions about the received data. Show that the vector $\mathbf{c}^*[n]$ that minimizes

$$\sum_{j=1}^{n}(\mathbf{c}^T\mathbf{y}[j])^2 \tag{6.131}$$

subject to $\mathbf{c}^T\mathbf{s}_1 = 1$ and $\|\mathbf{c}\|^2 = 1 + \chi$ has the form

$$\mathbf{c}^*[n] = \alpha[n]\hat{\mathbf{R}}_1^{-1}[n]\mathbf{s}_1, \tag{6.132}$$

where

$$\hat{\mathbf{R}}_1[n] = \sum_{j=1}^{n}\mathbf{y}[j]\mathbf{y}^T[j] + \nu\mathbf{I} \tag{6.133}$$

and

$$\alpha[n] = \left(\mathbf{s}_1^T\hat{\mathbf{R}}_1^{-1}[n]\mathbf{s}_1\right)^{-1}.$$

PROBLEM 6.46. The obvious drawback of the method in Problem 6.45 is that the matrix inverse in (6.133) has to be recomputed at each iteration. In this problem we exhibit the method of *recursive least squares*, whereby this drawback is eliminated. Modify the criterion in (6.131) to include a *forgetting factor*, λ, that weighs recent observations more heavily:

$$\sum_{j=1}^{n} \lambda^{n-j} (\mathbf{c}^T \mathbf{y}[j])^2 \qquad (6.134)$$

under the same constraints as in Problem 4.46. Let

$$\hat{\mathbf{R}}_\lambda^{-1}[n] = \sum_{j=1}^{n} \lambda^{n-j} \mathbf{y}[j] \mathbf{y}^T[j] + \nu \mathbf{I}.$$

(a) Show that a solution akin to (6.132) holds in this case.
(b) Show the following recursive equation:

$$\hat{\mathbf{R}}_\lambda^{-1}[n] = \lambda^{-1} \hat{\mathbf{R}}_\lambda^{-1}[n-1] - \lambda^{-1} \beta[n]$$
$$\times \hat{\mathbf{R}}_\lambda^{-1}[n-1] \mathbf{y}[n] \mathbf{y}^T[n] \hat{\mathbf{R}}_\lambda^{-1}[n-1]$$

with

$$\beta[n] = \left(\lambda + \mathbf{y}^T[n] \hat{\mathbf{R}}_\lambda^{-1}[n-1] \mathbf{y}[n] \right)^{-1}.$$

[*Hint*: Use the matrix inversion lemma (4.99)]

PROBLEM 6.47. Derive a recursive least squares adaptive algorithm for MMSE linear detection with training sequences.

PROBLEM 6.48. (Chen and Roy [51]) Derive a recursive least squares adaptive algorithm that converges to the decorrelating detector. [*Hint*: Incorporate a forgetting factor in (5.17).]

PROBLEM 6.49. Consider a synchronous two-user direct-sequence spread-spectrum system, such that the signature waveforms $\{s_1, s_2\}$ are not known to the receiver.

(a) Find the unit-energy linear transformation e_1^* that maximizes the output variance

$$E\left[\left(\langle \sum_{k=1}^{2} A_k b_k s_k + \sigma n, e_1 \rangle \right)^2 \right].$$

(b) Find the unit-energy linear transformation e_2^* that maximizes the output variance subject to the constraint

$$\langle e_1^*, e_2^* \rangle = 0.$$

(c) Find an adaptive algorithm that converges to the maximum variance signals e_1^* and e_2^*. What is the usefulness of such an algorithm?

PROBLEM 6.50. (Bar-Ness and Punt [23]) For the purposes of this problem, if $\mathbf{a}$ is a K-vector, then $\mathbf{a}_k$ denotes the $(K-1)$-vector where the kth component of $\mathbf{a}$ has been removed.

For any $\mathbf{w}$ and the usual matched filter bank output vector $\mathbf{y}$ denote

$$z_k = y_k - \mathbf{w}_k^T \mathbf{y}_k.$$

Consider the following nondecorrelating linear detector for user l:

$$\hat{b}_l = \text{sgn}(z_l)$$

where $\mathbf{w}_l$ is chosen so that

$$E\left[z_l \hat{b}_j^d\right] = 0, \quad j \neq l,$$

with the output of the decorrelating detector for user j denoted by

$$\hat{b}_j^d = \text{sgn}((\mathbf{R}^{-1}\mathbf{y})_j).$$

(a) Show that $\mathbf{w}_l \rightarrow [0 \cdots 0]$ as $\sigma \rightarrow \infty$. Therefore, this detector is indeed nondecorrelating.
(b) Show that as $\sigma \rightarrow 0$ this detector becomes the decorrelating detector.
(c) Find $\mathbf{w}_1$ in the case $K = 2$.
(d) Find the probability of error of this receiver in the two-user case.
(e) The *bootstrap* adaptive algorithm computes the sequence

$$\mathbf{w}_k[i + 1] = \mathbf{w}_k[i] + \mu z_k[i]\text{sgn}(\mathbf{z}_k[i]),$$

where the sgn of a vector is the vector of the signs of the individual components. Discuss the rationale for this adaptive rule.

PROBLEM 6.51. This problem considers the application of the Constant Modulus Algorithm (Johnson et al. [179]) to blind multiuser detection. Consider the following cost function:

$$\Xi(\mathbf{v}) = E[((\mathbf{v}^T\mathbf{r})^2 - 1)^2], \qquad (6.135)$$

where

$$\mathbf{r} = \mathbf{SAb} + \sigma\mathbf{m}.$$

(a) Obtain an adaptive linear detector by stochastic approximation of (6.135).

(b) Show that, in the absence of noise ($\sigma = 0$), $\mathbf{v}^*$ is a global minimum of $\Xi(\mathbf{v})$, if and only if there is a user k, such that

$$A_k \mathbf{s}_k^T \mathbf{v}^* = \pm 1, \qquad (6.136)$$

$$\mathbf{s}_j^T \mathbf{v}^* = 0, \quad j \neq k, \qquad (6.137)$$

that is, assuming the signature waveforms are linearly independent there are $2K$ global minima, placed at the decorrelating transformations.

(c) Show that, in the two-user case, the stationary points (zero-gradient points) of $\Xi(\mathbf{v})$ are (assuming $\sigma = 0$):

$\pm\left(\frac{1}{2(1+\rho)}, \frac{1}{2(1+\rho)}\right)$	saddle point
$\pm\left(\frac{1}{2(1-\rho)}, \frac{-1}{2(1-\rho)}\right)$	saddle point
$(0, 0)$	local maximum
$\pm\left(\frac{1}{1-\rho^2}, \frac{-\rho}{1-\rho^2}\right)$	global minimum
$\pm\left(\frac{-\rho}{1-\rho^2}, \frac{1}{1-\rho^2}\right)$	global minimum

CHAPTER SEVEN

DECISION-DRIVEN MULTIUSER DETECTORS

A number of works in multiuser detection have proposed nonlinear detectors that use decisions on the bits of interfering users in the demodulation of the bit of interest. Some of those solutions use final decisions, while other proposed solutions employ tentative decisions used only internally by the demodulator. Although the bit-error-rate analysis of decision-driven detectors does not have the rich structure of the detectors discussed in previous chapters, it is an instructive exercise in the analysis of nonlinear decision regions. The multiuser detectors in this chapter are bootstrapping techniques that require at least a modicum of reliability for their initial decisions. They are particularly suited to high signal-to-noise ratio channels with power imbalances.

7.1   SUCCESSIVE CANCELLATION

We encountered the technique of successive cancellation (also known as *stripping* or *successive decoding*[1]) in Section 1.4 (cf. Figure 1.8). This approach is based on a simple and natural idea: if a decision has been made about an interfering user's bit, then that interfering signal can be recreated at the receiver and subtracted from the received waveform. This will cancel the interfering signal provided that the decision was correct; otherwise it will double the contribution of the interferer. Once the subtraction has taken place, the receiver takes the optimistic view that the resulting

---

[1] The culinary term *onion peeling* is also used.

344

signal contains one fewer user and the process can be repeated with another interferer, until all but one user have been demodulated. In order to fully describe the receiver we just need to specify how the intermediate decisions are obtained. In its simplest form, successive cancellation uses decisions produced by single-user matched filters, which neglect the presence of interference. Since erroneous intermediate decisions affect the reliability of all successive decisions, the order in which users are demodulated affects performance.

A popular approach is to demodulate the users in the order of decreasing received powers. However, this is not necessarily best (Problem 7.7) since it fails to take into account the crosscorrelations among users. A sensible alternative is to order users according to

$$E\left[\left(\int_0^T y(t)s_k(t)\,dt\right)^2\right] = \sigma^2 + A_k^2 + \sum_{j\neq k} A_j^2\rho_{jk}^2, \qquad (7.1)$$

which can be estimated easily from the matched filter outputs.

As usual, let us consider the synchronous two-user channel first. Suppose that user 2 is demodulated by its matched filter:

$$\hat{b}_2 = \text{sgn}(y_2).$$

Remodulating the signal of user 2 with $\hat{b}_2$ we get $A_2\hat{b}_2 s_2(t)$, which upon subtraction from the received signal yields

$$\hat{y}(t) = y(t) - A_2\hat{b}_2 s_2(t) \qquad (7.2)$$

$$= A_1 b_1 s_1(t) + A_2(b_2 - \hat{b}_2)s_2(t) + n(t). \qquad (7.3)$$

Processing $\hat{y}$ with the matched filter for s_1, we obtain the decision:

$$\hat{b}_1 = \text{sgn}(\langle \hat{y}, s_1 \rangle) \qquad (7.4)$$

$$= \text{sgn}(y_1 - A_2\hat{b}_2\rho) \qquad (7.5)$$

$$= \text{sgn}(y_1 - A_2\rho\,\text{sgn}(y_2)) \qquad (7.6)$$

$$= \text{sgn}(A_1 b_1 + A_2(b_2 - \hat{b}_2)\rho + \sigma\,\langle n, s_1 \rangle). \qquad (7.7)$$

The two-user successive cancellation detector is depicted in Figure 7.1. Using (7.5) we obtain the equivalent implementation in Figure 7.2. It is interesting to compare the decision rule (7.5) with that of the decorrelating

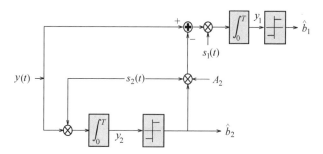

FIGURE 7.1.
Successive
cancellation for
two synchronous
users.

detector:

$$\hat{b}_1 = \text{sgn}(y_1 - \rho y_2).$$

In contrast to the successive canceler, the decorrelating detector truly succeeds in canceling the component of the decision statistic due to user 2. As
we saw in Chapter 5, such complete cancellation comes at the expense of
attenuating the component of the desired user.

The successive-cancellation decision regions for $\rho = 1/2$ are shown in
the cases:

- Figure 7.3: $A_1 = 1$; $A_2 = 1$.
- Figure 7.4: $A_1 = 0.5$; $A_2 = 1$.
- Figure 7.5: $A_1 = 2.5$; $A_2 = 1$.

In all cases, the decision regions are identical because the decision rule
does not depend on A_1. For comparison purposes the minimum-distance
decision region (which results in jointly optimum and maximum-likelihood
decisions) is shown in Figure 7.6 in the case $A_1 = A_2 = 1$.

We see in Figure 7.5 that if the weak user is detected first, then the
receiver may make errors even in the absence of background noise because
some hypotheses have migrated outside their own decision regions.

In K-user successive cancellation, the received signal is stripped away
of each of the interfering waveforms one at a time. When making a decision
about the kth user we assume that the decisions of users $k + 1, \ldots, K$ are

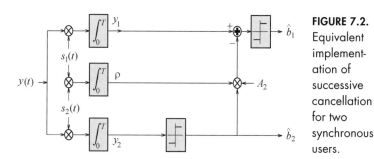

FIGURE 7.2.
Equivalent
implement-
ation of
successive
cancellation
for two
synchronous
users.

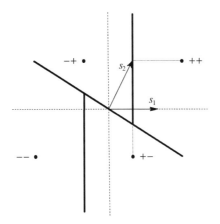

FIGURE 7.3.
Decision regions
of successive
cancellation with
$A_1 = 1$ and
$A_2 = 1$.

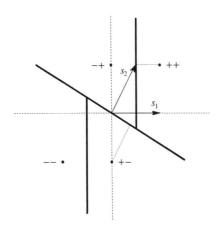

FIGURE 7.4.
Decision regions
of successive
cancellation with
$A_1 = 0.5$ and
$A_2 = 1$.

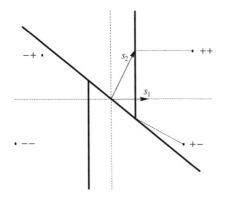

FIGURE 7.5.
Decision regions
of successive
cancellation with
$A_1 = 2.5$ and
$A_2 = 1$.

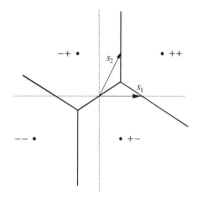

FIGURE 7.6.
Decision regions
of maximum-
likelihood
detection with
$A_1 = A_2 = 1$.

correct and we neglect the presence of users $1, \ldots, k - 1$. Therefore,

$$\hat{b}_k = \mathrm{sgn}\left(y_k - \sum_{j=k+1}^{K} A_j \rho_{jk} \hat{b}_j \right). \qquad (7.8)$$

Successive cancellation generalizes fairly easily to the asynchronous channel. Up to now, we had assumed when convenient that the users are numbered so that their offsets are increasing, and, thus, the jth user's bit that overlaps with $b_k[i]$ on the right side is $b_j[i]$ if $j > k$. However, in the current setting it is convenient to number the users in the (inverse) order they are canceled, which is, normally, dictated by their relative received powers or matched filter output powers (7.1). In order to describe the stripping detector in the asynchronous case it is useful to introduce the following notation:

$$\delta_{kj} = 1\{\tau_k < \tau_j\}.$$

Then $b_k[i]$ overlaps on the right side with $b_j[i - \delta_{kj} + 1]$ and overlaps on the left side with $b_j[i - \delta_{kj}]$ (cf. (2.6)). We can now generalize (7.8) to

$$\hat{b}_k[i] = \mathrm{sgn}\left(y_k[i] - \sum_{j=k+1}^{K} A_j (\rho_{jk} \hat{b}_j[i - \delta_{kj}] + \rho_{kj} \hat{b}_j[i - \delta_{kj} + 1]) \right).$$

It should be noted that selecting $\hat{b}_1[i]$ requires two consecutive decisions of user 2: $\hat{b}_2[i - \delta_{12}]$ and $\hat{b}_2[i - \delta_{12} + 1]$, which in turn require three consecutive decisions of user 3: $\hat{b}_3[i - \delta_{12} - \delta_{23}]$, $\hat{b}_3[i - \delta_{12} - \delta_{23} + 1]$, and $\hat{b}_3[i - \delta_{12} - \delta_{23} + 2]$. In general, in order to demodulate user 1 we need to store a sliding window of k decisions for the kth user. Figure 7.7 depicts a three-user case with $\delta_{12} = 1$, $\delta_{23} = 0$, and $\delta_{13} = 1$.

Regarding the practical implementation of successive cancellation we can point out the following features:

1. It requires knowledge of the received amplitudes. Any errors in the estimation of received amplitudes directly translate into noise for succeeding decisions.

348

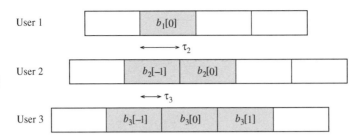

FIGURE 7.7.
Sliding window of decisions for demodulation of user 1.

2. Users weaker than the user (or users) of interest are neglected.

3. In contrast to the (nonadaptive) multiuser linear detectors, successive cancellation requires no arithmetic computations with the crosscorrelations beyond their product with the received amplitudes. (The optimum multiuser detector enjoys a similar advantage.)

4. The time complexity per bit is linear in the number of users.

5. The philosophy of successive cancellation is very general in scope: it applies not only to the basic CDMA model (where signals are linearly modulated) but to any multiple-access channel where the receiver observes the additive superposition of the transmitted signals.

6. The demodulation delay in successive cancellation grows linearly with the number of users.

A shortcoming of successive cancellation, which will be apparent in Section 7.2, is that its performance is asymmetric: equal-power users are demodulated with disparate reliability.

Several improvements to the basic successive cancellation structure have been proposed. One possible direction is to replace the "hard" intermediate decisions by "soft" decisions. This means that the inner sign function in (7.6) is replaced by a different nonlinearity. In fact, we have already found the optimum nonlinearity in (4.9); however, that approach is difficult to generalize to K users. Another possibility is to replace the sign by a hyperbolic tangent function (scaled with the signal-to-noise ratio). Such a function closely approximates the sign function when the signal-to-noise ratio is high and is justified on the basis that it minimizes mean-square error in the single-user channel (Problem 6.6). Intuitively, such a modification of the cancellation mechanism is appealing since it tends to deemphasize the effect of unreliable tentative decisions based on low-magnitude observables. Taking this approach to its ultimate conclusion we could make the nonlinearity "infinitely soft," in which case the successive canceler becomes a linear multiuser detector.

In view of the fact that tentative decisions are not completely reliable, another way to improve the quality of decision-driven multiuser detectors is not to attempt to cancel interference completely. This can be accomplished by lowering the cancellation weights: instead of subtracting $A_k \hat{b}_k s_k(t)$ from the received signal we can subtract $\hat{A}_k \hat{b}_k s_k(t)$, where $\hat{A}_k$ is chosen to minimize the energy of the cancellation error

$$
\begin{aligned}
E[\|A_k b_k s_k - \hat{A}_k \hat{b}_k s_k\|^2] &= E[(A_k b_k - \hat{A}_k \hat{b}_k)^2] \\
&= A_k^2 + \hat{A}_k^2 - 2\hat{A}_k A_k E[b_k \hat{b}_k] \\
&= A_k^2 + \hat{A}_k^2 - 2\hat{A}_k A_k(1 - 2P[b_k \neq \hat{b}_k]). \quad (7.9)
\end{aligned}
$$

The minimum of the right side of (7.9) is attained by

$$
\hat{A}_k = A_k(1 - 2P[b_k \neq \hat{b}_k]). \quad (7.10)
$$

If the detector does not have an accurate estimate of $P[b_k \neq \hat{b}_k]$, then $\hat{A}_k$ can be obtained by an adaptive least-mean-squares algorithm. Substituting (7.10) into (7.9) we obtain the minimum energy of the cancellation error:

$$
E[\|A_k b_k s_k - \hat{A}_k \hat{b}_k s_k\|^2] = 4A_k^2 P[b_k \neq \hat{b}_k](1 - P[b_k \neq \hat{b}_k]). \quad (7.11)
$$

The solution in (7.10) and (7.11) is intuitively reasonable: almost no cancellation is attempted for very weak interferers (for which full cancellation doubles the interference with significant probability), whereas strong interferers whose decisions are very reliable are greatly attenuated.

Successive cancellation has also been proposed for the noncoherent demodulation of m-ary orthogonal modulation (3.170):

$$
y(t) = \sum_{k=1}^{K} |A_k| e^{j\theta_k} s_k(t; b_k) + \sigma n(t), \quad (7.12)
$$

and

$$
\langle s_k(\cdot; b), s_k(\cdot; d) \rangle = \delta_{bd}.
$$

In principle, cancellation requires knowledge of $|A_k|$, θ_k, and $s_k(t; b_k)$ for all k. If that is indeed the case and the decisions $\{\hat{b}_{k+1}, \ldots, \hat{b}_K\}$ have been made, then $\hat{b}_k$ can be obtained by selecting the largest magnitude output from the bank of filters matched to $\{s_k(\cdot; b), b \in \mathcal{A}\}$ and driven by

$$
y^{(k)}(t) = y^{(k+1)}(t) - |A_{k+1}| e^{j\theta_{k+1}} s_{k+1}(t; \hat{b}_{k+1}). \quad (7.13)
$$

The obvious shortcoming of this demodulator is that it requires knowledge of amplitudes and phases of all but the last user. At the expense of an increase of cancellation error, it is possible to come up with a noncoherent demodulator by noticing that

$$\int_0^T y(t)s_{k+1}^*(t; \hat{b}_{k+1}) \, dt = |A_{k+1}| e^{j\theta_{k+1}} \qquad (7.14)$$

in the hypothetical case in which (a) $b_{k+1} = \hat{b}_{k+1}$, (b) no background noise is present, and (c) $s_{k+1}(t; \hat{b}_{k+1})$ is orthogonal to all other signals. This motivates the noncoherent successive cancellation receiver (cf. (7.13)):

$$y^{(k)}(t) = y^{(k+1)}(t) - \left(\int_0^T y^{(k+1)}(t)s_{k+1}^*(t; \hat{b}_{k+1}) \, dt \right) s_{k+1}(t; \hat{b}_{k+1}). \quad (7.15)$$

7.2 PERFORMANCE ANALYSIS OF SUCCESSIVE CANCELLATION

In the analysis of the linear multiuser detectors in Chapter 6, we saw that the interference term corrupting the decision statistic consists of the sum of a binomial random variable and an independent Gaussian random variable. This lead to a representation of the error probability as a weighted sum of Q-functions. In two-user successive cancellation, the interference term corrupting the decision statistic (7.7) consists of the sum of a random variable taking values on $\{-A_2\rho, 0, A_2\rho\}$ and a Gaussian random variable. Unfortunately, those two random variables are dependent because the noise realization dictates whether or not $b_2 = \hat{b}_2$. To make matters worse, the cancellation residual $A_1\rho(b_2 - \hat{b}_2)$ is, in general, dependent on b_1 (Problem 7.7). For this reason, rather than analyzing the conditional distribution of the decision statistic, it is preferable to make effective use of the geometry of the decision regions and analyze performance in the high signal-to-noise ratio region.

Before we derive the two-user asymptotic efficiency of successive cancellation, it is instructive to reflect on what we can expect to obtain. We saw in Figure 6.2 that in order to approach optimum asymptotic efficiency when

$$A_2 \gg A_1$$

it is necessary to recur to nonlinear detectors. We can anticipate that stripping will be optimal if $A_2 \gg A_1$ because, in that case,

$$A_2 - |\rho|A_1 \gg A_1,$$

and therefore the decisions of a conventional single-user matched filter for user 2 in the presence of user 1 will be much more reliable than those of user 1 with ideal canceling of user 2. So most of the errors in demodulating

351

user 1 will occur when the signal of user 2 has been perfectly canceled. At the other extreme:

$$A_2 \ll A_1,$$

the single-user matched filter for user 2 will make errors much more frequently than that of user 1. Then the most likely process by which the successive canceler makes an error for user 1 is by doubling the interference of user 2 rather than eliminating it. So we should expect that successive cancellation will not have the asymptotic optimality property (as $A_2/A_1 \to 0$) of the single-user matched filter detector.

In the previous paragraph, the assessment of the probability of rare events (bit-error-rate in the high signal-to-noise ratio region) was based on the intuition that although there are several ways in which an error can be explained (and all of them are improbable), the error probability is dominated by the probability of the most likely explanation. We have seen the same phenomenon in previous chapters: the error probability of linear multiuser receivers is an average of Q-functions, which, at high signal-to-noise ratios, is dominated by the one with the smallest argument; in the analysis of the optimum multiuser efficiency, performance is dominated by the minimum distance between multiuser signals that differ in the bit of interest. This is very much akin to the discipline of *large deviations* in probability theory, which analyzes the exponential rate of decrease of the probability of a rare event by focusing on the explanation of the event that is overwhelmingly more likely than any other. The main difference between the classical setting in large deviations and our problem is that the asymptotics of rare-event likelihoods are examined as a function of sample sizes in large deviations, and as a function of the background noise level in our case. Instead of adapting general large deviations results to our setting, we will give the following self-contained result, which is sufficient for our purposes.

Proposition 7.1 *Fix* $\mathbf{b} \in \{-1, +1\}^K$ *and* $\mathbf{b}' \in \{-1, +1\}^K$. *Let* $P[\mathbf{b} \to \mathbf{b}']$ *be the probability that the detector outputs* $\mathbf{b}'$ *given that* $\mathbf{b}$ *is transmitted (cf. (4.49)). Denote the multiuser signal modulated by* $\mathbf{b}$ *by*

$$S(\mathbf{b}) = \sum_{k=1}^{K} A_k b_k s_k,$$

let $D(\mathbf{b}')$ *be the decision region corresponding to* $\mathbf{b}'$ *in the K-dimensional space spanned by the signature waveforms, and denote the distance from* $S(\mathbf{b})$ *to* $D(\mathbf{b}')$ *by*

$$\Delta(\mathbf{b}, \mathbf{b}') = \min_{v \in D(\mathbf{b}')} \|v - S(\mathbf{b})\|. \tag{7.16}$$

If $D(\mathbf{b}')$ is a polytope that includes its boundary (and does not depend on σ), then

$$\lim_{\sigma \to 0} 2\sigma^2 \log 1/P[\mathbf{b} \to \mathbf{b}'] = \Delta^2(\mathbf{b}, \mathbf{b}'). \qquad (7.17)$$

Proof. If $\bar{n}$ denotes the projection of the received noise realization, n, on the K-dimensional space spanned by the signature waveforms, then

$$P[\mathbf{b} \to \mathbf{b}'] = P[S(\mathbf{b}) + \sigma\bar{n} \in D(\mathbf{b}')].$$

As customary in the proof of this kind of asymptotic result, we will prove $\geq$ and $\leq$ in (7.17) separately. To that end, we will simply find a superset and a subset, respectively, of $D(\mathbf{b}')$ whose conditional probabilities given that $\mathbf{b}$ is transmitted are easy to compute. We will assume $S(\mathbf{b}) \notin D(\mathbf{b}')$, for otherwise, both sides of (7.17) are obviously 0. Throughout this proof, we will use the notation

$$d(\mathbf{b}, \mathbf{b}') = v^* - S(\mathbf{b}),$$

where v^* denotes a signal that achieves the minimum in (7.16) and thus

$$\|d(\mathbf{b}, \mathbf{b}')\| = \Delta(\mathbf{b}, \mathbf{b}').$$

To show

$$\lim_{\sigma \to 0} 2\sigma^2 \log 1/P[\mathbf{b} \to \mathbf{b}'] \geq \Delta^2(\mathbf{b}, \mathbf{b}'), \qquad (7.18)$$

denote the half-space of all signals $S(\mathbf{b}) + v$, such that

$$\langle v, d(\mathbf{b}, \mathbf{b}') \rangle \geq \Delta^2(\mathbf{b}, \mathbf{b}'),$$

by $C(\mathbf{b}, \mathbf{b}')$. It is easy to see (cf. Figure 7.8 and Problem 7.8) that

$$D(\mathbf{b}') \subset C(\mathbf{b}, \mathbf{b}').$$

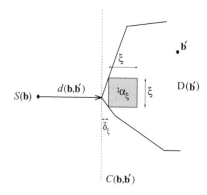

FIGURE 7.8. $H_\xi(\mathbf{b}, \mathbf{b}')$ is the shaded ξ-square normal to $d(\mathbf{b}, \mathbf{b}')$ and closest to $S(\mathbf{b})$ in $D(\mathbf{b}')$.

Accordingly, we can upper bound the desired probability by

$$P[\mathbf{b} \to \mathbf{b}'] \leq P[S(\mathbf{b}) + \sigma\bar{n} \in C(\mathbf{b}, \mathbf{b}')]$$

$$= P[\sigma \langle \bar{n}, d(\mathbf{b}, \mathbf{b}') \rangle \geq \Delta^2(\mathbf{b}, \mathbf{b}')]$$

$$= Q\left(\frac{\Delta(\mathbf{b}, \mathbf{b}')}{\sigma}\right)$$

$$\leq \frac{1}{2} \exp\left(\frac{-\Delta^2(\mathbf{b}, \mathbf{b}')}{2\sigma^2}\right). \qquad (7.19)$$

Upon taking $\lim_{\sigma \to 0} \sigma^2 \log 1/(\cdot)$ of both sides of (7.19) we get (7.18).

To show

$$\lim_{\sigma \to 0} 2\sigma^2 \log 1/P[\mathbf{b} \to \mathbf{b}'] \leq \Delta^2(\mathbf{b}, \mathbf{b}'), \qquad (7.20)$$

we will assume for the sake of concreteness that (as in Figure 7.3) the signals belong to a two-dimensional space. (The extension to the K-user case is straightforward.) We fix an arbitrary $\xi > 0$ and define

$$H_\xi(\mathbf{b}, \mathbf{b}') \subset D(\mathbf{b}')$$

so that

$$P[S(\mathbf{b}) + \sigma\bar{n} \in H_\xi(\mathbf{b}, \mathbf{b}')]$$

is easily computable. One way to accomplish this is to let $H_\xi(\mathbf{b}, \mathbf{b}')$ be a square with sides of length ξ normal to $d(\mathbf{b}, \mathbf{b}')$ and centered at

$$\left(1 + \frac{\delta_\xi + \xi/2}{\Delta(\mathbf{b}, \mathbf{b}')}\right) d(\mathbf{b}, \mathbf{b}') + \alpha_\xi q(\mathbf{b}, \mathbf{b}')$$

(Figure 7.8), where $q(\mathbf{b}, \mathbf{b}')$ is a unit-norm vector orthogonal to $d(\mathbf{b}, \mathbf{b}')$, $-\xi/2 \leq \alpha_\xi \leq \xi/2$, and δ_ξ is the smallest nonnegative scalar such that such a square fits inside $D(\mathbf{b}')$. Since $D(\mathbf{b}')$ is a polytope not only is it possible to find such a δ_ξ for all sufficiently small ξ, but $\delta_\xi \to 0$ as $\xi \to 0$.

The orthogonal components of the noise vector are independent, so we can write

$$P[\mathbf{b} \to \mathbf{b}'] \geq P[S(\mathbf{b}) + \sigma\bar{n} \in H_\xi(\mathbf{b}, \mathbf{b}')]$$

$$= \left[Q\left(\frac{\alpha_\xi - \xi/2}{\sigma}\right) - Q\left(\frac{\alpha_\xi + \xi/2}{\sigma}\right)\right]$$

$$\times \left[Q\left(\frac{\Delta(\mathbf{b}, \mathbf{b}') + \delta_\xi}{\sigma}\right) - Q\left(\frac{\Delta(\mathbf{b}, \mathbf{b}') + \delta_\xi + \xi}{\sigma}\right)\right]$$

$$\geq \left[\frac{1}{2} - Q\left(\frac{\xi}{\sigma}\right)\right]$$

$$\times \left[Q\left(\frac{\Delta(\mathbf{b}, \mathbf{b'}) + \delta_\xi}{\sigma}\right) - Q\left(\frac{\Delta(\mathbf{b}, \mathbf{b'}) + \delta_\xi + \xi}{\sigma}\right)\right],$$

(7.21)

where the inequality in (7.21) follows from (3.51)

Upon taking $\lim_{\sigma \to 0} 2\sigma^2 \log 1/(\cdot)$ of both sides of (7.21) and using (3.41), we obtain

$$\lim_{\sigma \to 0} 2\sigma^2 \log 1/P[\mathbf{b} \to \mathbf{b'}] \leq \left(\Delta(\mathbf{b}, \mathbf{b'}) + \delta_\xi\right)^2,$$

for sufficiently small ξ. But since $\delta_\xi \to 0$ as $\xi \to 0$, (7.20) must hold.

Returning to the analysis of the asymptotic efficiency of the two-user successive cancellation detector, we first note that, by symmetry, the answer will depend on the crosscorrelation ρ only through $|\rho|$. Therefore, it is advisable to assume $\rho > 0$ and refer to the decision regions displayed in Figures 7.3, 7.4, and 7.5. Furthermore, by symmetry, we can write the error probability of the successive cancellation detector for user 1 as

$$P_1^s(\sigma) = \frac{1}{2}P[++ \to -+] + \frac{1}{2}P[++ \to --]$$

$$+ \frac{1}{2}P[+- \to --] + \frac{1}{2}P[+- \to -+].$$

(7.22)

Applying Proposition 7.2 to (7.22) we obtain

$$\lim_{\sigma \to 0} 2\sigma^2 \log 1/P_1^s(\sigma) = \min\{\Delta^2(++, -+), \Delta^2(++, --),$$

$$\Delta^2(+-, --), \Delta^2(+-, -+)\}$$

$$= \min\{\Delta^2(++, -+), \Delta^2(+-, -+)\}.$$

(7.23)

To verify (7.23), note that regardless of the value of A_2/A_1 (cf. Figure 7.3 and Problem 7.5):

$$\Delta(++, -+) = \Delta(+-, --),$$

(7.24)

$$\Delta(++, --) \geq \Delta(+-, -+).$$

(7.25)

The distance from $S(++)$ to the region $D(-+)$ is A_1. Thus,

$$\Delta(++, -+) = A_1.$$

(7.26)

The computation of $\Delta^2(+-, -+)$ is more involved. Two cases may occur depending on the relative values of A_1 and A_2: in Figure 7.3 we see that the closest point to $S(+-)$ in $D(-+)$ is the vertex of the region, which will be denoted by v_{-+}, whereas in the case depicted in Figure 7.4, $d(+-, -+)$ is a multiple of s_2. To obtain v_{-+}, we just have to notice that it is orthogonal to s_2 and its inner product with s_1 yields $A_2\rho$:

$$v_{-+} = \gamma(s_1 - \rho s_2),$$

$$\langle v_{-+}, s_1 \rangle = A_2\rho,$$

where γ is a constant. Therefore,

$$v_{-+} = \frac{A_2\rho}{1 - \rho^2}(s_1 - \rho s_2),$$

and

$$\|v_{-+} - S(+-)\|^2 = \left\| \frac{A_2\rho}{1 - \rho^2}(s_1 - \rho s_2) - A_1 s_1 + A_2 s_2 \right\|^2$$

$$= A_1^2 - 4A_1 A_2 \rho + \frac{A_2^2}{1 - \rho^2}. \tag{7.27}$$

The distance from $S(+-)$ to the line that goes through the origin and is perpendicular to s_2 is equal to $A_2 - A_1\rho$. To see this, we can write $S(+-)$ as a multiple of s_2 plus an orthogonal component to s_2:

$$A_1 s_1 - A_2 s_2 = A_1(s_1 - \rho s_2) + (A_1\rho - A_2)s_2,$$

or, equivalently,

$$S(+-) + (A_2 - A_1\rho)s_2 = A_1(s_1 - \rho s_2).$$

Both cases involved in the computation of $\Delta^2(+-, -+)$ (Figures 7.3 and 7.4) are identical if

$$S(+-) + (A_2 - A_1\rho)s_2 = v_{-+},$$

which occurs when

$$\frac{A_2}{A_1} = \frac{1}{\rho} - \rho. \tag{7.28}$$

Summarizing these results, we get

$$\Delta^2(+-, -+) = \begin{cases} A_1^2 - 4A_1 A_2 \rho + \frac{A_2^2}{1-\rho^2}, & \text{if } \frac{A_2}{A_1} \leq \frac{1}{\rho} - \rho; \\ (A_2 - A_1\rho)^2, & \text{otherwise.} \end{cases} \tag{7.29}$$

But according to (7.23), (7.26), and the definition of asymptotic efficiency (3.113) we can write the asymptotic efficiency achieved by successive cancellation as

$$\eta_1^s = \min\left\{1, \frac{\Delta^2(+-,-+)}{A_1^2}\right\}. \tag{7.30}$$

Putting together (7.29) and (7.30) we readily obtain that, if $|\rho| \geq 1/2$, then

$$\eta_1^s = \begin{cases} 1 - 4\frac{A_2}{A_1}|\rho| + \frac{1}{1-\rho^2}\frac{A_2^2}{A_1^2}, & \text{if } \frac{A_2}{A_1} \leq \frac{1}{|\rho|} - |\rho|; \\ \left(\frac{A_2}{A_1} - |\rho|\right)^2, & \text{if } \frac{1}{|\rho|} - |\rho| \leq \frac{A_2}{A_1} \leq 1 + |\rho|; \\ 1, & \text{if } 1 + |\rho| < \frac{A_2}{A_1}; \end{cases} \tag{7.31}$$

whereas if $|\rho| < 1/2$, then

$$\eta_1^s = \begin{cases} 1 - 4\frac{A_2}{A_1}|\rho| + \frac{1}{1-\rho^2}\frac{A_2^2}{A_1^2}, & \text{if } \frac{A_2}{A_1} \leq 4|\rho|(1-\rho^2); \\ 1, & \text{otherwise.} \end{cases} \tag{7.32}$$

In Figure 7.9 we have compared the asymptotic efficiency of user 1 obtained for successive cancellation (7.31) with the optimum one and with that achieved by the single-user matched filter. As we had predicted, if $A_2 \ll A_1$, then successive cancellation is markedly inferior to the single-user matched filter, whereas if A_2/A_1 is sufficiently high, then the asymptotic efficiency of the successive canceler is equal to 1.

The asymptotic efficiency achieved by the single-user matched filter for user 1 is better than that of the successive canceler for

$$\frac{A_2}{A_1} \leq \psi(\rho) \overset{\text{def}}{=} \begin{cases} \frac{2|\rho|}{\frac{1}{1-\rho^2}-\rho^2}, & \text{if } 0 < |\rho| < \sqrt{\frac{3-\sqrt{5}}{2}}; \\ 1, & \text{otherwise} \end{cases}$$

(see Problem 7.9). Accordingly, if the successive canceler not only knows the value of A_2, but that of A_1, then it can simply switch to a single-user matched filter for user 1 when the asymptotic efficiency of the latter is higher

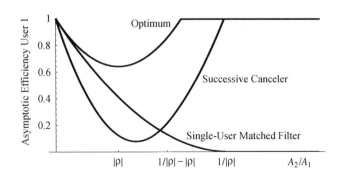

FIGURE 7.9.
Asymptotic multiuser efficiencies for two synchronous users; $|\rho| = 0.6$.

357

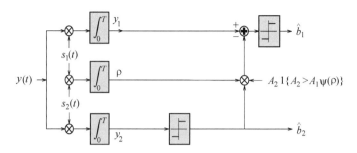

FIGURE 7.10.
Modified
successive
cancellation for
two synchronous
users.

(Figure 7.10). With this precaution, the near–far resistance of the successive canceler is given by

$$
\bar{\eta}_1^s = \begin{cases} \left(\frac{1-3\rho^2+3\rho^4}{1-\rho^2+\rho^4}\right)^2, & \text{if } 0 < |\rho| < \sqrt{\frac{3-\sqrt{5}}{2}}; \\ (1-|\rho|)^2, & \text{otherwise.} \end{cases} \tag{7.33}
$$

Figure 7.11 compares the near–far resistance of the successive canceler (7.33) with the optimum near–far resistance achieved by the multiuser detectors studied in Chapters 4, 5, and 6.

The power-tradeoff regions of successive cancellation are shown in Figures 7.12 and 7.13. In contrast to the power-tradeoff regions found in previous chapters, the regions in Figures 7.12 and 7.13 are asymmetric because each user is demodulated in a different way. Notice that for user 1 there is a window of feasible powers, whereas for user 2 only a lower threshold exists. Indeed, once the relative power of user 2 exceeds the value required for $\eta_1^s = 1$, then (for low bit-error-rates) the bit-error-rate of user 1 remains essentially unaffected by the power of user 2. Conversely, if the power of user 1 is too large, then it forces user 2 to employ much higher signal-to-noise ratio so that the single-user matched filter for user 2 can provide reliable demodulation. The disadvantage of using equal powers in successive cancellation is made evident in Figure 7.12: if $\rho = 0.5$, then $SNR_1 \simeq 12$ dB and $SNR_2 \simeq 15$ dB are sufficient to achieve bit-error-rates

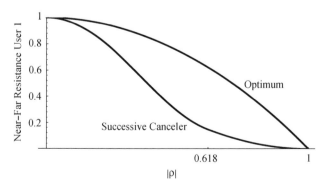

FIGURE 7.11.
Near–far
resistance for two
synchronous users
as a function of
crosscorrelation.

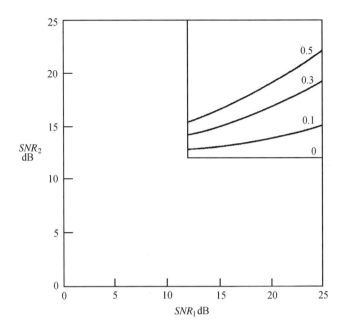

FIGURE 7.12.
Signal-to-noise ratios necessary for successive cancellation to achieve bit-error-rate not higher than 3×10^{-5} for both users. Shown for $|\rho| = $ 0.1, 0.3, 0.5.

not higher than 3×10^{-5}, whereas if one insists that both received powers be identical, then it is required that $SNR_1 = SNR_2 \simeq 18$ dB. This behavior should be contrasted to that of linear multiuser detectors (single-user matched filter, decorrelating, and MMSE) and of the optimum detector (for $\rho \leq 0.5$) for which equal received powers minimize the total power required to guarantee a certain bit-error-rate level.

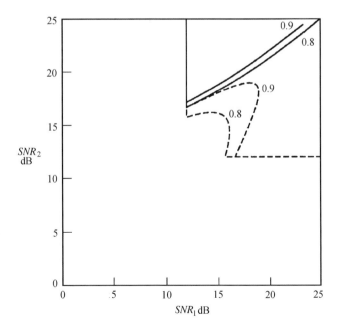

FIGURE 7.13.
Signal-to-noise ratios necessary for successive cancellation to achieve bit-error-rate not higher than 3×10^{-5} for both users. Shown for $|\rho| = 0.8, 0.9,$ and compared with the optimal regions (dashed).

359

The performance analysis of successive cancellation in the simple setting of the two-user synchronous channel has revealed an important shortcoming of this technique, namely, its mediocre performance (compared to the detectors in previous chapters) when the received signal-to-noise ratios are comparable.

Let us return now to the analysis of error probability for nonasymptotic bit-error-rate. From (3.70) and (7.8) we obtain

$$\hat{b}_k = \mathrm{sgn}\left(A_k b_k + n_k + \sum_{j=1}^{k-1} A_j \rho_{jk} b_j + \sum_{j=k+1}^{K} A_j \rho_{jk}(b_j - \hat{b}_j)\right). \quad (7.34)$$

Since the cancellation residuals depend on the other random variables inside the sign function in (7.34), an analytical exact evaluation of bit-error-rate is difficult. For very small K, it is possible to obtain exact results by integrating multidimensional normal distributions over the decision regions (cf. Problem 4.25 and Varanasi and Aazhang [474]). In general, however, simulations or analytical approximations appear to be inevitable. A popular approximation is based on the asymptotic ($K \to \infty$) results for the error probability of the single-user matched filter for direct-sequence spread-spectrum with spreading gain equal to N (Section 3.4), where performance is the same as that of a single-user system with an extra Gaussian noise source with zero mean and variance $1/N$ for every interferer. If, in addition, the correlation of the cancellation residuals with the other interfering quantities in (7.34) is neglected, then the probability of error of the successive canceler is approximated by the recursive formula

$$P_k^{sc}(\sigma) \approx Q\left(\frac{A_k}{\sqrt{\sigma^2 + \frac{1}{N}\sum_{j=1}^{k-1} A_j^2 + \frac{4}{N}\sum_{j=k+1}^{K} A_j^2 P_j^{sc}(\sigma)}}\right), \quad (7.35)$$

where we have used

$$E[(b_j - \hat{b}_j)^2] = 4P_j^{sc}(\sigma).$$

It is evident from (7.35) that equal received powers will result in a wide range of error probabilities for the various users. Furthermore, we have already seen that successive cancellation makes the most sense when the received powers are disparate. In some situations it may make sense to pose the problem of controlling the received powers so that the bit-error-rates for all users are equal. From (7.35) one can solve numerically for the values of $\sigma^{-2}(A_1^2, \ldots, A_K)$ that achieve the desired bit-error-rate level.

When that level is very small it is sensible to set up the received powers as

$$\frac{A_k^2}{\sigma^2} = \gamma \left(1 + \frac{\gamma}{N}\right)^{K-k} \tag{7.36}$$

because this guarantees that in the hypothetical scenario of perfect cancellation, the signal-to-noise ratio seen after each stage remains the same (Problem 7.2).

7.3 MULTISTAGE DETECTION

The order in which users are canceled greatly affects the performance of successive cancellation for a particular user. In this section, we explore a symmetrized version of successive cancellation, which mitigates some of the shortcomings of that technique.

7.3.1 CONVENTIONAL FIRST STAGE

The two-user successive cancellation receiver in Figure 7.2 outputs the conventional single-user matched filter decisions for user 2. This receiver can easily be symmetrized by using successive cancellation for user 2, as we can see in Figure 7.14, where we have depicted a *multistage* detector for two users.

The decision regions of the two-stage detector in Figure 7.14 are shown in Figure 7.15.

The term *multistage detection* is suggested by the fact that various decisions are produced at consecutive stages; in the first stage, the conventional bank of single-user matched filters is used, whereas in the second stage successive cancellation is used for both users. This approach can be further iterated, by cleaning up the original matched filter outputs with, hopefully, increasingly reliable tentative decisions of the interfering users (Figure 7.16). Some motivation for this approach can be given heuristically by considering

FIGURE 7.14.
Two-stage detector for two synchronous users.

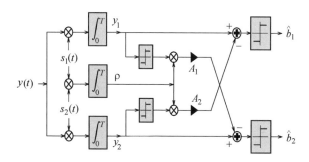

361

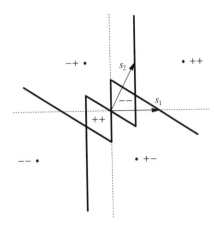

FIGURE 7.15.
Decision regions of two-stage detector with $A_1 = 1$, $A_2 = 1$.

the property satisfied by the jointly optimum decisions (Problem 4.1):

$$\hat{b}_k = \text{sgn}\left(y_k - \sum_{j \neq k} A_j \rho_{jk} \hat{b}_j \right). \tag{7.37}$$

Equation (7.37) states that the optimum decisions are a *fixed point* of the K-dimensional nonlinear transformation $\mathbf{p}[x_1, \dots, x_K]$ whose kth component is given by

$$p_k[x_1, \dots, x_{k-1}, x_{k+1}, \dots x_K] = \text{sgn}\left(y_k - \sum_{j \neq k} A_j \rho_{jk} x_j \right),$$

or in matrix notation (letting the sgn of a vector be the vector of signs)

$$\mathbf{p}[\mathbf{x}] = \text{sgn}(\mathbf{y} + \mathbf{A}[\mathbf{I} - \mathbf{R}]\mathbf{x}).$$

The decisions of the m-stage detector with a conventional first stage are the result of applying $(m - 1)$ times the transformation $\mathbf{p}$ to the tentative decisions supplied by the bank of conventional single-user receivers. This operation can be succinctly written as

$$[\hat{b}_1^m, \dots, \hat{b}_K^m]^T = \mathbf{p}^{(m)}[0, \dots, 0]. \tag{7.38}$$

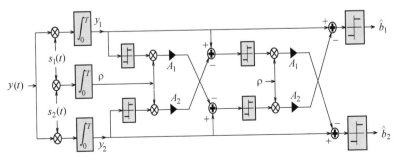

FIGURE 7.16.
Three-stage detector for two synchronous users.

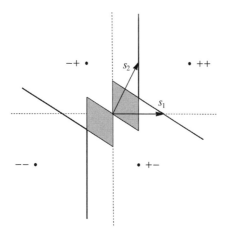

FIGURE 7.17.
Decision regions
of *m*-stage
detector with
shaded regions
leading to
limit-cycle
decisions.

Contrary to what might have been expected,[2] as $m \to \infty$, the decisions of the m-stage detector do not converge to the optimum ones. Figure 7.17 shows the two-user decision regions of the m-stage detector (Problem 7.10). For observations belonging to the shaded regions, the multistage detector enters a limit cycle, with decisions given by a succession of $\{++, --, ++, --, \cdots\}$. In fact, the decision regions of the $(m + 2)$-stage detector are identical to those of the m-stage detector with $m = 2, 3, \ldots$. This means that an additional stage may actually hurt performance.

Multistage detectors for asynchronous channels require that sliding windows of tentative decisions be stored at each stage. For example, in a K-user asynchronous channel, a two-stage detector has to store a window of K decisions for each user (see also Problem 7.7).

The results on the two-user asymptotic efficiency of successive cancellation apply to the two-stage detector verbatim since the decisions for user 1 are identical in both cases. To obtain the asymptotic efficiency of user 2, we just need to reverse the roles of users 1 and 2 in (7.31) and (7.32).

7.3.2   DECORRELATING FIRST STAGE

The multistage approach can be used with tentative decisions other than those provided by the bank of single-user matched filters. For example, a decorrelating detector can serve as the first stage of the receiver. The

---

[2] Classical *fixed-point theorems* of functional analysis give sufficient conditions (not satisfied by **p**) under which repeated applications of a function converge to its fixed point.

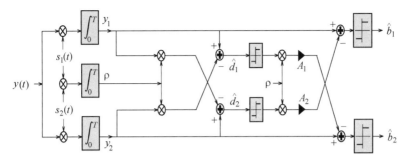

FIGURE 7.18.
Two-stage detector
with decorrelating
first stage.

decisions put out by the second stage are

$$\hat{b}_k = \text{sgn}\left(y_k - \sum_{j \neq k} A_j \rho_{jk} \hat{d}_j \right) \qquad (7.39)$$

with the tentative decisions made by the decorrelator:

$$\hat{d}_k = \text{sgn}((\mathbf{R}^{-1}\mathbf{y})_k).$$

In the two-user case depicted in Figure 7.18, we obtain

$$\hat{b}_1 = p_1[\hat{d}_2] = \text{sgn}(y_1 - A_2\rho \, \text{sgn}(y_2 - \rho y_1)), \qquad (7.40)$$

$$\hat{b}_2 = p_2[\hat{d}_1] = \text{sgn}(y_2 - A_1\rho \, \text{sgn}(y_1 - \rho y_2)), \qquad (7.41)$$

with the corresponding decision regions shown in Figure 7.19. The decision regions of the m-stage detector (Problem 7.7) with a decorrelating first
stage are shown in Figure 7.20, where we can see the presence of limit
cycles for certain observables as in the case of the conventional first stage
(Figure 7.17).

Despite the shape of the decision regions in Figure 7.19, it turns out
that the analysis of the two-stage detector with a decorrelating first stage
is easier than the analysis in Section 7.2. By symmetry we can write the

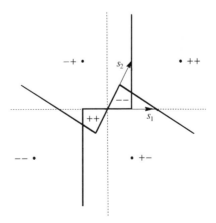

FIGURE 7.19.
Decision regions
of multistage
detector with a
decorrelating first
stage.

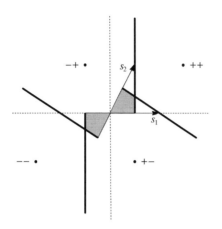

FIGURE 7.20.
Decision regions
of *m*-stage
detector
(decorrelating first
stage) with
shaded regions
leading to
limit-cycle
decisions.

bit-error-rate of the two-stage detector with a decorrelating first stage as

$$P_1^{dd}(\sigma) = \frac{1}{2} P[\hat{b}_1 = -1 \,|\, (b_1, b_2) = (+1, +1)]$$

$$+ \frac{1}{2} P[\hat{b}_1 = -1 \,|\, (b_1, b_2) = (+1, -1)]. \qquad (7.42)$$

We will now compute the first term in the right side of (7.42). Conditioning on $\hat{d}_2$, we have

$$P[\hat{b}_1 = -1 \,|\, (b_1, b_2) = (+1, +1)]$$

$$= P[\hat{b}_1 = -1 \,|\, (b_1, b_2, \hat{d}_2) = (+1, +1, +1)]$$

$$\times P[\hat{d}_2 = +1 \,|\, (b_1, b_2) = (+1, +1)]$$

$$+ P[\hat{b}_1 = -1 \,|\, (b_1, b_2, \hat{d}_2) = (+1, +1, -1)]$$

$$\times P[\hat{d}_2 = -1 \,|\, (b_1, b_2) = (+1, +1)]. \qquad (7.43)$$

But we saw in Chapter 5 that the decision of the decorrelating detector $\hat{d}_2$ is unaffected by the value of b_1. Thus,

$$P[\hat{d}_2 = +1 \,|\, (b_1, b_2) = (+1, +1)]$$

$$= 1 - P[\hat{d}_2 = -1 \,|\, b_2 = +1]$$

$$= 1 - Q\left(\frac{A_2\sqrt{1 - \rho^2}}{\sigma}\right). \qquad (7.44)$$

Furthermore, using (7.39) we get

$$P[\hat{b}_1 = -1 \,|\, (b_1, b_2, \hat{d}_2) = (+1, +1, +1)]$$

$$= P[A_1 + A_2\rho + n_1 - A_2\rho < 0 \,|\, A_2(1 - \rho^2) + (n_2 - \rho n_1) > 0]$$

$$= P[n_1 < -A_1]$$

$$= Q\left(\frac{A_1}{\sigma}\right), \tag{7.45}$$

where we used the independence of the zero-mean Gaussian random variables n_1 and $(n_2 - \rho n_1)$ (cf. (2.79)):

$$E[n_1(n_2 - \rho n_1)] = \sigma^2\rho - \rho\sigma^2 = 0.$$

Analogously,

$$P[\hat{b}_1 = -1 \,|\, (b_1, b_2, \hat{d}_2) = (+1, +1, -1)]$$

$$= P[A_1 + 2A_2\rho + n_1 < 0 \,|\, A_2(1 - \rho^2) + (n_2 - \rho n_1) < 0]$$

$$= P[n_1 < -A_1 - 2A_2\rho]$$

$$= Q\left(\frac{A_1 + 2\rho A_2}{\sigma}\right). \tag{7.46}$$

Putting together (7.43), (7.44), (7.45), and (7.46) we get

$$P[\hat{b}_1 = -1 \,|\, (b_1, b_2) = (+1, +1)]$$

$$= Q\left(\frac{A_1}{\sigma}\right)\left[1 - Q\left(\frac{A_2\sqrt{1 - \rho^2}}{\sigma}\right)\right]$$

$$+ Q\left(\frac{A_1 + 2\rho A_2}{\sigma}\right) Q\left(\frac{A_2\sqrt{1 - \rho^2}}{\sigma}\right). \tag{7.47}$$

Proceeding in an entirely similar way, we get

$$P[\hat{b}_1 = -1 \,|\, (b_1, b_2) = (+1, -1)]$$

$$= Q\left(\frac{A_2\sqrt{1 - \rho^2}}{\sigma}\right) Q\left(\frac{A_1 - 2A_2\rho}{\sigma}\right)$$

$$+ \left[1 - Q\left(\frac{A_2\sqrt{1 - \rho^2}}{\sigma}\right)\right] Q\left(\frac{A_1}{\sigma}\right). \tag{7.48}$$

It follows that (7.42), (7.47), and (7.48) result in the closed-form expression

$$P_1^{dd}(\sigma) = Q\left(\frac{A_1}{\sigma}\right)\left[1 - Q\left(\frac{A_2\sqrt{1-\rho^2}}{\sigma}\right)\right] + Q\left(\frac{A_2\sqrt{1-\rho^2}}{\sigma}\right)$$
$$\times \left[\frac{1}{2}Q\left(\frac{A_1 + 2A_2\rho}{\sigma}\right) + \frac{1}{2}Q\left(\frac{A_1 - 2A_2\rho}{\sigma}\right)\right], \quad (7.49)$$

which is one of the rare closed-form expressions one can obtain for the bit-error-rate of a nonlinear detector. Equation (7.49) has an obvious interpretation: the event $b_1 \neq \hat{b}_1$ can happen in two different ways depending on whether the decorrelator makes the right decision for user 2; if so, then the error probability is that of a single-user system; otherwise the error probability is that of a conventional receiver for user 1 with interfering user amplitude equal to $2A_2$.

To find the asymptotic multiuser efficiency we just need to use (7.49) to compute

$$\lim_{\sigma \to 0} \frac{2\sigma^2}{A_1^2} \log 1/P_1^{dd}(\sigma).$$

Making use of (3.43) we obtain

$$\eta_1^{dd} = \min\left\{1, \frac{A_2^2}{A_1^2}(1-\rho^2) + \max^2\left\{0, 1 - 2\frac{A_2}{A_1}|\rho|\right\}\right\}. \quad (7.50)$$

Figure 7.21 shows η_1^{dd} as a function of A_2/A_1 when $|\rho| = 0.6$. We see that $\eta_1^{dd} = 1$ for sufficiently large interference. Moreover, for sufficiently low interference, it does not behave as well as the conventional single-user matched filter detector (cf. Figure 7.9), and its near–far resistance is not optimal (i.e., it is worse than that of the simpler decorrelating detector). The near–far resistance of the two-stage detector with decorrelating first stage can be readily found by minimizing (7.50) with respect to A_2/A_1

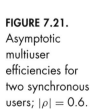

FIGURE 7.21. Asymptotic multiuser efficiencies for two synchronous users; $|\rho| = 0.6$.

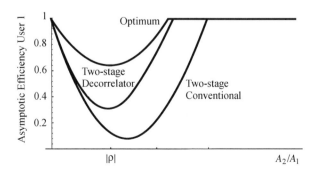

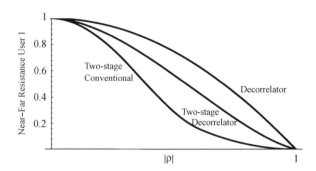

FIGURE 7.22. Near–far resistance of two-stage detector with conventional and decorrelating first stage for two synchronous users.

(Problem 7.13):

$$\bar{\eta}_1^{dd} = \frac{1 + 4|\rho| - \rho^2 - 12|\rho|^3 + 8\rho^4}{(1 + 2|\rho| - \rho^2)^2}. \tag{7.51}$$

Figure 7.22 compares (7.51) to the near–far resistances of the two-stage receiver with conventional first stage (Problem 7.9) and to the near–far resistance of the decorrelator (equal to $1 - \rho^2$). Although the decorrelating first stage is preferable to a conventional first stage, the second stage appended to the decorrelator actually decreases its near–far resistance. The reason for this behavior is that the reliability of the tentative decorrelating decision for user 2 may not be sufficiently high (for certain A_2/A_1 and ρ) compared to the reliability of the decorrelating decision for user 1, but yet A_2/A_1 is sufficiently high that subtracting wrong decisions from user 2 has a substantial negative effective on performance. In those cases, it is preferable to tune out user 2 rather than to attempt to cancel it. Consequently, it is sensible to switch off the second stage when the relative amplitude of the interferer is such that the asymptotic efficiency of the decorrelator $\eta_k^d = 1 - \rho^2$ is higher than η_k^{dd}.

The allowable power-tradeoff regions to obtain a guaranteed bit-error-rate for both users are shown in Figure 7.23. If $|\rho| \leq \frac{1}{3}$, then it can be shown (Problem 7.14) that the power-tradeoff region (for asymptotically large signal-to-noise ratio) is identical to the optimal one (i.e., there is no coupling between the energies). As the crosscorrelations are increased, we observe the interesting phenomenon that for sufficiently similar powers the curves coincide with those of the decorrelator, whereas for sufficiently imbalanced powers, the behavior is identical to the optimum detector.

The multistage detector with decorrelating first stage is one of those cases where the two-user analysis does not capture the full essence of the K-user analysis. The reason we were able to obtain a closed-form solution to the bit-error-rate in the two-user case was the independence between the noise components at the output of the matched filter for user 1 and at the output of the decorrelating detector for user 2. This property carries over to the

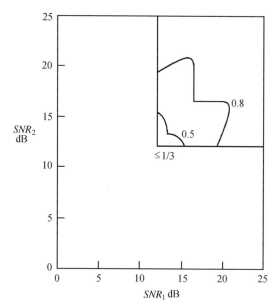

FIGURE 7.23. Signal-to-noise ratios necessary for the two-stage detector with decorrelating first stage to achieve bit-error-rate not higher than 3×10^{-5} for both users.

K-user case since every decorrelating linear transformation is orthogonal to all other signature waveforms. However, in the K-user case it is necessary to compute the joint probabilities of error of the decorrelating detector. That is, for every given pair of $\mathbf{b} \in \{-1, +1\}^K$ and $\mathbf{d} \in \{-1, +1\}^K$, we must find

$$P[\mathrm{sgn}(\mathbf{Ab} + \mathbf{R}^{-1}\mathbf{n}) = \mathbf{d}],$$

which involves multidimensional Gaussian integration since $\mathbf{R}^{-1}\mathbf{n}$ is a correlated vector. An upper bound on K-user asymptotic multiuser efficiency based on the two-user analysis is given in Problem 7.18.

7.4 CONTINUOUS-TIME TENTATIVE DECISIONS

If we receive

$$y(t) = Abs(t) + m(t), \quad t \in [0, T],$$

we can make tentative decisions for b not only at time T, but for all $t \in [0, T]$:

$$\hat{b}(t) = \mathrm{sgn}\left(\int_0^t s(\lambda) y(\lambda) \, d\lambda \right).$$

This principle can be employed to continuously cancel asynchronous multiuser interference with running tentative decisions

$$\hat{b}_k[i, t^+] \stackrel{\text{def}}{=} \text{sgn}\left(\int_{iT+\tau_k}^{t} s_k(\lambda - iT - \tau_k)\right.$$

$$\left. \times \left[\tilde{y}(\lambda) + \hat{A}_k(\lambda)\hat{b}_k[i, \lambda]s_k(\lambda - iT - \tau_k)\right] d\lambda\right), \quad (7.52)$$

where

$$\tilde{y}(t) = y(t) - \sum_{k=1}^{K}\sum_{i=-M}^{M} \hat{A}_k(t)\hat{b}_k[i, t]s_k(t - iT - \tau_k) \qquad (7.53)$$

and (cf. (7.10))

$$\hat{A}_k(t) = A_k(1 - 2P[\hat{b}_k[i, t] \neq b_k[i]]) \qquad (7.54)$$

with i chosen in (7.54) such that $iT + \tau_k \leq t \leq iT + \tau_k + T$. Alternatively, we can express the term in large brackets in (7.52) as

$$\tilde{y}(\lambda) + \hat{A}_k(\lambda)\hat{b}_k[i, \lambda]s_k(\lambda - iT - \tau_k)$$

$$= y(\lambda) - \sum_{j \neq k} \hat{A}_j(\lambda)\hat{b}_j[i, \lambda]s_j(\lambda - iT - \tau_j)$$

$$- \sum_{j \neq k} \hat{A}_j(\lambda)\hat{b}_j[i + 1, \lambda]s_j(\lambda - iT - T - \tau_j)$$

$$- \sum_{j \neq k} \hat{A}_j(\lambda)\hat{b}_j[i - 1, \lambda]s_j(\lambda - iT + T - \tau_j). \qquad (7.55)$$

The final decision for the ith bit of the kth user is $\hat{b}_k[i, iT + T + \tau_k]$. In contrast to the asynchronous decision-driven multiuser detectors we have seen in previous sections, there is no need to maintain a sliding window of tentative decisions (Figure 7.7).

Performing a rigorous analysis of this detector is even more difficult than those done previously in this chapter. An analysis of the random signature case that invokes several simplifying approximations (Abrams et al. [8]) leads to the conclusion that the continuous-time tentative decision detector achieves the performance of the decorrelator.

7.5   DECISION-FEEDBACK MULTIUSER DETECTION

In this section we consider multiuser detectors that combine several of the features of the decision-driven detectors we have seen in previous sections:

- As in successive cancellation, the intermediate decisions used in decision-feedback detectors are final (output) decisions.
- Decision-feedback detectors operate sequentially, demodulating one bit at a time.
- As in multistage detection with decorrelating first stage, both linear and nonlinear methods are used to combat multiuser interference.

The term *decision-feedback* is borrowed from an approach extensively used for the demodulation of single-user channels subject to intersymbol interference, whereby previous output ("final") decisions are used as surrogate transmitted data to cancel intersymbol interference.

As usual, our main emphasis will be on the synchronous setting, in which the basic principles of decision-feedback multiuser detection can be explained succinctly.

7.5.1 SYNCHRONOUS DECORRELATING DECISION-FEEDBACK

According to Proposition 2.2, the synchronous $K \times K$ crosscorrelation matrix can be factored as

$$\mathbf{R} = \mathbf{F}^T\mathbf{F}, \tag{7.56}$$

where $\mathbf{F}$ is a lower triangular matrix (i.e., $F_{kl} = 0$ if $k < l$). If $\mathbf{R}$ is nonsingular, so is $\mathbf{F}$, and we can process the vector of matched filter outputs by the upper triangular matrix

$$\mathbf{F}^{-T} \stackrel{\text{def}}{=} (\mathbf{F}^T)^{-1},$$

yielding the whitened matched filter outputs (cf. (2.87)):

$$\bar{\mathbf{y}} = \mathbf{F}^{-T}\mathbf{y} = \mathbf{FAb} + \bar{\mathbf{n}}, \tag{7.57}$$

where $\bar{\mathbf{n}}$ is a Gaussian K-vector with independent components, each with variance σ^2. Since $\mathbf{F}$ is lower triangular, the first whitened matched filter output is

$$\bar{y}_1 = F_{11}A_1b_1 + \bar{n}_1. \tag{7.58}$$

Thus, $\bar{y}_1$ contains no interference from other users and its signal-to-noise ratio is that of the decorrelating detector (5.46):

$$\left(\frac{A_1F_{11}}{\sigma}\right)^2 = \frac{A_1^2}{\sigma^2 R_{11}^+}.$$

371

Indeed, we saw in Chapter 5 that

$$\hat{b}_1 = \text{sgn}(\bar{y}_1) \tag{7.59}$$

is the decision of the decorrelator for user 1. For $k > 1$, $\bar{y}_k$ does contain interference from lower-numbered users:

$$\bar{y}_k = F_{kk}A_kb_k + \sum_{j=1}^{k-1} F_{kj}A_jb_j + \bar{n}_k. \tag{7.60}$$

In view of (7.60) and following the philosophy of successive cancellation (Section 7.1), we can demodulate the users sequentially (in the order $1, 2, \ldots, K$) by

$$\hat{b}_k = \text{sgn}\left(\bar{y}_k - \sum_{j=1}^{k-1} F_{kj}A_j\hat{b}_j\right), \tag{7.61}$$

which is the decorrelating decision-feedback detector.

In matrix notation, (7.61) becomes

$$\hat{\mathbf{b}} = \text{sgn}(\mathbf{F}^{-T}\mathbf{y} - (\mathbf{F} - \text{diag}\{\mathbf{F}\})\mathbf{A}\hat{\mathbf{b}}). \tag{7.62}$$

Reflecting on (7.61) we see that the first user is demodulated by its decorrelating detector whereas other users subtract a linear combination of previous decisions from the whitened matched filter outputs. We recall from Chapter 5 that the whitened matched filter output $\bar{y}_k$ is the output of a decorrelating linear transformation for s_k against $\{s_{k+1}, \ldots, s_K\}$. Accordingly, $\bar{y}_k$ contains no trace of users $k + 1, \ldots, K$. Furthermore, if the previous decision on user $j < k$ was correct, then its contribution to $\bar{y}_k$ (7.60) is perfectly canceled in the decision statistic (7.61).

Unlike successive cancellation, in the hypothetical absence of background noise ($\sigma = 0$), the decorrelating decision-feedback detector guarantees error-free demodulation, as we can see from (7.60) and (7.61).

The bit-error-rate analysis of the decorrelating decision-feedback detector benefits from the fact that the noise samples $(\bar{n}_1, \ldots, \bar{n}_K)$ affecting the observables in (7.60) are independent. This implies that the noise $\bar{n}_k$ affecting the decision statistic in (7.61),

$$z_k = \bar{y}_k - \sum_{j=1}^{k-1} F_{kj}A_j\hat{b}_j \tag{7.63}$$

$$= F_{kk}A_kb_k + \sum_{j=1}^{k-1} F_{kj}A_j(b_j - \hat{b}_j) + \bar{n}_k, \tag{7.64}$$

is independent of previous decisions. Accordingly, if we knew $\{(b_j - \hat{b}_j), j = 1, \ldots, k - 1\}$ in addition to z_k, the optimum decision for b_k would be

$$\operatorname{sgn}\left(z_k - \sum_{j=1}^{k-1} F_{kj} A_j (b_j - \hat{b}_j)\right),$$

immediately leading to a lower bound on the bit-error-rate of the decorrelating decision-feedback detector:

$$\mathsf{P}_k^{ddf}(\sigma) \geq Q\left(\frac{A_k F_{kk}}{\sigma}\right). \tag{7.65}$$

Thanks to the independence of the noise components that affect the observables of the decorrelating decision-feedback detector we will be able to obtain a closed-form expression for $\mathsf{P}_k^{ddf}(\sigma)$ for an arbitrary number of users and arbitrary crosscorrelation matrices. Recall that up to now we had been able to obtain exact expressions (for arbitrary K) only for linear multiuser detectors.

User 1 achieves the same error probability as the decorrelating detector,

$$\mathsf{P}_1^{ddf}(\sigma) = Q\left(\frac{A_1 F_{11}}{\sigma}\right) = \mathsf{P}_1^d(\sigma), \tag{7.66}$$

and user 2 achieves the same error probability as a multistage detector with decorrelating first stage (cf. (7.49)):

$$\mathsf{P}_2^{ddf}(\sigma) = \mathsf{P}_2^{dd}(\sigma) = Q\left(\frac{A_2}{\sigma}\right)\left[1 - Q\left(\frac{A_1\sqrt{1-\rho^2}}{\sigma}\right)\right]$$

$$+ Q\left(\frac{A_1\sqrt{1-\rho^2}}{\sigma}\right)\left[\frac{1}{2}Q\left(\frac{A_2 + 2A_1\rho}{\sigma}\right)\right.$$

$$+ \frac{1}{2}Q\left(\frac{A_2 - 2A_1\rho}{\sigma}\right)\right]. \tag{7.67}$$

By symmetry we can condition on $b_k = +1$:

$$\mathsf{P}_k^{ddf}(\sigma) = P[z_k < 0 \,|\, b_k = 1]$$

$$= 2^{1-k} \sum_{(b_1,\ldots,b_{k-1}) \in \{-1,1\}^{k-1}} P[z_k < 0 \,|\, (b_1,\ldots,b_{k-1},1)]$$

$$= 2^{1-k} \sum_{(b_1,\ldots,b_{k-1}) \in \{-1,1\}^{k-1}} \sum_{(\hat{b}_1,\ldots,\hat{b}_{k-1}) \in \{-1,1\}^{k-1}}$$

$$P[z_k < 0 \,|\, (b_1,\ldots,b_{k-1},1),(\hat{b}_1,\ldots,\hat{b}_{k-1})]$$

$$\times P[(\hat{b}_1,\ldots,\hat{b}_{k-1}) \,|\, (b_1,\ldots,b_{k-1},1)]$$

$$= 2^{1-k} \sum_{(b_1,\ldots,b_{k-1})\in\{-1,1\}^{k-1}} \sum_{(\hat{b}_1,\ldots,\hat{b}_{k-1})\in\{-1,1\}^{k-1}}$$

$$P[z_k < 0 \,|\, (b_1,\ldots,b_{k-1},1),(\hat{b}_1,\ldots,\hat{b}_{k-1})]$$

$$\times \prod_{j=1}^{k-1} P[\hat{b}_j \,|\, (b_1,\ldots,b_{j-1}),(\hat{b}_1,\ldots,\hat{b}_{j-1})]$$

$$= 2^{1-k} \sum_{(b_1,\ldots,b_{k-1})\in\{-1,1\}^{k-1}} \sum_{(\hat{b}_1,\ldots,\hat{b}_{k-1})\in\{-1,1\}^{k-1}}$$

$$\pi_k(b_1 - \hat{b}_1,\ldots,b_{k-1} - \hat{b}_{k-1})$$

$$\times \prod_{j=1}^{k-1} \left[1\{b_j = \hat{b}_j\} + (1 - 2\,1\{b_j = \hat{b}_j\}) \right.$$

$$\left. \times \pi_j(b_1 - \hat{b}_1,\ldots,b_{j-1} - \hat{b}_{j-1}) \right], \tag{7.68}$$

where

$$\pi_k(e_1,\ldots,e_{k-1}) = Q\left(F_{kk}\frac{A_k}{\sigma} + \sum_{l=1}^{k-1} F_{kl}\frac{A_l}{\sigma}e_l \right). \tag{7.69}$$

Note that the independence of the noise components is crucial to express the conditional probabilities in the left side of (7.68) in terms of (7.69). Although formidable looking, Equation (7.68) is an exact expression for the error probability of the decorrelating decision-feedback detector.

The probability that the detector makes no errors is easily obtained:

$$P(b_1 = \hat{b}_1,\ldots,b_K = \hat{b}_K) = \prod_{k=1}^{K} \left(1 - Q\left(\frac{A_k F_{kk}}{\sigma} \right) \right). \tag{7.70}$$

The explicit expression in (7.68) is a sum of products of Q-functions. Therefore, it is straightforward to find the asymptotic efficiency (Problem 7.20):

$$\eta_k^{ddf} = \min_{\epsilon\in\{-1,0,1\}^{k-1}} \max^2 \left\{ 0, F_{kk} + 2\sum_{l=1}^{k-1} \frac{A_l}{A_k} F_{kl}\epsilon_l \right\}$$

$$+ \sum_{i=1}^{k-1} v_{\epsilon_i}\left(F_{ii}\frac{A_i}{A_k} + 2\sum_{l=1}^{i-1} \frac{A_l}{A_k} F_{il}\epsilon_l \right), \tag{7.71}$$

with

$$v_\epsilon(x) \stackrel{\text{def}}{=} (1\{x < 0\}1\{\epsilon = 0\} + 1\{x > 0\}1\{\epsilon \neq 0\})\, x^2. \tag{7.72}$$

According to (7.65),

$$\eta_k^{ddf} \leq F_{kk}^2 \tag{7.73}$$

with equality if $k = 1$ (7.66). Furthermore, we can obtain the following expression for the worst asymptotic effective energy:

$$\min_{k=1,\ldots,K} A_k^2 \eta_k^{ddf} = -2 \lim_{\sigma \to 0} \sigma^2 \log P \left[\bigcup_{k=1}^{K} \{b_k \neq \hat{b}_k\} \right] \qquad (7.74)$$

$$= -2 \lim_{\sigma \to 0} \sigma^2 \log \left(1 - \prod_{k=1}^{K} \left[1 - Q \left(\frac{A_k F_{kk}}{\sigma} \right) \right] \right) \qquad (7.75)$$

$$= \min_{k=1,\ldots,K} A_k^2 F_{kk}^2, \qquad (7.76)$$

where (7.74), (7.75), and (7.76) come from (3.116), (7.70), and (3.44), respectively. Since (7.76) holds for all $(A_1, \ldots, A_K)$, the power-tradeoff region of the decorrelating decision-feedback detector is the same (asymptotically as $\sigma \to 0$) as that of a hypothetical multiuser detector that would attain (7.65) with equality, that is, to achieve error probability no worse than P for all users, the signal-to-noise ratio should satisfy, asymptotically,

$$\frac{A_k^2}{\sigma^2} \geq \frac{(Q^{-1}(P))^2}{F_{kk}^2}. \qquad (7.77)$$

Comparing this to the power-tradeoff region of the decorrelating detector (5.49):

$$\frac{A_k^2}{\sigma^2} \geq (Q^{-1}(P))^2 R_{kk}^+, \qquad (7.78)$$

and recalling (Problem 5.23) that $F_{kk}^2 R_{kk}^+ \geq 1$, we conclude that the power-tradeoff region of the decorrelating decision-feedback detector is better than (i.e., contains) the power-tradeoff region of the decorrelating detector. We emphasize that this conclusion is guaranteed to hold asymptotically as $\sigma \to 0$; however, we did not take the optimistic view that previous decisions are correct nor did we assume that users are decoded in any particular order.

In the two-user case, $F_{11}^2 = 1 - \rho^2$ and $F_{22} = 1$. Therefore, in that case, the power-tradeoff region of the decorrelating decision feedback detector is

$$\frac{A_1^2}{\sigma^2} \geq \frac{(Q^{-1}(P))^2}{1 - \rho^2}, \quad \frac{A_2^2}{\sigma^2} \geq (Q^{-1}(P))^2, \qquad (7.79)$$

whereas the decorrelating detector has power-tradeoff region

$$\frac{A_1^2}{\sigma^2} \geq \frac{(Q^{-1}(P))^2}{1 - \rho^2}, \quad \frac{A_2^2}{\sigma^2} \geq \frac{(Q^{-1}(P))^2}{1 - \rho^2}. \qquad (7.80)$$

According to (7.77), if users adjust their powers to minimize the total transmitted power required to achieve a certain common level of error probability, then each user should satisfy (7.77) with equality (i.e., the power transmitted should be $1/F_{kk}^2$ times the power required to achieve the target error probability in a single-user system). It is interesting to investigate the average per-user power penalty:

$$\zeta^{ddf} = \frac{1}{K} \sum_{k=1}^{K} \frac{1}{F_{kk}^2}, \qquad (7.81)$$

and contrast it to that of the decorrelator:

$$\zeta^d = \frac{1}{K} \sum_{k=1}^{K} R_{kk}^+. \qquad (7.82)$$

In particular, we can examine the asymptotic ($K \to \infty$) behavior of these quantities under the random spread-spectrum model with spreading factor equal to N and

$$\beta = \lim_{K \to \infty} \frac{K}{N} < 1.$$

According to (4.121),

$$\lim_{K \to \infty} E[\zeta^d] = \lim_{K \to \infty} \frac{1}{K} \sum_{k=1}^{K} E[R_{kk}^+] = \frac{1}{1 - \beta}. \qquad (7.83)$$

Now, the average power penalty of the decorrelating decision-feedback detector satisfies

$$E[\zeta^{ddf}] = \frac{1}{K} \sum_{k=1}^{K} E\left[\frac{1}{F_{kk}^2}\right] \qquad (7.84)$$

$$= \frac{1}{K} \sum_{k=1}^{K} E[\mathbf{R}^+[k, K]_{11}] \qquad (7.85)$$

$$\to \lim_{K \to \infty} \frac{1}{K} \sum_{k=1}^{K} \frac{1}{1 - \frac{K-k}{N}} \qquad (7.86)$$

$$= \lim_{K \to \infty} \frac{1}{K} \sum_{j=1}^{K} \frac{1}{1 - \frac{j}{N}} \qquad (7.87)$$

$$= \lim_{K \to \infty} \frac{N}{K} \sum_{j=1}^{K} \frac{1}{N - j} \qquad (7.88)$$

$$= \frac{1}{\beta} \lim_{K \to \infty} \sum_{j=1}^{K} \log\left(1 + \frac{1}{N - j}\right) \qquad (7.89)$$

$$= \frac{1}{\beta} \lim_{K \to \infty} \log\left(\prod_{j=1}^{K} \frac{N - j + 1}{N - j}\right) \qquad (7.90)$$

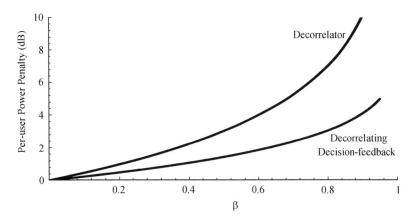

FIGURE 7.24.
Average per-user power penalty for random signature sequences.

$$= \frac{1}{\beta} \lim_{K \to \infty} \log\left(\frac{1}{1 - \frac{K}{N}}\right) \qquad (7.91)$$

$$= \frac{1}{\beta} \log\left(\frac{1}{1 - \beta}\right), \qquad (7.92)$$

where (7.85) follows from (2.137); (7.86) follows from (4.121) and the definition of $\mathbf{R}[k, K]$; and (7.89) is a standard property of the logarithmic function. The functions of β in (7.83) and (7.92) are compared in Figure 7.24.

7.5.2 SYNCHRONOUS MMSE DECISION-FEEDBACK

The demodulation of user k by the decorrelating decision-feedback detector assumes that the decisions of users $\{1, \ldots, k-1\}$ are correct and decorrelates against users $\{k+1, \ldots, K\}$. This is seen in Equation (7.62),

$$\hat{\mathbf{b}} = \text{sgn}(\mathbf{F}^{-T}\mathbf{y} - (\mathbf{F} - \text{diag}\{\mathbf{F}\})\mathbf{A}\hat{\mathbf{b}}), \qquad (7.93)$$

where the purpose of the upper triangular matrix $\mathbf{F}^{-T}$ is to decorrelate against "future" users and the purpose of the strictly lower triangular feedback matrix $(\mathbf{F} - \text{diag}\{\mathbf{F}\})$ is to cancel the interference from "previous" users. In view of the results in Chapter 6 it is natural to investigate the replacement of those linear transformations by nondecorrelating linear transformations.

A general form for a synchronous decision-feedback detector is

$$\hat{\mathbf{b}} = \text{sgn}(\mathbf{G}\mathbf{y} - \mathbf{B}\mathbf{A}\hat{\mathbf{b}}), \qquad (7.94)$$

where the $K \times K$ matrices $\mathbf{G}$ and $\mathbf{B}$ are to be specified. A basic constraint is that $\mathbf{B}$ be strictly lower triangular so that (7.94) can be implemented sequentially (i.e., $\hat{b}_k$ does not depend on $\hat{b}_j$, $j \geq k$). Note that the special case

$$G = I,$$

$$B_{ij} = \begin{cases} \rho_{ij}, & \text{if } i > j, \\ 0, & \text{if } i \leq j \end{cases}$$

corresponds to successive cancellation (7.8), and the special case

$$G = F^{-T},$$

$$B = F - \text{diag}\{F\}$$

corresponds to decorrelating decision-feedback.

One way to assess the quality of the decision statistics,

$$z = Gy - BA\hat{b},$$

is to consider the covariance matrix of the difference between the inputs to the decision device and the true transmitted bits weighted by their amplitudes:

$$\text{cov}\{z - Ab\} = \text{cov}\{Gy - BA\hat{b} - Ab\}$$

$$= \text{cov}\{GRAb + Gn - BA\hat{b} - Ab\}$$

$$= \sigma^2 GRG^T + \text{cov}\{(GR - I)Ab - BA\hat{b}\}$$

$$= \sigma^2 GRG^T + [GR - I]A^2[GR - I]^T$$

$$+ BAE[\hat{b}\hat{b}^T]AB^T$$

$$- BAE[\hat{b}b^T][GR - I]^T - [GR - I]AE[b\hat{b}^T]AB^T.$$

$$(7.95)$$

Now, to be able to give an expression for the error covariance matrix in terms of the design parameters, we would need expressions for $E[\hat{b}\hat{b}^T]$ and $E[b\hat{b}^T]$ as functions of G and B. Even if those expressions were known, which they are not, the complexity of the ensuing optimization problem would be unwieldy. A judicious alternative to this design problem is to substitute those matrices by the approximations

$$E[\hat{b}\hat{b}^T] \approx I$$

and

$$E[b\hat{b}^T] \approx I,$$

which are manifestly justified unless the error probabilities are high, and select $\mathbf{G}$ and $\mathbf{B}$ to minimize the trace of the ensuing covariance matrix (7.95):

$$\sigma^2 \mathbf{G} \mathbf{R} \mathbf{G}^T + [\mathbf{G}\mathbf{R} - \mathbf{B} - \mathbf{I}] \mathbf{A}^2 [\mathbf{G}\mathbf{R} - \mathbf{B} - \mathbf{I}]^T.$$

We will proceed in two stages. First we optimize the choice of $\mathbf{G}$ for arbitrary $\mathbf{B}$, and then we select the optimum strictly lower triangular matrix $\mathbf{B}$. The solution to this problem will provide the matrices that minimize the mean-square errors between the inputs to the thresholds and the true data if the fed-back decisions were correct.

Using the definition

$$\bar{\mathbf{G}} \stackrel{\text{def}}{=} [\mathbf{B} + \mathbf{I}](\sigma^2 \mathbf{A}^{-2} + \mathbf{R})^{-1}, \tag{7.96}$$

it is straightforward to check (cf. (6.24)) that

$$\sigma^2 \mathbf{G} \mathbf{R} \mathbf{G}^T + [\mathbf{G}\mathbf{R} - \mathbf{B} - \mathbf{I}] \mathbf{A}^2 [\mathbf{G}\mathbf{R} - \mathbf{B} - \mathbf{I}]^T$$

$$= \sigma^2 [\mathbf{B} + \mathbf{I}](\sigma^2 \mathbf{A}^{-2} + \mathbf{R})^{-1} [\mathbf{B} + \mathbf{I}]^T$$

$$+ (\mathbf{G} - \bar{\mathbf{G}})(\mathbf{R} \mathbf{A}^2 \mathbf{R} + \sigma^2 \mathbf{R})(\mathbf{G} - \bar{\mathbf{G}})^T. \tag{7.97}$$

Because of the nonnegative definiteness of $(\mathbf{R} \mathbf{A}^2 \mathbf{R} + \sigma^2 \mathbf{R})$, the matrix $\mathbf{G} = \bar{\mathbf{G}}$ minimizes the trace of (7.97).

Now let us proceed to minimize

$$\sigma^2 \, \text{trace}\{[\mathbf{B} + \mathbf{I}](\sigma^2 \mathbf{A}^{-2} + \mathbf{R})^{-1} [\mathbf{B} + \mathbf{I}]^T\}$$

with respect to strictly lower triangular $\mathbf{B}$. To that end, it is convenient to introduce the Cholesky factorization of the nonsingular matrix

$$\sigma^2 \mathbf{A}^{-2} + \mathbf{R} = \mathbf{F}_\sigma^T \mathbf{F}_\sigma \tag{7.98}$$

and to define the lower triangular matrix

$$\mathbf{L} \stackrel{\text{def}}{=} [\mathbf{B} + \mathbf{I}] \mathbf{F}_\sigma^{-1}.$$

Then

$$\text{trace}\{[\mathbf{B} + \mathbf{I}](\sigma^2 \mathbf{A}^{-2} + \mathbf{R})^{-1} [\mathbf{B} + \mathbf{I}]^T\}$$

$$= \text{trace}\{\mathbf{L} \mathbf{L}^T\}$$

$$= \sum_{k=1}^{K} \sum_{l=1}^{k} L_{kl}^2$$

$$\geq \sum_{k=1}^{K} L_{kk}^2$$

$$= \sum_{k=1}^{K} \left(\mathbf{F}_\sigma^{-1}\right)_{kk}^2 \tag{7.99}$$

$$= \sum_{k=1}^{K} \frac{1}{\left(\mathbf{F}_\sigma\right)_{kk}^2}, \tag{7.100}$$

where we used the lower triangularity of $[\mathbf{B} + \mathbf{I}]$, $\mathbf{F}_\sigma^{-1}$, and $\mathbf{F}_\sigma$ and the fact that the diagonal elements of the product of two lower triangular matrices are the product of the respective diagonal elements.

Moreover, letting

$$\tilde{\mathbf{F}}_\sigma \stackrel{\text{def}}{=} \mathrm{diag}^{-1}\{\mathbf{F}_\sigma\}\mathbf{F}_\sigma$$

we can express

$$[\mathbf{B} + \mathbf{I}]\left(\mathbf{F}_\sigma^T \mathbf{F}_\sigma\right)^{-1}[\mathbf{B} + \mathbf{I}]^T = [\mathbf{B} + \mathbf{I}]\left(\tilde{\mathbf{F}}_\sigma^T \mathrm{diag}^2\{\mathbf{F}_\sigma\}\tilde{\mathbf{F}}_\sigma\right)^{-1}[\mathbf{B} + \mathbf{I}]^T$$

$$= [\mathbf{B} + \mathbf{I}](\tilde{\mathbf{F}}_\sigma)^{-1}\mathrm{diag}^{-2}\{\mathbf{F}_\sigma\}(\tilde{\mathbf{F}}_\sigma)^{-T}[\mathbf{B} + \mathbf{I}]^T$$

$$= \mathrm{diag}^{-2}\{\mathbf{F}_\sigma\}, \tag{7.101}$$

where (7.101) holds if we choose the strictly lower triangular matrix

$$\bar{\mathbf{B}} = \mathrm{diag}^{-1}\{\mathbf{F}_\sigma\}\mathbf{F}_\sigma - \mathbf{I}. \tag{7.102}$$

But the trace of the matrix in (7.101) coincides with the lower bound in (7.100). Thus, we have shown that (7.102) is the optimum strictly lower triangular matrix. Substituting (7.102) into (7.96) we obtain

$$\bar{\mathbf{G}} = \mathrm{diag}^{-1}\{\mathbf{F}_\sigma\}\mathbf{F}_\sigma(\sigma^2\mathbf{A}^{-2} + \mathbf{R})^{-1} \tag{7.103}$$

$$= \mathrm{diag}^{-1}\{\mathbf{F}_\sigma\}\mathbf{F}_\sigma\mathbf{F}_\sigma^{-1}\mathbf{F}_\sigma^{-T} \tag{7.104}$$

$$= \mathrm{diag}^{-1}\{\mathbf{F}_\sigma\}\mathbf{F}_\sigma^{-T}. \tag{7.105}$$

With those choices of $\bar{\mathbf{G}}$ and $\bar{\mathbf{B}}$, the decision-feedback detector becomes

$$\hat{\mathbf{b}} = \mathrm{sgn}\left(\mathrm{diag}^{-1}\{\mathbf{F}_\sigma\}\mathbf{F}_\sigma^{-T}\mathbf{y} - (\mathrm{diag}^{-1}\{\mathbf{F}_\sigma\}\mathbf{F}_\sigma - \mathbf{I})\mathbf{A}\hat{\mathbf{b}}\right)$$

$$= \mathrm{sgn}\left(\mathrm{diag}^{-1}\{\mathbf{F}_\sigma\}[\mathbf{F}_\sigma^{-T}\mathbf{y} - (\mathbf{F}_\sigma - \mathrm{diag}\{\mathbf{F}_\sigma\})\mathbf{A}\hat{\mathbf{b}}]\right) \tag{7.106}$$

$$= \mathrm{sgn}\left(\mathbf{F}_\sigma^{-T}\mathbf{y} - (\mathbf{F}_\sigma - \mathrm{diag}\{\mathbf{F}_\sigma\})\mathbf{A}\hat{\mathbf{b}}\right). \tag{7.107}$$

We conclude that the same detector is obtained with

$$\mathbf{G} = \mathbf{F}_\sigma^{-T}$$

and

$$\mathbf{B} = \mathbf{F}_\sigma - \mathrm{diag}\{\mathbf{F}_\sigma\},$$

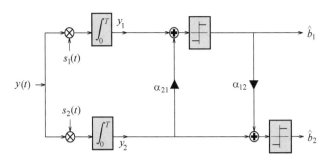

FIGURE 7.25.
Two-user
synchronous
decision-feedback
detector.

which is the MMSE decision-feedback detector. Note that in the special case of $\sigma = 0$ and invertible $\mathbf{R}$, the MMSE decision-feedback detector becomes the decorrelating decision-feedback detector.

Even though we placed no such restriction at the outset, the optimum matrix $\mathbf{G}$ is upper triangular, as in the case of the decorrelating decision-feedback detector. This means that in the demodulation of user k, the mitigation of interferers $k+1, \ldots, K$ is purely linear, whereas the mitigation of interferers $1, \ldots, k-1$ is by nonlinear cancellation. In Figure 7.25 we have depicted the general form of a two-user decision-feedback detector where user 1 is demodulated before user 2. For the MMSE decision-feedback detector the coefficients in Figure 7.25 are equal to

$$\alpha_{21} = \frac{-\rho}{1 + \sigma^2 / A_2^2}, \tag{7.108}$$

$$\alpha_{12} = -\rho A_1. \tag{7.109}$$

Comparing to Figure 6.4 we see that user 1's decision is the same as that of a linear MMSE detector. This observation holds for any number of users because $\mathbf{F}_\sigma^{-1}$ is lower triangular and its diagonal elements are positive. Thus,

$$\mathrm{sgn}\big((\mathbf{F}_\sigma^{-T} \mathbf{y})_1\big) = \mathrm{sgn}\big((\mathbf{F}_\sigma^{-1} \mathbf{F}_\sigma^{-T} \mathbf{y})_1\big)$$

$$= \mathrm{sgn}(((\sigma^2 \mathbf{A}^{-2} + \mathbf{R})^{-1} \mathbf{y})_1). \tag{7.110}$$

In addition to the fact that it does not require the crosscorrelation matrix to be invertible, one of the main advantages of the MMSE decision-feedback detector over the decorrelating decision-feedback detector is that adaptive training-sequence-based versions are easy to obtain.

Unlike the decorrelating decision-feedback detector, the MMSE decision-feedback detector does not lend itself to an exact analysis of bit-error-rate since the noise vector in $\mathbf{F}_\sigma^{-T} \mathbf{y}$ no longer has independent components. Naturally, for high signal-to-noise ratios the bit-error-rate will be close to the decorrelating decision-feedback detector. In particular, for low bit-error-rates the power-tradeoff region of the MMSE decision-feedback detector is guaranteed to contain the power-tradeoff region of the MMSE linear detector.

7.5.3   ASYNCHRONOUS DECISION-FEEDBACK

The generalization of the decision-feedback detectors we have seen in Sections 7.5.1 and 7.5.2 to asynchronous channels is conceptually straightforward.

We can start by viewing the asynchronous channel with a finite number of bits per frame as a synchronous channel with as many "users" as bits (cf. Section 2.2). A natural way to order the "users" for decision-feedback is by their time of arrival. But this is neither mandatory nor necessarily optimal; consider the case in which bit $b_j[i]$ occurs slightly after $b_l[i]$ but $A_j \gg A_l$. Conversely, ordering by received powers is not necessarily optimal. Depending on whether a decorrelating decision-feedback detector or an MMSE decision-feedback detector is implemented, the feedforward and feedback matrices can be obtained by factoring either the crosscorrelation matrix or its sum with the diagonal matrix of noise-to-signal ratios. However, this approach is hardly implementable because the size of the matrices is proportional to the frame length. Fortunately, as we saw in Chapters 5 and 6, those linear transformations become time-invariant K-dimensional linear filters as the frame length grows. The role of the crosscorrelation matrix $\mathbf{R}$ is now played by the transfer function (2.112):

$$\mathbf{S}(z) = \mathbf{R}^T[1]z + \mathbf{R}[0] + \mathbf{R}[1]z^{-1}, \tag{7.111}$$

which according to Proposition 2.3 can be factored as

$$\mathbf{S}(z) = [\mathbf{F}[0] + \mathbf{F}[1]z]^T [\mathbf{F}[0] + \mathbf{F}[1]z^{-1}], \tag{7.112}$$

where $\mathbf{F}[0]$ is lower triangular and $\mathbf{F}[1]$ is upper triangular with zero diagonal.

For the asynchronous decorrelating decision-feedback receiver the feedforward and feedback filters have $K \times K$ transfer functions:

$$\mathbf{G}(z) = [\mathbf{F}[0] + \mathbf{F}[1]z]^{-T}, \tag{7.113}$$

$$\mathbf{B}(z) = \mathbf{F}[0] - \text{diag}\,\mathbf{F}[0] + \mathbf{F}[1]z^{-1}. \tag{7.114}$$

As in the synchronous case, $\mathbf{G}(z)$ acts as a linear decorelating transformation for those bits that have not yet been demodulated. If the decoding order is determined by the time of arrival, then $\mathbf{G}(z)$ is a purely noncausal transformation (including at each time the bit of interest). Conversely, in

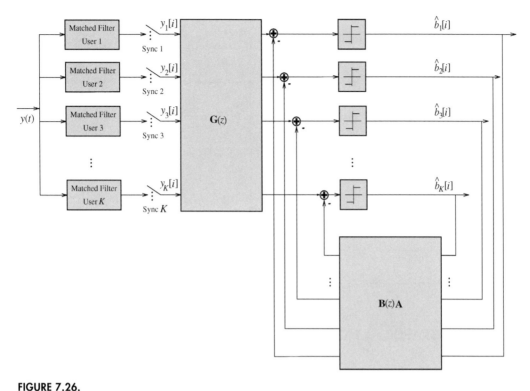

FIGURE 7.26.

Asynchronous
decision-feedback
multiuser detector.

that case, the feedback filter is causal. Other asynchronous decorrelating decision-feedback receivers can be envisioned. For example, the decorrelation against future bits need not have an infinite impulse response if the one-shot decorrelator approach of Section 5.3 is used (Problem 7.27).

The asynchronous MMSE decision-feedback multiuser detector (Figure 7.26) uses feedforward and feedback filters with transfer functions

$$\mathbf{G}(z) = [\mathbf{F}_\sigma[0] + \mathbf{F}_\sigma[1]z]^{-T} \tag{7.115}$$

and

$$\mathbf{B}(z) = \mathbf{F}_\sigma[0] - \mathrm{diag}\,\mathbf{F}_\sigma[0] + \mathbf{F}_\sigma[1]z^{-1}, \tag{7.116}$$

where we have factored

$$\mathbf{S}(z) + \sigma^2\mathbf{A}^{-2} = [\mathbf{F}_\sigma[0] + \mathbf{F}_\sigma[1]z]^T[\mathbf{F}_\sigma[0] + \mathbf{F}_\sigma[1]z^{-1}]. \tag{7.117}$$

A comparison of the requirements of various multiuser detectors is shown in Figure 7.27. The MMSE detector is the only detector for which the requirements of both the adaptive and nonadaptive versions are shown in Figure 7.27.

	Single-User Matched Filter	Maximum Likelihood	Minimum Bit-Error-Rate	Decorrelator	MMSE	Adaptive MMSE	Blind MMSE	Decision Driven
Signature Waveform of Desired User.	■	■	■	■	■		■	■
Timing of Desired User.	■	■	■	■	■	■	■	■
Received Amplitudes.		■	■		■			■
Noise Level.			■		■			
Signature Waveforms of Interfering Users.		■	■	■	■			■
Timing of Interfering Users.		■	■	■	■			■
Training Sequence of Desired User's Data.						■		

FIGURE 7.27.
Comparison of requirements for multiuser detectors.

7.6 BIBLIOGRAPHICAL NOTES

The idea of successive decoding in multiaccess communication goes back to the information theoretic study of the Gaussian multiaccess channel (Cover [61]), where it was devised as a technique to show the achievability of the capacity region. The order in which users are decoded depends on their transmission rate (with the user transmitting at full single-user capacity being decoded last). In that context, information is encoded with vanishing blockwise probability of error and successive decoding is asymptotically optimal (in contrast to our setting). In the analysis of asymptotic efficiency of the demodulators in this chapter (as well as optimum asymptotic efficiency) we found that strong interference is equivalent to no interference. An information theoretic study of the Gaussian interference channel (Carleial [45]) revealed a similar property.

In the context of multiuser detection of CDMA channels (without error control decoding), successive cancellation has been examined in Kohno [213], Dent et al. [65], Yoon et al. [555, 556], Patel and Holtzman [323], Holtzman [149], and Divsalar et al. [67]. In those works, cancellation residuals are approximated as white Gaussian independent sources. The use of matched-filter outputs for estimating the cancellation coefficients and the order of cancellation (cf. (7.1) and (7.15)) is proposed in Patel and Holtzman [323] and [324]. Other implementation issues related to the determination of the cancellation order and power control are examined in Kubota et al. [223], Pedersen et al.

384

[328], and Seskar et al. [403]. Analyses of successive cancellation in fading channels have been reported in Yoon et al. [556], Patel and Holtzman [325], Wijk et al. [537], Johansson and Svensson [177], and Mazzinni [271]. Successive cancellation of nonlinearly modulated signals is considered in Ewerbring et al. [83]. Successive cancellation has been applied to array receivers in Kohno et al. [218] and Ghazi-Moghadam and Kaveh [111].

Prior to the foregoing works on successive cancellation, the multistage detectors were proposed in Varanasi and Aazhang [472, 474] and Kohno et al. [217] for asynchronous channels and in Varanasi and Aazhang [473, 476] for synchronous channels. Multistage detectors with conventional front-end and multipath combining have been explored in Yoon et al. [556] and Fawer and Aazhang [87]. Further results on the bit-error-rate of multistage detectors with conventional front-ends are reported in Divsalar and Simon [66] and Hottinen et al. [163]. The two-user asymptotic efficiency of asynchronous multistage detection with soft decisions is found in Zhang and Brady [560]. The effect of receiver mismatch (timing, amplitude, and phase) on the performance of the multistage detector with conventional first stage is explored in Gray et al. [126]. Multistage detectors with multiple-antenna reception are considered in Kandala et al. [199] and Hottinen [162]. The latter paper and Hottinen and Pehkonen [164] consider multistage detection for multirate CDMA. An adaptive form of the multistage detector with decorrelating first stage is analyzed in Siveski et al. [411] and Zhu et al. [564]. The differentially-coherent version of the multistage detector with decorrelating first stage is obtained in Hegarty and Vojcic [140]. Multistage noncoherent detection for nonlinear modulation is explored in Hegarty and Vojcic [142], Halford and Brandt-Pearce [135], and Visotsky and Madhow [506]; in the latter reference, tentative decisions supplied by a successive canceler are used to select a decorrelating final stage. Multistage coherent demodulation of m-ary spread-spectrum multiaccess is explored in Tachikawa [438]. Algebraic methods for multistage decoding of noncoherent frequency-hopped spread-spectrum multiaccess date back to Timor [446]. Multistage detection has been explored for optical CDMA in Brandt-Pearce and Aazhang [33].

The continuous-time tentative decision detector of Section 7.4 was proposed in Abrams et al. [8].

The decorrelating decision-feedback detector for synchronous channels is due to Duel-Hallen [73]. The optimal allocation of powers to minimize total power (cf. (7.77)) is found in Varanasi [469]. Further results on the analysis of the decorrelating decision-feedback detector (cf. Problem 7.21) were given in Varanasi [471]. The replacement of the decorrelating transformation with a linear transformation that maximizes asymptotic efficiency (Section 6.1) has been explored in Varanasi [470, 471]. A detector that works with the observables (7.63) of the decorrelating decision-feedback detector is proposed in Wei and Schlegel [527]. The use of maximum-likelihood sequence detection in the decorrelating decision-feedback structure has been studied in Sauer-Greff and Kennedy [382] for multiuser channels with intersymbol interference. The differentially-coherent version of decorrelating decision-feedback detection is considered in Wu and Duel-Hallen [546]. An adaptive synchronous decorrelating decision-feedback structure is presented in Chen and Roy [51].

The first decision-driven multiuser detector reported in the literature appears to be a four-user MMSE decision-feedback detector (Kavehrad and Salz [203]) designed to cancel cross-polarization interference.

Various forms of asynchronous decision-feedback multiuser detectors have been proposed and analyzed (Xie et al. [550]; Abdulrahman and Falconer [4]; Duel-Hallen [72, 74]; Verdú [498]; Abdulrahman et al. [5]; Rapajic and Vucetic [352]; Petersen and Falconer [330]; Yang and Roy [551]; Wu and Duel-Hallen [545]; Tidestav et al. [444]; Varanasi [469]). Some of those works consider detectors that instead of operating with the sufficient statistics provided by a bank of matched filters operate with the fractionally sampled output of the matched filter of the user of interest. Adaptive implementations are reviewed in Falconer et al. [84]. Joint antenna-diversity reception and decision-feedback multiuser detection is considered in Jung et al. [184], Jung and Blanz [183], and Gray et al. [127]. The MMSE decision-feedback receiver has been shown to minimize geometric MSE (i.e., not only the trace but the determinant of the error covariance matrix) in Yang and Roy [551]. Algorithms for recursive Cholesky factorization of asynchronous crosscorrelation matrices are studied in Kaleh [195], Wei and Rasmussen [525], and Alexander and Rasmussen [15].

The replacement of the sign nonlinearities by soft tentative decisions has been explored in Zhang and Brady [559], Brady and Catipovic [32], Chen et al. [53], Divsalar and Simon [66], Vanghi and Vojcic [465], and Nelson and Poor [300]. The notion of partial interference cancellation that deemphasizes less reliable decisions was introduced in Divsalar and Simon [66] and Abrams et al. [8].

386

Decision-driven multiuser detectors are naturally coupled with error-control codes. At the expense of increased delay, the reliability of decisions is improved by error-control decoding. An illustrative set of references on this research direction includes Hoeher [146], El-Ezabi and Duel-Hallen [77], Shaheen and Gupta [404], Fawer and Aazhang [87], Vojcic [512], Hafeez and Stark [131], Giallorenzi and Wilson [115], Sanada and Wang [379], Saifudding et al. [374], Saifudding [373], and Sanada and Nakagawa [378].

Nonlinear multiuser detectors that fall outside the scope of Chapters 4 and 7 are pursued in Verdú [487], Poor and Verdú [337], Miyajima and Hasegawa [283], Douglas and Cichocki [70], Paris et al. [318], and Aazhang et al. [1].

7.7 PROBLEMS

PROBLEM 7.1. In this problem we show that successive cancellation in the order of decreasing powers is not necessarily optimum. Consider a three-user synchronous channel with crosscorrelation matrix

$$\mathbf{R} = \begin{bmatrix} 1 & \rho_{12} & 0.8 \\ \rho_{12} & 1 & 0.4 \\ 0.8 & 0.4 & 1 \end{bmatrix}$$

and

$$A_1 = 1.0,$$
$$A_2 = 1.1,$$
$$A_3 = 1.2.$$

(a) Show that in the absence of background noise the successive canceler that demodulates the users in the order $\{3, 2, 1\}$ has bit-error-rates:

$$P_1^s(0) = \frac{1}{4},$$
$$P_2^s(0) = \frac{1}{4}1\{\rho_{12} < -0.14\},$$
$$P_3^s(0) = \frac{1}{4}.$$

(b) Show that in the absence of background noise the successive canceler that demodulates the users in the order $\{2, 3, 1\}$ recovers the data error-free if $\rho_{12} < 0.62$.

PROBLEM 7.2. (Viterbi [509]) Assume a hypothetical perfect successive cancellation receiver that after cancelling the ith user obtains a signal free of the contributions of users $1, \ldots, i-1$. Further, assume that the contribution of the background noise after each cancellation is equal to σ^2. Using the infinite-user approximation to the error probability of the single-user matched filter with random signature waveforms, show that for the signal-to-noise ratio seen at each stage to be equal, the received powers must satisfy

$$\frac{A_k^2}{\sigma^2} = \gamma \left(1 + \frac{\gamma}{N} \right)^{K-k}. \tag{7.118}$$

PROBLEM 7.3. (Viterbi [509]) The signal-to-noise ratio in (7.35) is upper bounded by

$$\gamma_k = \frac{A_k^2}{\sigma^2 + \frac{1}{N} \sum_{j=1}^{k-1} A_j^2}.$$

Show that

$$\sum_{k=1}^{K} \gamma_k \geq N \log \left(1 + \frac{1}{N} \sum_{k=1}^{K} \frac{A_k^2}{\sigma^2} \right).$$

PROBLEM 7.4. Express the multistage detectors with conventional and decorrelating first stages in terms of the independent-noise L-dimensional model (2.97):

$$\mathbf{r} = \mathbf{SAb} + \sigma \mathbf{m}.$$

PROBLEM 7.5. Verify equations (7.24) and (7.25), which hold for $\rho > 0$. [*Hint:* Show

$$\Delta(+-, -+) \leq \|S(+-)\| < \|S(++)\| \leq \Delta(++, --).]$$

PROBLEM 7.6. (Patel and Holtzman [324]) Assume that the received amplitudes are independent identically distributed with common cumulative distribution function F_A (density function f_A), and label the users so that

$$A_K \leq \cdots \leq A_1.$$

Show that the density function of A_k is

$$f_{A_k}(x) = \frac{k!}{(K-k)!(k-1)!} F_A^{K-k}(x)[1 - F_A(x)]^{k-1} f_A(x).$$

PROBLEM 7.7. Consider a two-user synchronous CDMA channel using the signature waveforms in Figure 7.28. Suppose that $\hat{b}_2$ is the decision of the conventional single-user matched filter detector

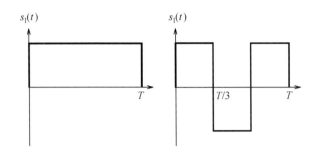

FIGURE 7.28.
Direct-sequence
signature
waveforms with
$N = 3$.

for user 2. Recall that a successive canceler demodulator for user 1 outputs

$$\hat{b}_1 = \text{sgn}(A_1 b_1 + A_2(b_2 - \hat{b}_2)\rho + \sigma \langle n, s_1 \rangle).$$

(a) Find the conditional distribution of the random variable $b_2 - \hat{b}_2$ given $b_1 = +1$.

(b) Find the conditional distribution of the random variable $b_2 - \hat{b}_2$ given $b_1 = -1$.

(c) Is $P[\langle n, s_1 \rangle \geq 0 \,|\, b_2 - \hat{b}_2 = 2]$ greater than, equal to, or less than $1/2 = P[\langle n, s_1 \rangle \geq 0]$?

(d) Find the variance of the random variable

$$A_2(b_2 - \hat{b}_2)\rho + \sigma \langle n, s_1 \rangle.$$

PROBLEM 7.8. Let s and D be an element and a convex subset, respectively, of a signal space such that $s \notin D$. Let d be such that $s + d \in D$ and

$$\min_{v \in D} \|v - s\| = \|d\|,$$

and denote by C the subset of all signals $s + v$ such that

$$\langle v, d \rangle \geq \|d\|^2.$$

Show that $D \subset C$. [*Hint:* The convexity of D must be used.]

PROBLEM 7.9. Consider two synchronous users such that user 1 is demodulated by cancellation of user 2 using the decisions of a conventional single-user matched filter for user 2.

(a) Show that the conventional matched filter receiver achieves higher asymptotic efficiency than the successive canceler if

$$\frac{A_2}{A_1} \leq \begin{cases} \frac{2|\rho|}{\frac{1}{1-\rho^2} - \rho^2}, & \text{if } 0 < |\rho| < \sqrt{\frac{3-\sqrt{5}}{2}}; \\ 1, & \text{otherwise.} \end{cases} \quad (7.119)$$

(b) Show that a successive cancellation receiver that switches to a single-user matched filter when (7.119) holds achieves the

389

following near–far resistance:

$$\bar{\eta}_1^s = \begin{cases} \left(\frac{1-3\rho^2+3\rho^4}{1-\rho^2+\rho^4}\right)^2, & \text{if } 0 < |\rho| < \sqrt{\frac{3-\sqrt{5}}{2}}; \\ (1-|\rho|)^2, & \text{otherwise.} \end{cases}$$

(c) Show that the successive cancellation receiver that does not switch to the single-user matched filter achieves near–far resistance:

$$\bar{\eta}_1^s = \begin{cases} (1-2\rho^2)^2, & \text{if } |\rho| < \frac{1}{\sqrt{2}}; \\ 0, & \text{otherwise.} \end{cases}$$

PROBLEM 7.10.

(a) Show that the decision regions shown in Figure 7.17 for a synchronous channel with $\rho > 0$ correspond to those of an m-stage detector, $m = 2, 3, \ldots$, with conventional first stage. The shaded regions in Figure 7.17 are those for which the detector enters a limit cycle. [*Hint:* Partition the space in sixteen regions according as

$$y_1 \in \{(-\infty, -A_2\rho], (-A_2\rho, 0), [0, A_2\rho), [A_2\rho, \infty)\},$$

$$y_2 \in \{(-\infty, -A_1\rho], (-A_1\rho, 0), [0, A_1\rho), [A_1\rho, \infty)\},$$

and find the sequence of decisions in each of those regions.]
(b) Repeat (a) for an m-stage detector with decorrelating first stage, (i.e., verify Figure 7.20).

PROBLEM 7.11. Find the length of the sliding window of tentative decisions that must be stored at the jth stage of an m-stage detector for a K-user asynchronous channel.

PROBLEM 7.12. Recall (Problem 4.47) that the single-user matched filter is asymptotically optimal in the sense that

$$\lim_{\sigma \to \infty} \frac{\sigma}{A_k} \left(P_k^c(\sigma) - P_k(\sigma) \right) = 0.$$

For a two-user synchronous two-stage detector with decorrelating first stage, find

$$\lim_{\sigma \to \infty} \frac{\sigma}{A_1} \left(P_1^{dd}(\sigma) - P_1(\sigma) \right).$$

PROBLEM 7.13. Show that the near–far resistance of the synchronous two-user two-stage detector with decorrelating first stage is given by

$$\bar{\eta}_1^{dd} = \frac{1 + 4|\rho| - \rho^2 - 12|\rho|^3 + 8\rho^4}{(1 + 2|\rho| - \rho^2)^2}.$$

PROBLEM 7.14. Consider two synchronous users with

$$|\rho| \leq \frac{1}{3}.$$

Show that the power-tradeoff region of the two-stage detector with decorrelating first stage is asymptotically optimal for large signal-to-noise ratios.

PROBLEM 7.15. Find the covariance matrix

$$E[\hat{\mathbf{b}}\hat{\mathbf{b}}^T]$$

for the two-user synchronous decorrelating decision-feedback detector.

PROBLEM 7.16. Find the range of values of A_2/A_1 for which the decorrelating detector gives better two-user asymptotic efficiency η_1 than the two-stage detector with decorrelating first stage.

PROBLEM 7.17.

(a) Show that if the initial decisions fed to the multistage detector are maximum-likelihood decisions, the decisions do not change after m stages.

(b) Suppose that the initial decisions fed to the multistage detector are the minimum bit-error-rate decisions. Sketch the decision regions of the two-stage detector in the two-user case. Conclude that for every $\sigma > 0$, the noise realization may be such that the decisions of the m-stage detector do not converge with m.

PROBLEM 7.18. (Vanghi and Vojcic [465]) Show that the asymptotic multiuser efficiency of the two-stage detector with decorrelating first stage satisfies

$$\eta_k^{dd} \leq \min_{i \neq k} \left\{ \frac{A_i^2}{A_k^2} \frac{1}{R_{ii}^+} + \max^2 \left\{ 0, 1 - 2\,|\rho_{ik}| \frac{A_i}{A_k} \right\} \right\}.$$

PROBLEM 7.19. (Vanghi and Vojcic [465]) Consider a two-user

391

synchronous channel and the following decision for user 1:

$$\hat{b}_1 = \text{sgn}(y_1 - f_\gamma(\rho y_2)),$$

$$f_\gamma(x) = \begin{cases} 1, & x > \gamma; \\ \frac{x}{\gamma}, & -\gamma \leq x \leq \gamma; \\ -1, & x < -\gamma; \end{cases}$$

$$\gamma = \frac{|\rho|}{1 - \rho^2} \left(\frac{A_1}{A_2} - |\rho| \right)^+.$$

(a) Sketch the decision regions of this detector in the case $A_1 = A_2$.

(b) Show that this detector achieves optimum asymptotic multiuser efficiency for all A_1/A_2 if $|\rho| \leq \frac{1}{\sqrt{2}}$.

PROBLEM 7.20. Verify that

$$2 \lim_{\sigma \to 0} \sigma^2 \log \left(1\{\epsilon = 0\} + [1 - 2 \ 1\{\epsilon = 0\}]Q\left(\frac{x}{\sigma}\right) \right) = -\nu_\epsilon(x),$$
$$(7.120)$$

with (cf. (7.72)):

$$\nu_\epsilon(x) \stackrel{\text{def}}{=} (1\{x < 0\}1\{\epsilon = 0\} + 1\{x > 0\}1\{\epsilon \neq 0\}) \ x^2 \quad (7.121)$$

PROBLEM 7.21.

(a) Generalize (7.76) to show

$$\min_{k=1,\ldots,j} A_k^2 \eta_k^{ddf} = \min_{k=1,\ldots,j} A_k^2 F_{kk}^2 \qquad (7.122)$$

for $j = 1, \ldots, K$.

(b) (Varanasi [471]) Show that if

$$A_K^2 F_{KK}^2 \leq \cdots \leq A_1^2 F_{11}^2, \qquad (7.123)$$

then

$$\eta_k^{ddf} = F_{kk}^2.$$

(c) Show that if all users have the same power, then (7.123) can only be satisfied if the signature waveforms are orthogonal.

PROBLEM 7.22. Verify the identity in (7.97):

$$\sigma^2 \mathbf{GRG}^T + [\mathbf{GR} - \mathbf{B} - \mathbf{I}]\mathbf{A}^2[\mathbf{GR} - \mathbf{B} - \mathbf{I}]^T$$

$$= \sigma^2[\mathbf{B} + \mathbf{I}](\sigma^2\mathbf{A}^{-2} + \mathbf{R})^{-1}[\mathbf{B} + \mathbf{I}]^T$$

$$+ (\mathbf{G} - \hat{\mathbf{G}})(\mathbf{RA}^2\mathbf{R} + \sigma^2\mathbf{R})(\mathbf{G} - \hat{\mathbf{G}})^T,$$

where

$$\hat{\mathbf{G}} \stackrel{\text{def}}{=} [\mathbf{B} + \mathbf{I}](\sigma^2\mathbf{A}^{-2} + \mathbf{R})^{-1}.$$

PROBLEM 7.23. Show that under the random signature sequence model, F_{11} converges in mean-square sense to $\sqrt{1-\beta}$, where

$$\beta = \lim_{K\to\infty} \frac{K}{N}.$$

PROBLEM 7.24. Consider the synchronous two-user two-stage detector with conventional first-stage.

(a) Find the worst asymptotic effective energy.
(b) Plot the power-tradeoff region to achieve bit-error-rates no larger than 3×10^{-5} for both users.

PROBLEM 7.25. Verify that the coefficients in Figure 7.25 are given by (7.108) and (7.109) in the case of an MMSE decision-feedback detector.

PROBLEM 7.26. Determine the matrices $\mathbf{F}_\sigma[0], \mathbf{F}_\sigma[1]$ (cf. (7.117)) in the two-user case. [*Hint:* Recall Problem 2.50.]

PROBLEM 7.27. (Verdú [498]) Using the one-shot decorrelator, design a two-user asynchronous decorrelating decision-feedback detector such that the decision $\hat{b}_k[i]$ is independent of $\{y(t), t > iT+T-\tau_k\}$. Find the asymptotic efficiency of the resulting detector.

PROBLEM 7.28. Let $Y = Ab + N$ where N is a zero-mean Gaussian random variable, A is a positive constant and b is equally likely to be -1 and $+1$. Show that

$$\frac{E[(Y-Ab)^2]}{E[(A \operatorname{sgn}(Y) - Ab)^2]} \geq \frac{3}{2}.$$

BIBLIOGRAPHY

[1] B. Aazhang, B. Paris, and G. Orsak. "Neural networks for multiuser detection in code division multiple access communications." *IEEE Trans. Communications,* 40:1212–1222, July 1992.

[2] B. Aazhang and H. V. Poor. "Performance of DS/SSMA communications in impulsive channels – Part I: Linear correlation receivers." *IEEE Trans. Communications*, 35:1179–1188, Nov. 1987.

[3] B. Aazhang and H. V. Poor. "Performance of DS/SSMA communications in impulsive channels – Part II: Hard-limiting correlation receivers." *IEEE Trans. Communications*, 36:88–96, Jan. 1988.

[4] M. Abdulrahman and D. Falconer. "Cyclostationary crosstalk suppression by decision feedback equalization on digital subscriber loops." *IEEE Journal on Selected Areas in Communications*, 10, no. 3:640–649, Apr. 1992.

[5] M. Abdulrahman, A. Sheikh, and D. Falconer. "Decision feedback equalization for CDMA in indoor wireless communications." *IEEE Journal on Selected Areas in Communications*, 12:698–706, May 1994.

[6] K. Abed-Meraiua, P. Loubaton, and E. Moulines. "Subspace blind indentification of multichannel FIR with unknown spatial covariance." *IEEE Signal Processing Letters*, 4, no. 5:135–137, May 1997.

[7] K. Abend and B. D. Fritchman. "Statistical detection for communication channels with intersymbol interference." *Proc. IEEE*, 58:779–785, May 1970.

[8] B. Abrams, A. Zeger, and T. Jones. "Efficiently structured CDMA receiver with near-far immunity." *IEEE Trans. Vehicular Technology*, 44, no. 1:1–13, 1995.

[9] N. Abramson. "Development of the ALOHANET." *IEEE Trans. Information Theory*, 31:119–123, Mar. 1985.

[10] N. Abramson, editor. *Multiple Access Communications: Foundations for Emerging Technologies.* IEEE Press, New York, 1993.

[11] I. Acar. *Performances of linear multiuser receivers for spread-spectrum communication systems.* PhD thesis, University of Massachusetts, Amherst, MA, Sept. 1996.

[12] I. Acar and S. Tantaratana. "Performance analysis of the decorrelating detector for DS spread-spectrum multiple-access communication systems." *Proc. 33rd Allerton Conf. Communications, Control and Computing*, pp. 1073–1082, Oct. 1995.

[13] P. Agashe and B. Woerner. "Interference cancellation for a multicellular CDMA environment." *Wireless Personal Communications*, pp. 1–14, nos. 1–2, 1996.

[14] A. Akki. "Statistical properties of mobile-to-mobile land communication channels." *IEEE Trans. Vehicular Technology*, 43, no. 4:826–831, Nov. 1994.

[15] P. Alexander and L. Rasmussen. "On the windowed Cholesky factorisation of the time-varying asynchronous CDMA channel." Submitted to *IEEE Trans. Communications*, 1996.

[16] P. Alexander, L. Rasmussen, and C. Schlegel. "A linear receiver for the coded asynchronous multiuser CDMA channel." *Proc. 33rd Allerton Conf. Communications, Control and Computing*, 3:29–38, Oct. 1995.

[17] P. Alexander, L. Rasmussen, and C. Schlegel. "A linear receiver for coded multiuser CDMA." *IEEE Trans. Communications*, 45, no. 5:605–610, May 1997.

[18] L. R. Bahl, J. Cocke, F. Jelinek, and J. Raviv. "Optimal decoding of linear codes for minimizing symbol error rate." *IEEE Trans. Information Theory*, 20:284–287, Mar. 1974.

[19] Z. D. Bai and Y. Q. Yin. "Limit of the smallest eigenvalue of a large dimensional sample covariance matrix." *Annals of Probability*, 21:1275–1294, 1993.

[20] C. A. Baird. "Recursive minimum variance estimation for adaptive sensor arrays." *Proc. 1972 IEEE International Conf. Cybernetics and Society*, pp. 412–414, Oct. 9–12, 1972.

[21] P. Balaban and J. Salz. "Optimum diversity combining and equalization in digital data transmission with applications to cellular mobile radio." *IEEE Trans. Communications*, 40:885–894, May 1992.

[22] Y. Bar-Ness, W. Chen, and E. Panayiric. "Eigenanalysis for interference cancellation with minimum redundancy array structure." *IEEE Trans. Aerospace and Electronic Systems*, 33, no. 3:977–988, July 1997.

[23] Y. Bar-Ness and J. Punt. "Adaptive 'bootstrap' CDMA multiuser detector." *Wireless Personal Communications*, pp. 55–71, nos. 1–2, 1996.

[24] Y. Bar-Ness and N. Sezgin. "Maximum signal-to-noise ratio data combining for one-shot asynchronous multiuser CDMA detector." *Proc. 6th IEEE Int. Symp. Personal, Indoor and Mobile Radio Communications*, 1:188–192, Sept. 1995.

[25] A. Barbosa and S. Miller. "Adaptive detection of DS/CDMA signals in fading channels." *IEEE Trans. Communications*, 46, no. 1:115–124, Jan. 1998.

[26] A. Batra and J. R. Barry. "Blind cancellation of co-channel interference." *Proc. 1995 IEEE Global Telecommunications Conf.*, pp. 157–162, Nov. 1995.

[27] S. Bensley and B. Aazhang. "Subspace-based channel estimation for code division multiple access communication systems." *IEEE Trans. Communications*, 44:1009–1020, Aug. 1996.

[28] X. Bernstein and A. Haimovich. "Space-time optimum combining for CDMA communications." *Wireless Personal Communications*, pp. 73–89, nos. 1–2, 1996.

396

[29] D. Bertsekas and R. Gallager. *Data Networks,* 2nd ed., Prentice-Hall, Englewood Cliffs, NJ, 1992.

[30] G. Bienvenue and L. Kopp. "Principe de la goniometrie passive adaptative." *Proc. 7éme Colloque GRETSI,* pp. 106/1–106/10, 1979.

[31] J. Blanz, A. Klein, M. Nasshan, and A. Steil. "Performance of a cellular hybrid C/TDMA mobile radio system applying joint detection and coherent receiver antenna diversity." *IEEE Journal on Selected Areas in Communications,* 12:568–579, May 1994.

[32] D. Brady and J. Catipovic. "Adaptive multiuser detection for underwater acoustical channels." *IEEE Journal of Oceanic Engineering,* 19:158–165, Apr. 1994.

[33] M. Brandt-Pearce and B. Aazhang. "Multiuser detection for optical code division multiple access systems." *IEEE Trans. Communications,* 42:1801–1810, Feb./Mar./Apr. 1994.

[34] M. Brandt-Pearce and B. Aazhang. "Performance analysis of single-user and multiuser detectors for optical code division multiple access communication systems." *IEEE Trans. Communications,* 43:435–444, Feb. 1995.

[35] A. Bravo. "Limited linear cancellation of multiuser interference in DS/CDMA asynchronous systems." *IEEE Trans. Communications,* 45, no. 11:1435–1443, Nov. 1997.

[36] R. Buehrer, A. Kaul, S. Striglis, and B. Woerner. "Analysis of DS-CDMA parallel interference cancellation with phase and timing errors." *IEEE Journal on Selected Areas in Communications,* 14, no. 8:1522–1435, Oct. 1996.

[37] R. Buehrer and B. Woerner. "Analysis of adaptive multistage interference cancellation for CDMA using an improved Gaussian approximation." *IEEE Trans. Communications,* 44, no. 10:1308–1321, Oct. 1996.

[38] S. Bulumulla and S. Venkatesh. "The quantized-input decorrelating detector." *Proc. 1996 Conf. on Information Sciences and Systems,* pp. 596–598, 1996.

[39] A. G. Burr. "Block codes on a dispersive channel." *IEE Proc. Part I,* 138:95–104, Apr. 1991.

[40] A. G. Burr. "Performance of linear separation of CDMA signals with FEC coding." *Proc. 1997 IEEE Int. Symp. on Information Theory,* p. 354, July 1997.

[41] G. Caire, G. Taricco, J. Ventura-Traveset, and E. Biglieri. "A multiuser approach to narrowband cellular communication." *IEEE Trans. Information Theory,* 43:1503–1517, Sept. 1997.

[42] R. Cameron and B. Woerner. "Performance analysis of CDMA with imperfect power control." *IEEE Trans. Communications,* 44:777–781, July 1996.

[43] G. Carayannis, D. Manolakis, and N. Kalouptsidis. "A fast sequential algorithm for least-squares filtering and prediction." *IEEE Trans. Acoustics, Speech, Signal Processing,* 31:1394–1402, Dec. 1983.

[44] J-F. Cardoso. "Blind signal separation: a review." *Proc. IEEE. Special issue on blind identification and estimation,* to appear, 1998.

[45] A. Carleial. *On the capacity of multiple-terminal communication networks.* PhD thesis, Stanford University, 1975.

[46] A. Carleial. "A case where interference does not reduce capacity." *IEEE Trans. Information Theory*, 21:569–570, Sept. 1975.

[47] L. Castedo, C. Escudero, and A. Dapena. "A blind signal separation method for multiuser communications." *IEEE Trans. Signal Processing*, 45:1343–1347, May 1997.

[48] J. Catipovic, D. Brady, and A. Etchemendy. "Development of underwater acoustic modems and networks." *Oceanography Magazine*, 6:112–119, 1993.

[49] R. Chang. "Synthesis of band-limited orthogonal signals for multichannel data transmission." *Bell System Technical Journal*, 45:1775–1796, Dec. 1966.

[50] R. Chang and J. Hancock. "On receiver structures for channels having memory." *IEEE Trans. Information Theory*, 12(4):463–468, Oct. 1966.

[51] D. Chen and S. Roy. "An adaptive multiuser receiver for CDMA systems." *IEEE Journal on Selected Areas in Communications*, 12, no. 5:808–816, June 1994.

[52] D. Chen, B. Sheu, and W. Young. "A CDMA communication detector with robust near-far resistance using paralleled array processors." *IEEE Trans. Circuits and Systems for Video Techn.*, 7, no. 4:654–662, Aug. 1997.

[53] D. Chen, Z. Siveski, and Y. Bar-Ness. "Synchronous multiuser CDMA detector with soft decision adaptive canceler." *Proc. 1994 Conf. on Information Sciences and Systems*, pp. 139–143, 1994.

[54] J. Chen and U. Mitra. "MMSE receivers for dual-rate DS/CDMA signals: Random signature sequence analysis." *Proc. 1997 IEEE Global Telecommunications Conf.*, pp. 139–143, Nov. 1997.

[55] R. Cheng and S. Verdú. "Capacity of RMS bandlimited Gaussian multiple-access channels." *IEEE Trans. Information Theory*, 37:453–465, May 1991.

[56] R. Cheng and S. Verdú. "The effect of asynchronism on the capacity of Gaussian multiple-access channels." *IEEE Trans. Information Theory*, 38:2–13, Jan. 1992.

[57] J. Cioffi, M. Eyuboglu, G. Dudevoir, and G. D. Forney. "MMSE decision-feedback equalizers and coding – Part I: Equalization results." *IEEE Trans. Communications*, 43, no. 10:2582–2594, 1995.

[58] J. Cioffi and T. Kailath. "Fast recursive-least-squares transversal filters for adaptive filtering." *IEEE Trans. Acoustics, Speech, Signal Processing*, 32:304–337, Apr. 1984.

[59] R. H. Clarke. "A statistical theory of mobile-radio reception." *Bell System Technical Journal*, 47:957–1000, July–Aug. 1968.

[60] R. Conot. *Thomas A. Edison: A Streak of Luck*. Da Capo, New York, 1979.

[61] T. Cover. "Some advances in broadcast channels." In *Advances in Communication Systems*, pp. 229–260. Academic, New York, 1975.

[62] S. Das, J. Cavallaro, and B. Aazhang. "Computationally efficient multiuser detectors." *Proc. 8th IEEE Int. Symp. Personal, Indoor and Mobile Radio Communications*, pp. 62–67, Sept. 1997.

[63] V. DaSilva and E. Sousa. "Multicarrier orthogonal CDMA signals for quasi-synchronous communications systems." *IEEE Journal on Selected Areas in Communications*, 12, no. 5:842–852, May 1994.

[64] M. Davis, A. Monk, and L. Milstein. "A noise whitening approach to multiple-access noise rejection – Part II: Implementation issues." *IEEE Journal on Selected Areas in Communications*, 14, no. 8:1488–1499, Oct. 1996.

[65] P. Dent, B. Gudmundson, and M. Ewerbring. "CDMA-IC: A novel code division multiple access scheme based on interference cancellation." *Proc. 3rd IEEE Int. Symp. Personal, Indoor and Mobile Radio Communications*, pp. 98–102, Sept. 1992.

[66] D. Divsalar and M. Simon. "Improved CDMA performance using parallel interference cancellation." *Tech. Rep. JPL Publication 95-21*, Jet Propulsion Laboratory, Oct. 1995. See also *Proc. 1994 IEEE Military Communications Conf.*, pp. 911–917.

[67] D. Divsalar, M. Simon, and D. Raphaeli. "Improved parallel interference cancellation for CDMA." *IEEE Trans. Communications*, 46, no. 2:258–268, Feb. 1998.

[68] R. C. Dixon. *Spread Spectrum Systems with Commercial Applications*, Wiley, New York, 1994.

[69] P. M. Dollard. "On the time-bandwidth concentration of signal functions forming given geometric vector configurations." *IEEE Trans. Information Theory*, 10:328–338, Oct. 1964.

[70] S. Douglas and A. Cichocki. "Neural networks for blind decorrelation of signals." *IEEE Trans. Communications*, 45, no. 11:2829–2842, Nov. 1997.

[71] Z. Drezner. "Computation of the multivariate normal integral." *ACM Trans. Mathematical Software*, 18:470–480, Dec. 1992.

[72] A. Duel-Hallen. "Equalizers for multiple input/multiple output channels and PAM systems with cyclostationary input sequences." *IEEE Journal on Selected Areas in Communications*, 10, no. 3:630–649, Apr. 1992.

[73] A. Duel-Hallen. "Decorrelating decision-feedback multiuser detector for synchronous CDMA." *IEEE Trans. Communications*, 41:285–290, Feb. 1993.

[74] A. Duel-Hallen. "A family of multiuser decision-feedback detectors for asynchronous code-division multiple-access channels." *IEEE Trans. Communications*, 43:421–434, Feb./Mar./Apr. 1995.

[75] A. Duel-Hallen, J. Holtzman, and Z. Zvonar. "Multiuser detection for CDMA systems." *IEEE Personal Communications*, 2, no. 2:46–58, Apr. 1995.

[76] A. Dutta and S. Kiaei. "Adaptive multiuser detector for asynchronous DS-CDMA in Rayleigh fading." *IEEE Trans. Circuits and Systems*, 44, no. 6:135–137, June 1997.

[77] A. El-Ezabi and A. Duel-Hallen. "Combined error correction and multiuser detection for synchronous CDMA channels." *Proc. 33rd Allerton Conf. on Communications, Control and Computing*, pp. 1–10, Oct. 1995.

[78] J. Epstein, P. Epstein, and H. Koch. "Screenplay." In H. Koch, ed., *Casablanca: Script and Legend*, pp. 23–228. The Overlook Press, Woodstock, NY, 1992.

[79] E. Esteves and R. Scholtz. "Bit error probability of linear multiuser detectors in the presence of unknown multiple access interference." *Proc. 1997 IEEE Global Telecommunications Conf.*, pp. 599–603, Nov. 1997.

[80] W. Van Etten. "An optimum linear receiver for multiple channel digital transmission systems." *IEEE Trans. Communications*, 23:828–834, Aug. 1975.

[81] W. Van Etten. "Maximum likelihood receiver for multiple channel transmission systems." *IEEE Trans. Communications*, 24(2):276–283, Feb. 1976.

[82] European Telecommunications Standardization Institute, Sophia Antipolis, France. *Group Speciale Mobile or Global System for Mobile Communication (GSM) Recommendation*, 1988.

[83] M. Ewerbring, B. Gudmundson, G. Larsson, and P. Teder. "CDMA with interference cancellation: A technique for high capacity wireless systems. *Proc. 1993 IEEE Int. Conf. on Communications*, pp. 1901–1906, May 1993.

[84] D. Falconer, M. Abdulrahman, N. Lo, and B. Petersen. "Advances in equalization and diversity for portable wireless systems." *Digital Signal Processing*, 3:148–162, 1993.

[85] D. Falconer, F. Adachi, and B. Gudmundson. "Time division multiple access methods for wireless personal communications." *IEEE Communications Magazine*, 33:50–57, 1995.

[86] D. Falconer and L. Ljung. "Application of fast Kalman estimation to adaptive equalization." *IEEE Trans. Communications*, 26:1439–1446, Oct. 1978.

[87] U. Fawer and B. Aazhang. "A multiuser receiver for code division multiple access communications over multipath channels." *IEEE Trans. Communications*, 43:1556–1565, Feb. 1995.

[88] U. Fawer and B. Aazhang. "Multiuser receivers for code-division multiple-access systems with trellis-based modulation." *IEEE Journal on Selected Areas in Communications*, 14, no. 8:1602–1609, Oct. 1996.

[89] K. Feher. "Modems for emerging digital cellular-mobile radio systems." *IEEE Trans. Vehicular Technology*, 40, no. 2:355–365, May 1991.

[90] T. Felhauer, A. Klein, and P. Baier. "A low-cost method for CDMA and other applications to separate nonorthogonal signals." *IEEE Trans. Communications*, 42, no. 2/3/4:881–883, Feb./Mar./Apr. 1994.

[91] W. Feller. *An Introduction to Probability Theory and Its Applications*. vol. 1. Wiley, New York, 1968.

[92] J. Fonollosa, J. Fonollosa, Z. Zvonar, and J. Catalá. "Blind multiuser detection with array observations." *Wireless Personal Communications*, 6(1–2):179–196, Jan. 1998.

[93] J. Fonollosa, J. Fonollosa, Z. Zvonar, and J. Vidal. "Blind multiuser identification and detection in CDMA systems." *Proc. 1995 IEEE Int. Conf. on Acoustic, Speech and Signal Processing*, pp. 1876–1879, May 1995.

[94] J. Fonollosa, J. Fonollosa, Z. Zvonar, and J. Vidal. "Blind multiuser deconvolution in fading and dispersive channels." *Proc. 1995 IEEE Int. Symp. on Information Theory*, p. 384, Sept. 1995.

[95] G. D. Forney. "Lower bounds on error probability in the presence of large intersymbol interference." *IEEE Trans. Communications*, 20:76–77, Feb. 1972.

400

[96] G. D. Forney. "Maximum likelihood sequence estimation of digital sequences in the presence of intersymbol interference." *IEEE Trans. Information Theory*, 18(3):363–378, May 1972.

[97] G. D. Forney. "The Viterbi algorithm." *Proc. IEEE*, 61(3):268–278, Mar. 1973.

[98] O. Frost. "An algorithm for linearly constrained adaptive linear processing." *Proc. IEEE*, 60:926–935, Aug. 1972.

[99] K. Fukawa and H. Suzuki. "Orthogonalizing matched filter (OMF) detection for DS-CDMA mobile radio systems." *Proc. 1994 IEEE Global Telecommunications Conf.*, 1:385–389, Nov. 28–Dec. 1 1994.

[100] R. Gallager. "A perspective on multiaccess channels." *IEEE Trans. Information Theory*, 31:124–142, Mar. 1985.

[101] A. El Gamal and T. Cover. "Multiple user information theory." *Proc. IEEE*, 68:1463–1483, Dec. 1980.

[102] M. Garey and D. Johnson. *Computers and Intractability: A Guide to the Theory of NP-completeness*. Freeman, San Francisco, 1979.

[103] R. De Gaudenzi, T. Garde, F. Giannetti, and M. Luise. "A performance comparison of orthogonal code-division multiple-access techniques for mobile satellite communications." *IEEE Journal on Selected Areas in Communications*, 13:325–331, Feb. 1995.

[104] R. De Gaudenzi, F. Giannetti, and M. Luise. "Advances in satellite CDMA transmission for mobile and personal communications." *Proc. IEEE*, 84:18–39, Jan. 1996.

[105] R. De Gaudenzi, F. Giannetti, and M. Luise. "Design of a low-complexity adaptive interference-mitigating detector for DS/SS receivers in CDMA radio networks." *IEEE Trans. Communications*, 46, no. 1:125–134, Jan. 1998.

[106] E. Geraniotis. "Performance of noncoherent direct-sequence spread-spectrum multiple-access communications." *IEEE Journal on Selected Areas in Communications*, 3, no. 5:687–694, Sept. 1985.

[107] E. Geraniotis. "Direct-sequence spread-spectrum multiple-access communications over nonselective and frequency-selective Rician fading channels." *IEEE Trans. Communications*, 34, no. 8:756–764, Aug. 1986.

[108] E. Geraniotis and M. Pursley. "Error probability for direct-sequence spread-spectrum multiple-access communications – Part II: Approximations." *IEEE Trans. Communications*, 30(5):985–995, May 1982.

[109] E. Geraniotis and M. Pursley. "Performance of coherent direct-sequence spread-spectrum communications over specular multipath fading channels." *IEEE Trans. Communications*, 33, no. 6:502–508, June 1985.

[110] I. Ghauri and R. Iltis. "Capacity of the linear decorrelating detector for QS-CDMA." *IEEE Trans. Communications*, 45:1039–1042, Sept. 1997.

[111] V. Ghazi-Moghadam and M. Kaveh. "Interference cancellation using antenna arrays." *Proc. 1995 IEEE Conf. on Personal, Indoor, Mobile Radio Communications*, pp. 936–939, 1995.

[112] V. Ghazi-Moghadam, L. Nelson, and M. Kaveh. "Parallel interference cancellation for CDMA systems." *Proc. 33rd Allerton Conf. on Communications, Control and Computing*, pp. 216–224, Oct. 1995.

[113] T. Giallorenzi and S. Wilson. "Multiuser ML sequence estimator for

convolutionally coded asynchronous DS-CDMA systems." *IEEE Trans. Communications*, 44:997–1008, Aug. 1996.

[114] T. Giallorenzi and S. Wilson. "Trellis-based multiuser receivers for convolutionally coded CDMA systems." *Proc. 31st Allerton Conf. on Communications, Control and Computing*, pp. 839–848, Oct. 1993.

[115] T. Giallorenzi and S. Wilson. "Suboptimum multiuser receivers for convolutionally coded asynchronous DS-CDMA systems." *IEEE Trans. Communications*, 44:1183–1196, Sept. 1996.

[116] K. Gilhousen, I. Jacobs, R. Padovani, A. J. Viterbi, L. Weaver, and C. Wheatley. "On the capacity of a cellular CDMA system." *IEEE Trans. Vehicular Technology*, 40:303–312, May 1991.

[117] R. Gitlin, H. Meadors, and S. Weinstein. "The tap-leakage algorithm for stable operation of a digitally implemented fractionally spaced adaptive equalizer." *Bell System Technical Journal*, 61:1817–1939, Oct. 1982.

[118] R. Gitlin and S. Weinstein. "Fractionally spaced equalization: An improved digital transversal equalizer." *Bell System Technical Journal*, 60:275–296, Feb. 1981.

[119] S. Glisic and B. Vucetic. *Spread Spectrum CDMA Systems for Wireless Communications*. Artech House, Boston, 1997.

[120] D. Goeckel and W. Stark. "Performance of coded direct-sequence systems with Rake reception in a multipath fading environment." *European Trans. Telecommunications*, 6, no. 1:41–51, Jan.– Feb. 1995.

[121] D. Goeckel and W. Stark. "Throughput optimization in multiple-access communication systems with decorrelator reception." *Proc. 1996 Int. Symp. on Information Theory and Its Applications*, 2:653–656, Sept. 1996.

[122] M. J. E. Golay. "An approach to multiple access satellite communications through the use of network synchronized orthogonal signals." In *Multiple Access to a Communication Satellite with a Hard-Limiting Repeater. Vol. II*, volume R-108, pp. 107–114. Institute for Defense Analyses, Jan. 1965.

[123] R. Gold. "Optimum binary sequences for spread spectrum multiplexing." *IEEE Trans. Information Theory*, 14:154–156, Oct. 1967.

[124] S. Golomb. *Shift Register Sequences*. Holden-Day, San Francisco, 1967.

[125] I. Gradshteyn and I. Ryzhik. *Tables of Integrals, Series, and Products*. Academic Press, San Diego, CA, 1980.

[126] S. Gray, M. Kocic, and D. Brady. "Multiuser detection in mismatched multiple-access channels." *IEEE Trans. Communications*, 43:3080–3089, Dec. 1995.

[127] S. Gray, J. Preisig, and D. Brady. "Multiuser detection in a horizontal underwater acoustic channel using array observations." *IEEE Trans. Signal Processing*, 45:148–160, Jan. 1997.

[128] G. Grimmett and D. Stirzaker. *Probability and Random Processes*. 2nd ed. Oxford Science Publications, Clarendon Press, Oxford, UK, 1992.

[129] M. Gudmundson. "Correlation model for shadowing fading in mobile radio systems." *Electronics Letters*, 27:2145–2146, Nov. 7, 1991.

[130] L. Gyorfi. "Adaptive linear procedures under general conditions." *IEEE Trans. Information Theory*, 40:262–267, Mar. 1984.

[131] A. Hafeez and W. Stark. "Combined decision-feedback multiuser

detection/soft-decision decoding for CDMA channels." *Proc. 1996 IEEE Vehicular Technology Conf.*, pp. 382–386, May 1996.

[132] A. Hafeez and W. Stark. "Soft-output multiuser estimation for asynchronous CDMA channels." *Proc. 1997 IEEE Vehicular Technology Conf.*, pp. 465–469, May 1997.

[133] F. Hagmanns and V. Hespelt. "On the detection of bandlimited direct-sequence spread-spectrum signals transmitted via fading multipath channels." *IEEE Journal on Selected Areas in Communications*, 12, no. 5:891–899, June 1994.

[134] A. M. Haimovich and Y. Bar-Ness. "An eigenanalysis interference canceller." *IEEE Trans. Signal Processing*, 39(1):76–84, Jan. 1991.

[135] K. Halford and M. Brandt-Pearce. "Performance of a multistage multiuser detector for FHMA." *Proc. 1996 Conf. on Information Sciences and Systems*, pp. 605–610, Mar. 1996.

[136] K. Halford and M. Brandt-Pearce. "New-users identification in a CDMA system." *IEEE Trans. Communications*, 46, no. 1:144–155, Jan. 1998.

[137] C. Haro, J. Fonollosa, Z. Zvonar, and J. Fonollosa. "Probabilistic algorithms for blind adaptive multiuser detection." Submitted to *IEEE Trans. Signal Processing*, 1997.

[138] J. Hayes, T. Cover, and J. Riera. "Optimal sequence detection and optimal symbol-by-symbol detection: Similar algorithms." *IEEE Trans. Communications*, 30(1):152–157, Jan. 1982.

[139] J. Hayes and D. Sherman. "A study of data multiplexing techniques and delay performance." *Bell System Technical Journal*, 51:1983–2011, Nov. 1972.

[140] C. Hegarty and B. Vojcic. "Two-stage multiuser detection for noncoherent CDMA." *Proc. 33rd Allerton Conf. on Communications, Control and Computing*, pp. 1063–1072, Oct. 1995.

[141] C. Hegarty and B. Vojcic. "Noncoherent multiuser detection of M-ary orthogonal signals using a decorrelator." *Proc. 1996 IEEE Military Communications Conf.*, pp. 903–907, Oct. 1996.

[142] C. Hegarty and B. Vojcic. "Noncoherent two-stage multiuser detection of M-ary orthogonal signals." *Wireless Networks*, to appear, 1998.

[143] T. Helleseth and P. V. Kumar. "Pseudonoise sequences." In J. D. Gibson, editor, *The Mobile Communications Handbook*. Chapter 8, pp. 110–122, CRC Press, Boca Raton, FL, 1996.

[144] F. Van Hesswyk, D. Falconer, and A. Sheikh. "A delay independent decorrelating detector for quasi-synchronous CDMA." *IEEE Journal on Selected Areas in Communications*, 14, no. 8:1619–1626, Oct. 1996.

[145] P. Hoeher. "A statistical discrete-time model for the WSSUS multipath channel." *IEEE Trans. Vehicular Technology*, 41, no. 4:461–468, Nov. 1992.

[146] P. Hoeher. "On channel coding and multiuser detection for DS/CDMA." *Proc. 2nd Int. Conf. on Universal Personal Communications*, pp. 641–646, 1993. See also ITG-Fachbericht 124 Mobile Kommunikation, pp. 55–66, Sept. 1993.

[147] J. Holtzman. "A simple accurate method to calculate spread-spectrum multiple access error probabilities." *IEEE Trans. Communications*, 40:461–464, Mar. 1992.

[148] J. Holtzman. "On calculating DS/SSMA error probabilities." *Proc. 1992 Int. Symp. on Spread Spectrum Techniques and Applications*, pp. 23–26, Nov.–Dec. 1992.

[149] J. Holtzman. "DS/CDMA successive interference cancellation." In S. G. Glisic and P. Leppänen, editors, *Code Division Multiple Access Communications*, pp. 161–182. Kluwer Academic, Dordrecht, The Netherlands, 1995.

[150] E. Hong, K. Kim, and K. Whang. "Performance evaluation of DS-CDMA system with m-ary orthogonal signaling." *IEEE Trans. Vehicular Technology*, 45, no. 1:57–63, Jan. 1996.

[151] M. L. Honig. "Rapid detection and suppression of multi-user interference in DS-CDMA." *Proc. 1995 IEEE Int. Conf. Acoustics, Speech and Signal Processing*, pp. 1057–1060, May 1995.

[152] M. L. Honig. "Orthogonally anchored interference suppression using the Sato cost criterion." *Proc. 1995 IEEE Int. Symp. on Information Theory*, p. 314, Sept. 1995.

[153] M. L. Honig. "Performance of adaptive interference suppression for DS-CDMA with a time-varying user population." *Proc. 4th Int. Symp. on Spread Spectrum Techniques and Applications*, pp. 267–271, Sept. 1996.

[154] M. L. Honig and U. Madhow. "Hybrid intra-cell TDMA/inter-cell CDMA with inter-cell interference suppression for wireless networks." *Proc. 1993 IEEE Vehicular Technology Conf.*, pp. 309–312, 1993.

[155] M. L. Honig, U. Madhow, and S. Verdú. "Blind adaptive interference suppression for near-far resistant CDMA." *Proc. 1994 IEEE Global Telecommunications Conf.*, 1:379–384, Dec. 1994.

[156] M. L. Honig, U. Madhow, and S. Verdú. "Blind adaptive multiuser detection." *IEEE Trans. Information Theory*, 41:944–960, July 1995.

[157] M. L. Honig and H. V. Poor. "Adaptive interference suppression in wireless communication systems. In H. V. Poor and G. Wornell, editors, *Wireless Communications: Signal Processing Perspectives*. Chapter 2, Prentice Hall, Upper Saddle River, NJ, 1998.

[158] M. L. Honig, M. Sensha, S. Miller, and L. Milstein. "Performance of adaptive linear interference suppression for DS-CDMA in the presence of flat Rayleigh fading." *Proc. 47th IEEE Vehicular Technology Conf.*, pp. 2191–2195, May 1997.

[159] M. L. Honig and W. Veerakachen. "Performance variability of linear multiuser detection for DS-CDMA." *Proc. 46th IEEE Vehicular Technology Conf.*, pp. 372–376, May 1996.

[160] D. Horwood and R. Gagliardi. "Signal design for digital multiple access communications." *IEEE Trans. Communications*, 23:378–383, Mar. 1975.

[161] S. Hosur, A. Tewfik, and V. Moghadam. "Adaptive multiuser receiver schemes for antenna arrays." *Proc. 6th IEEE Int. Symp. on Personal, Indoor and Mobile Radio Communications*, 3:940–944, Sept. 1995.

[162] A. Hottinen. "Multiuser detection for multirate CDMA communications." *Proc. 1996 IEEE Int. Communications Conf.*, pp. 1819–1823, 1996.

[163] A. Hottinen, H. Holma, and A. Toskala. "Performance of multistage multiuser detection in a fading multipath channel." *Proc. 6th IEEE*

404

Int. Symp. on Personal, Indoor and Mobile Radio Communications, 3:960–964, Sept. 1995.

[164] A. Hottinen and K. Pehkonen. "A flexible multirate CDMA concept with multiuser detection." *Proc. 1996 IEEE Int. Symp. on Spread Spectrum Techniques and Applications*, pp. 556–560, Sept. 1996.

[165] H. Huang. *Combined multipath processing, array processing, and multiuser detection for DS-CDMA channels.* PhD dissertation, Princeton University, Jan. 1996.

[166] H. Huang and S. Schwartz. "A comparative analysis of linear multiuser detectors for fading multipath channels." *Proc. 1994 IEEE Global Telecommunications Conf.*, pp. 11–21, Dec. 1994.

[167] H. Huang and S. Verdú. "Linear differentially coherent multiuser detection for multipath channels." *Wireless Personal Communications*, 6(1–2):113–136, Jan. 1998.

[168] C-L. I, C. Webb, H. Huang, S. Brink, S. Nanda, and R. Gitlin. "IS-95 enhancements for multimedia services." *Bell Labs Technical Journal*, 1:60–87, Autumn 1996.

[169] R. Iltis. "Demodulation and code acquisition using decorrelator detectors for QS-CDMA." *IEEE Trans. Communications*, 44:1553–1560, Nov. 1996.

[170] R. Iltis. "Performance of constrained and unconstrained adaptive multiuser detectors for quasi-synchronous CDMA." *IEEE Trans. Communications*, 46, no. 1:135–143, Jan. 1998.

[171] R. Iltis and L. Mailaender. "Multiuser code acquisition using parallel decorrelators." *Proc. 1994 Conf. on Information Sciences and Systems*, pp. 109–114, Mar. 1994.

[172] R. Iltis and L. Mailaender. "An adaptive multiuser detector with joint amplitude and delay estimation." *IEEE Journal on Selected Areas in Communications*, 12, no. 5:774–785, June 1994.

[173] R. Iltis and L. Mailaender. "Multiuser detection of quasi-synchronous CDMA signals using linear decorrelators." *IEEE Trans. Communications*, 44:1561–1571, Nov. 1996.

[174] W. Jakes. *Microwave Mobile Communications.* Wiley, New York, 1974.

[175] L. Jalloul and J. Holtzman. "Performance analysis of DS/CDMA with noncoherent *m*-ary orthogonal modulation in multipath fading channels." *IEEE Journal on Selected Areas in Communications*, 12, no. 5:862–870, June 1994.

[176] A. Johansson and A. Svensson. "Successive interference cancellation schemes in multi-rate DS/CDMA systems." *Workshop Records, 5th WINLAB Workshop on 3rd Generation Wireless Information Networks*, pp. 155–174, 1995.

[177] A. Johansson and A. Svensson. "Multi-stage interference cancellation in multi-rate DS/CDMA systems." *Proc. 6th IEEE Int. Symp. on Personal, Indoor and Mobile Radio Communications*, 3:965–969, Sept. 1995.

[178] C. Johnson. "Admissibility in blind adaptive equalization." *IEEE Control Systems Magazine*, 11:3–15, Jan. 1991.

[179] C. Johnson, P. Schniter, T. Engres, J. Behm, R. Casas, V. Brown, and C. Berg. "Blind equalization using the constant modulus criterion:

405

A review." *Proc. IEEE, Special issue on Blind Identification and Estimation*, to appear, 1998.

[180] D. Johnson and D. Dudgeon. *Array Signal Processing: Concepts and Techniques*. Prentice-Hall, 1993.

[181] P. Jung and P. Alexander. "A unified approach to multiuser detectors for CDMA and their geometrical interpretations." *IEEE Journal on Selected Areas in Communications*, 14, no. 8:1595–1601, Oct. 1996.

[182] P. Jung, P. Baier, and A. Steil. "Advantages of CDMA and spread spectrum techniques over FDMA and TDMA in cellular mobile radio applications." *IEEE Trans. Vehicular Technology*, 42, no. 3:357–364, Aug. 1993.

[183] P. Jung and J. Blanz. "Joint detection with coherent receiver antenna diversity in CDMA mobile radio systems." *IEEE Trans. Vehicular Technology*, 44, no. 1:76–88, 1995.

[184] P. Jung, J. Blanz, M. Nasshan, and P. Baier. "Simulation of the uplink of JD-CDMA mobile radio systems with coherent receiver antenna diversity." *Wireless Personal Communications*, 1:61–89, 1994.

[185] P. Jung, A. Klein, M. Nasshan, and A. Steil. "Performance of a cellular hybrid C/TDMA mobile radio system applying joint detection and coherent receiver antenna diversity." *IEEE Journal on Selected Areas in Communications*, 12:568–579, May 1994.

[186] M. Juntti. "Linear multiuser detector update in synchronous dynamic CDMA systems." *Proc. 6th IEEE Int. Symp. on Personal, Indoor and Mobile Radio Communications*, 3:980–984, Sept. 1995.

[187] M. Juntti. *Multiuser demodulation for CDMA systems in fading channels*. PhD thesis, University of Oulu, Oulu, Finland, 1997.

[188] M. Juntti and B. Aazhang. "Finite memory-length linear multiuser detection for asynchronous CDMA communications." *IEEE Trans. Communications*, 45, no. 5:611–622, May 1997.

[189] V. Kaasila and A. Mammela. "Bit error probability of a matched filter in a Rayleigh fading multipath channel." *IEEE Trans. Communications*, 42:826–828, Feb. 1994.

[190] S. R. Kadaba, S. B. Gelfand, M. P. Fitz, and R. L. Kashyap. "Bayesian techniques for interference suppression in TDMA wireless communication." *Wireless Networks*, to appear, 1998.

[191] J. Kahn, J. Komlós, and E. Szemerédi. "On the probability that a random ±-matrix is singular." *Journal of the American Mathematical Society*, 8:223–240, Jan. 1995.

[192] J. Kaiser, J. W. Schwartz, and J. M. Aein. "Multiple access to a communication satellite with a hard-limiting repeater. vol. I: Modulation techniques and their applications." *Tech. Rep.* R-108, Institute for Defense Analyses, Jan. 1965.

[193] A. Kajiwara and M. Nakagawa. "Crosscorrelation cancellation in SS/DS block demodulator." *IEICE Trans.*, pp. 2596–2602, Sept. 1991.

[194] A. Kajiwara and M. Nakagawa. "Microcellular CDMA system with a linear multiuser interference canceler." *IEEE Journal on Selected Areas in Communications*, 12:605–611, May 1994.

[195] G. K. Kaleh. "Channel equalization for block transmission systems." *IEEE Journal on Selected Areas in Communications*, 13:110–121, Jan. 1995.

[196] R. Kamel and Y. Bar-Ness. "Anchored blind equalization using the constant modulus algorithm." *IEEE Trans. Circuits and Systems–II: Analog and Digital Signal Processing*, 44:397–403, May 1997.

[197] K. Kammerlander. "Benefits of combined TDMA/CDMA operation for third generation mobile radio systems." *Proc. 1996 IEEE Int. Symp. on Spread Spectrum Techniques and Applications*, p. 507, Sept. 1996.

[198] S. Kandala, E. Sousa, and S. Pasupathy. "Decorrelators for multi-sensor systems in CDMA networks." *European Trans. Telecommunications*, 6, no. 1:29–40, 1995.

[199] S. Kandala, E. Sousa, and S. Pasupathy. "Multi-user multi-sensor detectors for CDMA networks." *IEEE Trans. Communications*, 43:946–957, Feb. 1995.

[200] T. Kasami. "Weight distribution formula for some class of cyclic codes." Tech. Rep. R-285, Coordinated Science Laboratory, University of Illinois, Urbana, IL, Apr. 1966.

[201] T. K. Kashihara. "Adaptive cancellation of mutual interference in spread spectrum multiple access." *Proc. 1980 IEEE Int. Communications Conf.*, pp. 44.4.1–5, 1980.

[202] M. Kavehrad. "Performance of nondiversity receivers for spread spectrum indoor wireless communications." *AT&T Technical Journal*, 64:1181–1985, July–Aug. 1995.

[203] M. Kavehrad and J. Salz. "Cross-polarization cancellation and equalization in digital transmission over dually polarized multipath fading channels." *AT&T Technical Journal*, 64:2211–2245, Dec. 1985.

[204] N. Kawabe, T. Kato, A. Kawahashi, T. Sato, and A. Fukasawa. "Advanced CDMA scheme based on interference cancellation." *Proc. IEEE 43rd Vehicular Technology Conf.*, pp. 448–451, 1993.

[205] T. Kawahara and T. Matsumoto. "Joint decorrelating multiuser detection and channel estimation in asynchronous CDMA mobile communications channels." *IEEE Trans. Vehicular Technology*, 44:506–515, Aug. 1995.

[206] A. Kaye and D. George. "Transmission of multiplexed PAM signals over multiple channel and diversity systems." *IEEE Trans. Communication Technology*, 18:520–525, Oct. 1970.

[207] P. Kempf. "A nonorthogonal synchronous DS-CDMA case, where successive cancellation and maximum-likelihood multiuser detectors are equivalent." *Proc. 1995 IEEE Int. Symp. on Information Theory*, p. 321, Sept. 1995.

[208] K. Kim, Y. Kim, J. Lee, I. Song, and S. Kim. "Suboptimum multiuser detection of DS/CDMA systems using antenna arrays in asynchronous channels." *Proc. 1997 IEEE Global Telecommunications Conf.*, Nov. 1997.

[209] A. Klein and P. Baier. "Linear unbiased data estimation in mobile radio systems applying CDMA." *IEEE Journal on Selected Areas in Communications*, 11:1058–1066, Sept. 1993.

[210] A. Klein, G. K. Kaleh, and P. Baier. "Zero forcing and minimum mean-square-error equalization for multiuser detection in code-division multiple-access channels." *IEEE Trans. Vehicular Technology*, 45(2):276–287, May 1996.

[211] L. Kleinrock and F. Tobagi. "Packet switching in radio channels: Part 1: CSMA modes and their throughput-delay characteristics." *IEEE Trans. Communications*, 23:1400–1416, 1975.

[212] H. Kobayashi. "Correlative level coding and maximum-likelihood decoding." *IEEE Trans. Information Theory*, 17:586–594, Jan. 1971.

[213] R. Kohno. "Pseudo-noise sequences and interference cancellation techniques for spread spectrum systems-spread spectrum theory and techniques in Japan." *IEICE Trans.*, E.74:1083–1092, May 1991.

[214] R. Kohno, M. Hatori, and H. Imai. "Cancellation techniques of co-channel interference in asynchronous spread spectrum multiple access systems." *Electronics and Communications*, 66-A:20–29, 1983.

[215] R. Kohno, H. Imai, and M. Hatori. "Optimum receiver using canceller of co-channel interference in SSMA." *1982 IECE National Conf. (Spring)*, pp. 7–201, 1982. In Japanese.

[216] R. Kohno, H. Imai, and M. Hatori. "Cancellation techniques of co-channel interference and application of Viterbi algorithm in asynchronous spread spectrum multiple access systems." *Proc. 1982 Symp. on Information Theory and Its Applications*, pp. 659–666, Oct. 1982. In Japanese.

[217] R. Kohno, H. Imai, M. Hatori, and S. Pasupathy. "An adaptive canceller of cochannel interference for spread-spectrum multiple-access communication networks in a power line." *IEEE Journal on Selected Areas in Communications*, 8, no. 4:691–699, Apr. 1990.

[218] R. Kohno, H. Imai, M. Hatori, and S. Pasupathy. "Combination of an adaptive array antenna and a canceller of interference for direct sequence spread spectrum multiple access system." *IEEE Journal on Selected Areas in Communications*, 8:675–681, Apr. 1990.

[219] R. Kohno, N. Ishii, and M. Nagatsuka. "A spatially and temporally optimal multi-user receiver using an array antenna for DS/CDMA." *Proc. 6th IEEE Int. Symp. on Personal, Indoor and Mobile Radio Communications*, 3:950–954, Sept. 1995.

[220] R. Kohno, R. Meidan, and L. Milstein. "Spread spectrum access methods in wireless communications." *IEEE Communications Magazine*, 33, no. 1:58–67, Jan. 1995.

[221] R. Kohno, P. Rapajic, and B. Vucetic. "An overview of adaptive techniques for interference minimization in CDMA systems." *Wireless Personal Communications*, 1, no. 1:3–21, 1994.

[222] J. Komlós. "On the determinant of (0,1) matrices." *Studia Sci. Math. Hungar.*, 2:7–21, 1967.

[223] S. Kubota, S. Kato, and K. Feher. "Inter-channel interference cancellation technique for CDMA mobile/personal communication systems." *Proc. 2nd IEEE Int. Symp. on Spread Spectrum Techniques and Applications*, pp. 112–117, 1992.

[224] E. Kudoh and T. Matsumoto. "Effects of power control error on the system user capacity of DS/SSMA cellular mobile radios." *IEICE Trans. Communications Japan*, E75-B:524–529, June 1992.

[225] P. Kumar and J. Holtzman. "Power control for a spread spectrum system with multiuser receivers." *Proc. 6th IEEE Int. Symp. on Personal, Indoor and Mobile Radio Communications*, 3:955–959, Sept. 1995.

408

[226] D. Laforgia, A. Luvison, and V. Zingarelli. "Bit error rate evaluation for spread-spectrum multiple access systems." *IEEE Trans. Communications*, 32:660–669, June 1984.

[227] P. Lancaster and M. Tismenetsky. *The Theory of Matrices*. 2nd ed. Academic Press, Orlando, FL, 1985.

[228] H. J. Landau and H. O. Pollak. "Prolate spheroidal wave functions, Fourier analysis and uncertainty–III: The dimension of the space of essentially time- and band-limited signals." *Bell System Technical Journal*, 41:1295–1336, 1962.

[229] J. Laster and J. Reed. "Interference rejection in digital wireless communications." *IEEE Signal Processing Magazine*, 14:37–62, May 1997.

[230] R. Learned, A. Willsky, and D. Boroson. "Low complexity optimal joint detection for oversaturated multiple access communications." *IEEE Trans. Signal Processing*, 45, no. 1:113–123, Jan. 1997.

[231] E. Lee. "Rapid converging adaptive interference suppression for direct sequence CDMA systems." *Proc. 1993 IEEE Global Telecommunications Conf.*, pp. 1683–1687, Nov. 1993.

[232] K. Lee. "Orthogonalization based adaptive interference suppression for direct-sequence code-division multiple-access systems." *IEEE Trans. Communications*, 44:1082–1085, Sept. 1996.

[233] W. Lee and R. Pickholtz. "Maximum likelihood multiuser detection with use of linear antenna arrays." *Proc. 1995 Asilomar Conf. on Signals, Systems, and Computers*, 1995.

[234] W. Lee and R. Pickholtz. "Convergence analysis of a CM array for CDMA systems." *Wireless Personal Communications*, pp. 37–53, Nos. 1–2, 1996.

[235] W. Lee, B. Vojcic, and R. Pickholtz. "Constant modulus algorithm for blind multiuser detection." *Proc. 1996 IEEE Int. Symp. on Spread Spectrum Techniques and Applications*, pp. 1262–1266, Sept. 1996.

[236] J. Lehnert and M. Pursley. "Error probability for binary direct-sequence spread-spectrum communications with random signature sequences." *IEEE Trans. Communications*, 35:87–98, 1987.

[237] K. B. Letaief. "On the performance of spread-spectrum multiple-access communications in multipath fading channels." *Proc. 6th IEEE Int. Symp. on Personal, Indoor and Mobile Radio Communications*, 2:692–696, Sept. 1995.

[238] K. B. Letaief. "Efficient evaluation of the error probabilities of spread-spectrum multiple-access communications." *IEEE Trans. Communications*, 45, no. 2:239–246, Feb. 1997.

[239] A. J. Levy. "Fast error rate evaluation in the presence of intersymbol interference." *IEEE Trans. Communications*, 45, no. 2:479–481, May 1985.

[240] J-W. Liang, J-T. Chen, and A. Paulraj. "A two-stage hybrid approach for CCI/ISI reduction with space-time processing." *IEEE Communications Letters*, 1:163–165, Nov. 1997.

[241] T. Lim and L. Rasmussen. "Adaptive symbol and parameter estimation in asynchronous multiuser CDMA detectors." *IEEE Trans. Communications*, 45, no. 2:213–220, Feb. 1997.

[242] T. Lim and S. Roy. "Adaptive detectors for multiuser CDMA." *Wireless Networks*, to appear, 1998.

[243] A. Lindsey. "Wavelet packet modulation for orthogonally multiplexed communication." *IEEE Trans. Signal Processing*, 45:1336–1339, May 1997.

[244] W. Lindsey. "Error probabilities for Rician fading multichannel reception of binary and N-ary signals." *IEEE Trans. Information Theory*, 10:339–350, Oct. 1964.

[245] H. Liu and Z. Siveski. "Differentially coherent decorrelating detector for CDMA single-path time-varying Rayleigh fading channels." Submitted to *IEEE Trans. Communications*, 1998; see also *Proc. 1996 Conf. on Information Sciences and Systems*, pp. 86–89, Princeton, NJ, Mar. 1996.

[246] H. Liu and G. Xu. "A subspace method for signature waveform estimation in synchronous CDMA systems." *IEEE Trans. Communications*, 44, no. 10:1346–1354, Oct. 1996.

[247] H. Liu and G. Xu. "Smart antennas in wireless systems: uplink multiuser blind channel and sequence detection." *IEEE Trans. Communications*, 45, no. 2:187–199, Feb. 1997.

[248] H. Liu and M. Zoltowski. "Blind equalization in antenna array CDMA systems." *IEEE Trans. Signal Processing*, 45, no. 1:161–172, Jan. 1997.

[249] Q. Liu, R. Scholtz, and Z. Zhang. "Complexity of Verdú optimum multiuser detection algorithm in random access multichannel CDMA systems." Submitted to *IEEE Trans. Communications*, 1997.

[250] B. Long, J. Hu, and P. Zhang. "Method to improve Gaussian approximation accuracy for calculation of spread-spectrum multiple-access error probabilities." *Electronics Letters*, 31, no. 7:529–531, 1995.

[251] C. Loo. "A statistical model for a land-mobile satellite link." *IEEE Trans. Vehicular Technology*, 34, no. 3:122–127, Aug. 1985.

[252] R. Lupas. "Near-far resistant linear multiuser detection." PhD thesis, Princeton University, Princeton, NJ, 1989.

[253] R. Lupas and S. Verdú. "Optimum near-far resistance of linear detectors for code-division multiple-access channels." *Abstr. 1988 IEEE Int. Symp. Information Theory*, p. 14, June 1988.

[254] R. Lupas and S. Verdú. "Linear multiuser detectors for synchronous code-division multiple-access channels." *IEEE Trans. Information Theory*, 35:123–136, Jan. 1989.

[255] R. Lupas and S. Verdú. "Near-far resistance of multiuser detectors in asynchronous channels." *IEEE Trans. Communications*, 38:496–508, Apr. 1990.

[256] R. Lupas-Golaszewski and S. Verdú. "Asymptotic efficiency of linear multiuser detectors." *Proc. 25th IEEE Conf. on Decision and Control*, pp. 2094–2100, Dec. 1986.

[257] D. Lyu, I. Song, Y. Han, and H. Kim. "Analysis of the performance of an asynchronous multiple-chip-rate DS/CDMA system." *Int. J. Electron. Communication*, AEU-51:213–218, July 1997.

[258] U. Madhow. "Blind adaptive interference suppression for acquisition and demodulation of direct-sequence CDMA signals." *Proc. 1995 Conf. Information Sciences and Systems*, pp. 180–185, Mar. 1995.

[259] U. Madhow. "MMSE interference suppression for joint acquisition and demodulation of direct-sequence CDMA signals." *Preprint*, 1997.

[260] U. Madhow. "Blind adaptive interference suppression for the near-far

resistant acquisition and demodulation of direct-sequence CDMA signals." *IEEE Trans. Signal Processing*, 45, no. 1:124–136, Jan. 1997.

[261] U. Madhow. "Blind adaptive interference suppression for direct-sequence CDMA." *IEEE Proc.*, to appear, 1998.

[262] U. Madhow and M. L. Honig. "Performance analysis of MMSE detectors for direct sequence CDMA assuming random signature sequences." Submitted to *IEEE Trans. Information Theory*.

[263] U. Madhow and M. L. Honig. "MMSE interference suppression for direct-sequence spread spectrum CDMA." *IEEE Trans. Communications*, 42:3178–3188, Dec. 1994.

[264] N. Mandayam and B. Aazhang. "Gradient estimation for sensitivity analysis and adaptive multiuser interference rejection in code division multiple access systems." *IEEE Trans. Communications*, 45:848–858, July 1997.

[265] N. Mandayam and S. Verdú. "Analysis of an approximate decorrelating detector." *Wireless Personal Communications*, 6(1–2):97–111, Jan. 1998.

[266] J. Marcum. "A statistical theory of target detection by pulsed radar." *IRE Trans. Information Theory*, 6:59–268, Apr. 1960. Reprinted from RAND Research Memo. RM-754, Dec. 1947.

[267] J. L. Massey and T. Mittelholzer. "Welch's bound and sequence sets for code-division multiple-access systems." In R. Capocelli, A. De Santis, and U. Vaccaro, editors, *Sequences II, Methods in Communication, Security, and Computer Science*. Springer-Verlag, New York, 1993.

[268] K. Mayyas and T. Aboulnasr. "Leaky LMS algorithm: MSE analysis for Gaussian data." *IEEE Trans. Signal Processing*, 45:927–934, Apr. 1997.

[269] J. Mazo. "Faster-than-Nyquist-signaling." *Bell System Technical Journal*, 54:1451–1462, Oct. 1975.

[270] J. Mazo. "Some theoretical observations on spread-spectrum communications." *Bell System Technical Journal*, 58:2013–2023, Nov. 1979.

[271] G. Mazzinni. "Equal BER with successive interference cancellation DS-CDMA systems on AWGN and Ricean channels." *Proc. 6th IEEE Int. Symp. on Personal, Indoor and Mobile Radio Communications*, 2:727–736, Sept. 1995.

[272] A. McDowell and J. Lehnert. "General techniques for analyzing DS/SSMA interference illustrated by evaluating a noncoherent system." *IEEE Trans. Communications*, 44:1730–1737, Dec. 1996.

[273] S. Miller. "An adaptive direct-sequence code-division multiple-access receiver for multi-user interference rejection." *IEEE Trans. Communications*, 43:1746–1755, Apr. 1995.

[274] S. Miller. "Training analysis of adaptive interference suppression for direct-sequence code-division multiple-access systems." *IEEE Trans. Communications*, 44:488–495, Apr. 1996.

[275] S. Miller, M. L. Honig, M. Shensa, and L. Milstein. "MMSE reception of DS-CDMA for frequency-selective fading channels." *Proc. 1997 IEEE Int. Symp. on Information Theory*, p. 51, July 1997.

[276] S. Y. Miller. *Detection and estimation in multiple-access channels*. PhD thesis, Princeton University, Princeton, NJ, Oct. 1989.

[277] S. Y. Miller and S. Schwartz. "Integrated spatial-temporal detectors for asynchronous Gaussian multiple-access channels." *IEEE Trans. Communications*, 43:396–411, Feb. 1995.

[278] L. Milstein. "Interference rejection techniques in spread spectrum communications." *Proc. IEEE*, 76:657–671, June 1988.

[279] U. Mitra and H. V. Poor. "Adaptive receiver algorithms for near-far resistant CDMA." *IEEE Trans. Communications*, pp. 1713–1724, Feb. 1995.

[280] U. Mitra and H. V. Poor. "Adaptive decorrelating detectors for CDMA systems." *Wireless Personal Communications*, pp. 415–440, 1996.

[281] U. Mitra and H. V. Poor. "Analysis of an adaptive decorrelating detector for synchronous CDMA channels." *IEEE Trans. Communications*, pp. 257–268, Feb. 1996.

[282] U. Mitra and H. V. Poor. "Activity detection in a multi-user environment." *Wireless Personal Communications*, pp. 149–174, nos. 1–2, 1996.

[283] T. Miyajima and T. Hasegawa. "Multiuser detection using a recurrent neural network in code-division multiple-access communications." *Technical Report of IEICE*, SST92-84:63–68, 1993.

[284] T. Miyajima, S. Miura, and K. Yamanaka. "Blind multiuser receivers for DS/CDMA systems." *Proc. 1997 IEEE Global Telecommunications Conf.*, pp. 118–122, Nov. 1997.

[285] A. Monk, M. Davis, L. Milstein, and C. Helstrom. "A noise-whitening approach to multiple access noise rejection – Part I: Theory and background." *IEEE Journal on Selected Areas in Communications*, 12, no. 5:817–827, June 1994.

[286] T. Moon, Z. Xie, C. K. Rushforth, and R. Short. "Parameter estimation in a multi-user communication system." *IEEE Trans. Communications*, 42:2553–2560, Aug. 1994.

[287] R. Morrow and J. Lehnert. "Bit-to-bit error dependence in slotted DS/SSMA packet systems with random signature sequences." *IEEE Trans. Communications*, 37:1052–1061, Oct. 1989.

[288] S. Moshavi. "Multiuser detection for DS-CDMA communications." *IEEE Communications Magazine*, 34:124–137, Oct. 1996.

[289] S. Moshavi, E. Kanterakis, and D. Schilling. "Multistage linear receivers for DS-CDMA systems." *Int. Journal of Wireless Information Networks*, 3:1–18, Jan. 1996.

[290] R. Mowbray, P. Grant, and R. Pringle. "New antimultipath technique for spread spectrum receivers." *Electronics Letters*, 29, no. 5:456–457, 1993.

[291] R. Mowbray, R. Pringle, and P. Grant. "Increased CDMA system capacity through adaptive cochannel interference regeneration and cancellation." *IEE Proc. I*, 139, no. 5:515–524, Oct. 1992.

[292] R. Müller, P. Schramm, and J. Huber. "Spectral efficiency of CDMA systems with linear interference suppression." *IEEE Workshop Communications Engineering*, Ulm, Germany, 1997.

[293] O. Muñoz-Medina and J. Fernández-Rubio. "Cancellation of external and multiple access interference in CDMA systems using antenna arrays." *Signal Processing*, 61:113–129, 1997.

[294] J. Musser and J. Daigle. "Derivation of asynchronous code division

multiple access (CDMA) throughput." In R. Pickholtz, editor, *Local Area and Multiple Access Networks*, pp. 167–183. Computer Science Press, Rockville, MD, 1986.

[295] M. Mydlow, S. Basavaraju, and A. Duel-Hallen. "Decorrelating detector with diversity combining for single user frequency-selective Rayleigh fading multipath channels." *Wireless Personal Communications*, pp. 175–193, nos. 1–2, 1996.

[296] T. Myers and M. Magaña. "An adaptive implementation of the "one-shot" decorrelating detector for CDMA communications." *IEEE Trans. Circuits and Systems: Analog and Digital Signal Processing*, 44:762–765, Sept. 1997.

[297] R. Van Nee, H. Misser, and R. Prasad. "Direct-sequence spread spectrum in a shadowed Rician fading land-mobile satellite channel." *IEEE Journal on Selected Areas in Communications*, 10:350–357, Feb. 1992.

[298] L. Nelson and H. V. Poor. "Performance of multiuser detection for optical CDMA – Part II: Asymptotic analysis." *IEEE Trans. Communications*, 43:3015–3024, Dec. 1995.

[299] L. Nelson and H. V. Poor. "Performance of multiuser detection for optical CDMA – Part I: Error probabilities." *IEEE Trans. Communications*, 43:2803–2811, Nov. 1995.

[300] L. Nelson and H. V. Poor. "Iterative multi-user receivers for the synchronous CDMA channel: An EM-based approach." *IEEE Trans. Communications*, 44:1700–1710, Dec. 1996.

[301] E. W. Ng and M. Geller. "A table of integrals of the error functions." *Journal of Research of the National Bureau of Standards-B. Mathematical Sciences*, 73B:1–20, Jan.–Mar. 1969.

[302] H. Nichols, A. Giordano, and J. Proakis. "MLD and MSE algorithms for adaptive detection of digital signals in the presence of interchannel interference." *IEEE Trans. Information Theory*, 23:563–575, Sept. 1977.

[303] D. North. "Analysis of the factors which determine signal/noise discrimination in radar." *Technical Report PTR-6-C*. RCA, Princeton, NJ, June 1943. Reprinted in *Proc. IEEE*, vol. 51, July 1963.

[304] A. Nuttall and F. Amoroso. "Minimum Gabor bandwidth of m orthogonal signals." *IEEE Trans. Information Theory*, 11:440–444, July 1965.

[305] H. Oda and Y. Sato. "A method of multidimensional blind equalization." *IEEE Int. Symp. on Information Theory*, p. 327, Jan. 1993.

[306] T. Ojanpera, K. Rikkinen, H. Hakkinen, K. Pehkonen, A. Hottinen, and J. Lilleberg. "Design of a 3rd generation multirate CDMA system with multiuser detection, MUD-CDMA." *Proc. 1996 IEEE Int. Symp. on Spread Spectrum Techniques and Applications*, pp. 334–338, Sept. 1996.

[307] J. Olsen, R. Scholtz, and L. Welch. "Bent-function sequences." *IEEE Trans. Information Theory*, 28:858–864, Nov. 1982.

[308] J. Omura. "Optimal receiver design for convolutional codes and channels with memory via control theoretical concepts." *Information Sciences*, 3:243–266, 1971.

[309] I. Oppermann and M. Latva-aho. "Adaptive LMMSE receiver for wideband CDMA systems" *Proc. 1997 IEEE Global Telecommunications Conf.*, pp. 133–138, Nov. 1997.

[310] I. Oppermann, B. Vucetic, and P. Rapajic. "Capacity of digital

413

cellular CDMA system with adaptive receiver." *1995 IEEE Int. Symp. Information Theory*, p. 110, Sept. 1995.

[311] P. Orten and T. Ottosson. "Robustness of DS-CDMA multiuser detectors." *Proc. 1997 IEEE Global Telecommunications Conf.*, pp. 144–148, Nov. 1997.

[312] D. Pados and S. Batalama. "Low-complexity blind detection of DS/CDMA signals: Auxiliary-vector receivers." *IEEE Trans. Communications*, 45, no. 12:1586–1594, Dec. 1997.

[313] C. Papadias and A. Paulraj. "A constant modulus algorithm for multiuser signal separation in presence of delay spread using antenna arrays." *IEEE Signal Processing Letters*, 4:178–181, June 1997.

[314] C. Papadimitrou and K. Steiglitz. *Combinatorial Optimization: Algorithms and Complexity*. Prentice-Hall, Englewood Cliffs, NJ, 1982.

[315] E. Papproth and G. Kaleh. "Near-far resistant channel estimation for the DS-CDMA uplink." *Proc. 1995 IEEE Int. Symp. Personal, Indoor and Mobile Radio Communications*, 2:758–762, Sept. 1995.

[316] B. Paris. "Asymptotic properties of self-adaptive maximum-likelihood sequence estimation." *Proc. 1993 Conf. Information Sciences and Systems*, pp. 161–166, Mar. 1993.

[317] B. Paris. "Finite precision decorrelating receivers for multiuser CDMA communication systems." *IEEE Trans. Communications*, 44:496–507, Apr. 1996.

[318] B. Paris, G. Orsak, M. Varanasi, and B. Aazhang. "Neural net receivers in multiple-access communications." In *Advances in Neural Network Information Processing Systems*. Morgan Kaufmann, San Mateo, CA, 1988.

[319] S. Park and J. Doherty. "Generalized projection algorithm for blind interference suppression in DS/CDMA communications." *IEEE Trans. Circuits and Systems*, 44, no. 6:453–460, June 1997.

[320] S. Parkvall, E. Strom, and B. Ottersten. "The impact of timing errors on the performance of linear DS-CDMA receivers." *IEEE Journal on Selected Areas in Communications*, 14, no. 8:1660–1668, Oct. 1996.

[321] D. Parsavand and M. Varanasi. "RMS bandwidth constrained signature waveforms that maximize the total capacity of PAM-synchronous CDMA channels." *IEEE Trans. Communications*, 44:65–75, Jan. 1996.

[322] J. D. Parsons. "The Mobile Radio Propagation Channel." Wiley, New York, 1992.

[323] P. Patel and J. Holtzman. "Analysis of a DS/CDMA successive interference cancellation scheme using correlations." *Proc. 1993 IEEE Global Telecommunications Conf.*, pp. 76–80, Nov. 29–Dec. 2, 1993.

[324] P. Patel and J. Holtzman. "Analysis of a simple successive interference cancellation scheme in DS/CDMA system." *IEEE Journal on Selected Areas in Communications*, 12:796–807, June 1994.

[325] P. Patel and J. Holtzman. "Analysis of successive interference cancellation in m-ary orthogonal DS-CDMA system with single path Rayleigh fading." *Proc. 1994 Int. Zurich Seminar on Digital Communications*, pp. 150–161, Mar. 1994.

[326] A. Paulraj and C. Papadias. "Space-time processing for wireless communications." *IEEE Signal Processing Magazine*, 14:49–83, Nov. 1997.

414

[327] A. Paulraj, R. Roy, and T. Kailath. "A subspace rotation approach to signal parameter estimation. " *Proc. IEEE*, 74:1044–1045, July 1986.

[328] K. Pedersen, T. Kolding, I. Seskar, and J. Holtzman. "Practical implementation of successive interference cancellation in DS/CDMA systems." *Proc. 1996 Int. Conf. on Universal Personal Communications*, pp. 321–325, Sept. 1996.

[329] B. Petersen and D. Falconer. "Minimum mean square equalization in cyclostationary and stationary interference – analysis and subscriber line calculations." *IEEE Journal on Selected Areas in Communications*, 9, no. 6:931–940, 1991.

[330] B. Petersen and D. Falconer. "Suppression of adjacent-channel, cochannel, and intersymbol interference by equalizers and linear combiners." *IEEE Trans. Communications*, 42:3109–3118, Dec. 1994.

[331] W. Peterson. *Error-Correcting Codes*. Wiley, New York, 1961.

[332] M. Petrich. "On the number of orthogonal signals which can be placed in a WT-product." *SIAM Journal*, pp. 936–940, Dec. 1963.

[333] R. Pickholtz, D. Schilling, and L. Milstein. "Theory of spread-spectrum communications – a tutorial." *IEEE Trans. Communications*, 30:855–884, May 1982.

[334] H. V. Poor. "Signal detection in multiple access channels." U.S. Army research proposal, University of Illinois, Urbana, 1980.

[335] H. V. Poor. "On parameter estimation in DS/SSMA formats." *Proc. 1988 Int. Conf. Advances in Communications and Control Systems*, 1:98–109, Oct. 1988.

[336] H. V. Poor. *An Introduction to Signal Detection and Estimation.* Springer, New York, 2nd ed., 1994.

[337] H. V. Poor and S. Verdú. "Single-user detectors for multiuser channels." *IEEE Trans. Communications*, 36:50–60, Jan. 1988.

[338] H. V. Poor and S. Verdú. "Probability of error in MMSE multiuser detection." *IEEE Trans. Information Theory*, 43:858–871, May 1997.

[339] H. V. Poor and X. Wang. "Code-aided interference suppression for DS/CDMA communications – Part I: Interference suppression capability." *IEEE Trans. Communications*, 45:1101–1111, Sept. 1997.

[340] H. V. Poor and X. Wang. "Code-aided interference suppression for DS/CDMA communications – Part II: Parallel blind adaptive implementations." *IEEE Trans. Communications*, 45:1112–1122, Sept. 1997.

[341] R. Prasad. *CDMA for Wireless Personal Communications*. Artech House, Boston, MA, 1996.

[342] R. Prasad and S. Hara. "An overview of multicarrier CDMA." *Proc. 1996 IEEE Int. Symp. on Spread Spectrum Techniques and Applications*, 1:107–114, Sept. 1996.

[343] R. Price. "A conversation with Claude Shannon." *IEEE Communications Magazine*, 22:123–126, May 1984.

[344] R. Price and P. Green. "A communication technique for multipath channels." *Proc. IRE*, 46:555–570, Mar. 1958.

[345] J. Proakis. *Digital Communications*. McGraw Hill, New York, 3rd ed., 1995.

[346] I. Psaromiligkos and S. Batalama. "On the unimodality of the bit error

rate function of linear receivers for DS-CDMA systems." *Proc. 1997 IEEE Global Telecommunications Conf.*, pp. 17–22, Nov. 1997.

[347] M. Pursley. "Performance evaluation for phase-coded spread-spectrum multiple-access communication – Part I: System analysis." *IEEE Trans. Communications*, 25:795–799, Aug. 1977.

[348] M. Pursley, D. Sarwate, and W. Stark. "Error probability for direct-sequence spread-spectrum multiple-access communications – Part I: Upper and lower bounds." *IEEE Trans. Communications*, 30(5):975–984, May 1982.

[349] C. Rader. "VLSI systolic arrays for adaptive nulling." *IEEE Signal Processing Magazine*, 13:29–49, July 1996.

[350] P. Ranta, Z. Honkasalo, and J. Tapaninen. "TDMA cellular network application of an interference cancellation technique." *Proc. 45th IEEE Vechicular Technology Conf.*, 1:296–300, July 1995.

[351] P. Ranta, A. Hottinen, and Z. Honkasalo. "Co-channel interference cancelling receiver for TDMA mobile systems." *Proc. 1995 IEEE Int. Conf. Communications*, 1:17–21, June 1995.

[352] P. Rapajic and B. Vucetic. "Adaptive receiver structures for asynchronous CDMA systems." *IEEE Journal on Selected Areas in Communications*, 12:685–697, May 1994.

[353] R. Rapajic and B. Vucetic. "Application of fast adaptive algorithms in asynchronous CDMA systems." *Int. Symp. on Information Theory and Its Applications*, pp. 97–102, Nov. 1994.

[354] R. Rapajic and B. Vucetic. "Narrow-band and multiple access interference rejection by adaptive single user receiver in asynchronous CDMA systems." *Int. Symp. on Information Theory and Its Applications*, pp. 73–78, Nov. 1994.

[355] P. Rapajic and B. Vucetic. "Linear adaptive transmitter-receiver structures for asynchronous CDMA systems." *European Trans. Telecommunications*, pp. 21–27, Jan./Feb. 1995.

[356] T. Rappaport and L. Milstein. "Effects of radio propagation path loss on DS-CDMA cellular frequency reuse efficiency for the reverse channel." *IEEE Trans. Vehicular Technology*, 41, no. 3:231–242, Aug. 1992.

[357] E. Del Re and L. Ronga. "Blind adaptive MUD with silence listening." *Signal Processing*, 61:91–100, Sept. 1997.

[358] S. Redl, M. Weber, and M. Oliphant. *An Introduction to GSM.* Artech House, Boston/London, 1995.

[359] J. Riba, J. Goldberg, and G. Vázquez. "Robust data detection in asynchronous DS-CDMA in the presence of timing uncertainty." *Proc. 8th IEEE Signal Processing Workshop on Statistical Signal and Array Processing*, pp. 521–524, 1996.

[360] B. Rimoldi and R. Urbanke. "A rate-splitting approach to the Gaussian multiple-access channel." *IEEE Trans. Information Theory*, 42:364–375, Mar. 1996.

[361] H. Robbins and S. Monro. "A stochastic approximation method." *Annals of Mathematical Statistics*, pp. 400–407, 1951.

[362] J. Ross and D. Taylor. "Multi-user signalling in the symbol-synchronous AWGN channel." *IEEE Trans. Information Theory*, 41:1174–1178, July 1995.

416

[363] S. Roy. "Subspace blind adaptive detection for multiuser CDMA." Submitted to *IEEE Trans. Communications*, 1997.

[364] M. Rupf and J. L. Massey. "Optimum sequence multi-sets for multiple-access channels." *IEEE Trans. Information Theory*, 40, no. 4:1261–1266, July 1994.

[365] M. Rupf, F. Tarkoy, and J. L. Massey. "User-separating demodulation for code-division multiple-access systems." *IEEE Journal on Selected Areas in Communications*, 12, no. 5:786–795, June 1994.

[366] J. Ruprecht, F. Neeser, and M. Hufschmid. "Code time division multiple access: An indoor cellular system." *Proc. 42nd IEEE Vehicular Technology Conf.*, pp. 2–4, 1992.

[367] L. Rusch and H. V. Poor. "Narrowband interference suppression in CDMA spread-spectrum communications." *IEEE Trans. Communications*, 42:1969–1979, Feb./Mar./Apr. 1994.

[368] L. Rusch and H. V. Poor. "Narrowband interference suppression in spread spectrum CDMA." *IEEE Personal Communications Magazine*, 1, no. 3:14–27, 1994.

[369] L. A. Rusch and H. V. Poor. "Effects of laser phasedrift on coherent optical CDMA." *IEEE Journal on Selected Areas in Communications*, 13:577–591, Apr. 1995.

[370] A. Russ and M. Varanasi. "Minimum error probability and suboptimum noncoherent multiuser detection for nonlinear nonorthogonal synchronous signaling over a Rayleigh fading channel." *Proc. 1997 IEEE, Personal, Indoor, Mobile Radio Communications Conf.*, pp. 58–62, Sept. 1997.

[371] A. Russ and M. Varanasi. "Noncoherent multiuser detection for nonlinear modulation: An asymptotic analysis of minimum probability of error for the Rayleigh fading channel." *Proc. 1997 IEEE Global Telecommunications Conf.*, pp. 195–199, Nov. 1997.

[372] J. Sadowsky and R. Bahr. "Direct sequence spread spectrum multiple access communication with random signature sequences." *IEEE Trans. Information Theory*, 37:514–527, May 1991.

[373] A. Saifudding. *Integrated Techniques for Efficient and Reliable Code Division Multiple Access*. PhD thesis, Yokohama National University, Yokohama, Japan, Dec. 1995.

[374] A. Saifudding, R. Kohno, and H. Imai. "Integrated receiver structure of staged decoder and CCI canceller for CDMA with multilevel coded modulation." *European Trans. Telecommunications*, vol. 6, no. 1:9–19, Jan. 1995.

[375] B. Saltzberg. "Intersymbol interference error bounds with application to ideal bandlimited signaling." *IEEE Trans. Information Theory*, 14:563–568, July 1968.

[376] J. Salz. "Digital transmission over cross-coupled linear channels." *AT&T Technical Journal*, 64, no. 6:1147–1159, 1985.

[377] Y. Sanada, A. Kajiwara, and M. Nakagawa. "Adaptive Rake receiver for mobile communications." *IEICE Trans. Communications*, E76-B:1002–1007, Aug. 1993.

[378] Y. Sanada and M. Nakagawa. "A multiuser interference cancellation technique utilizing convolutional codes and orthogonal multicarrier

modulation for wireless indoor communications." *IEEE Journal on Selected Areas in Communications*, 14, no. 8:1500–1509, Oct. 1996.

[379] Y. Sanada and Q. Wang. "A co-channel interference cancellation technique using orthogonal convolutional codes." *IEEE Trans. Communications*, 44:549–556, May 1996.

[380] C. Sankaran and A. Ephremides. "Optimum multiuser detection with polynomial complexity." *1998 IEEE Int. Symp. on Information Theory*, Aug. 1998.

[381] D. Sarwate and M. Pursley. "Crosscorrelation properties of pseudorandom and related sequences." *Proc. IEEE*, 68:593–620, May 1980.

[382] W. Sauer-Greff and R. Kennedy. "Suboptimal MLSE for distorted multiple-access channels using the M-algorithm." *Aachener Kolloquium Signaltheorie Mobile Kommunikationssysteme*, pp. 267–269, Mar. 1994.

[383] J. Savage. "Signal detection in the presence of multiple-access noise." *IEEE Trans. Information Theory*, 20:42–49, Jan. 1974.

[384] A. Sayeed and B. Aazhang. "Exploiting Doppler diversity in mobile wireless communications." *Proc. 1997 Conf. on Information Sciences and Systems*, pp. 287–292, Mar. 1997.

[385] L. Scharf and B. Friedlander. "Matched subspace detectors." *IEEE Trans. Signal Processing*, 42, no 8:2146–2157, Aug. 1994.

[386] C. Schlegel. "Error probability calculation for multibeam Rayleigh channels." *IEEE Trans. Communications*, 44:290–293, Mar. 1996.

[387] C. Schlegel and V. Mathews. "Computationally efficient multiuser detection for coded CDMA." *Proc. 1997 IEEE Int. Conf. Universal and Personal Communications*, Oct. 1997.

[388] C. Schlegel, S. Roy, P. Alexander, and Z. Xiang. "Multi-user projection receivers." *IEEE Journal on Selected Areas in Communications*, 14:1610–1618, Oct. 1996.

[389] C. Schlegel and L. Wei. "A simple way to compute the minimum distance in multiuser CDMA systems." *IEEE Trans. Communications*, 45, no. 5:532–535, May 1997.

[390] C. Schlegel and Z. Xiang. "A new projection receiver for coded synchronous multi-user CDMA systems." *Proc. 1995 IEEE Int. Sym. on Information Theory*, p. 318, Sept. 1995.

[391] R. O. Schmidt. "Multiple emitter location and signal parameter estimation." *Proc. RADC Spectrum Estimation Workshop*, pp. 243–258, 1979.

[392] K. Schneider. "Optimum detection of code division multiplexed signals." *IEEE Trans. Aerospace and Electronic Systems*, AES-15(1):181–185, Jan. 1979.

[393] K. Schneider. "Crosstalk resistant receiver for m-ary multiplexed communications." *IEEE Trans. Aerospace and Electronic Systems*, AES-16:426–433, July 1980.

[394] M. Schnell. "Interference calculations for MC-SSMA systems in mobile communications." *Proc. 6th IEEE Int. Symp. on Personal, Indoor and Mobile Radio Communications*, 1:158–163, Sept. 1995.

[395] J. Schodorf and D. Williams. "A constrained optimization approach to multiuser detection." *IEEE Trans. Signal Processing*, 45, no. 1:258–262, Jan. 1997.

418

[396] J. Schodorf and D. Williams. "Array processing techniques for multiuser detection." *IEEE Trans. Communications*, 45, no. 11:1375–1378, Nov. 1997.

[397] R. Scholtz. "The origins of spread-spectrum communications." *IEEE Trans. Communications*, 30:822–854, May 1982. See also Jan. 1983, pp. 82–97.

[398] R. Scholtz. "The evolution of spread-spectrum multiple-access communications." In S. G. Glisic and P. Leppänen, ed. *Code Division Multiple Access Communications*, pp. 3–28. Kluwer, Dordrecht, The Netherlands, 1995.

[399] P. Schramm, R. Müller, and J. Huber. "Spectral efficiency of multiuser systems based on CDMA with linear MMSE interference suppression." *Proc. 1997 IEEE Int. Symp. on Information Theory*, p. 357, June–July 1997.

[400] A. Sendonaris and B. Aazhang. "An optimal one-shot multiuser detector for flat Rayleigh fading channels." *Proc. 35th Allerton Conf. on Communications, Control and Computing*, Oct. 1997.

[401] N. Seshadri. "Joint data and channel estimation using blind trellis search techniques." *IEEE Trans. Communications*, 42:313–323, Feb./Mar./Apr. 1994.

[402] I. Seskar and N. Mandayam. "A software radio architecture for linear multiuser detection." *Proc. 1998 Conf. Information Sciences and Systems*, Mar. 1998.

[403] I. Seskar, K. Pedersen, T. Kolding, and J. Holtzman. "Implementation aspects of successive interference cancellation in DS/CDMA systems." *Wireless Networks*, to appear, 1998.

[404] K. Shaheen and S. Gupta. "Adaptive combination of cancelling co-channel interference and decoding of error-correcting codes for DS-SS CDMA mobile communication system." *Proc. 6th IEEE Int. Symp. on Personal, Indoor and Mobile Radio Communications*, 2:737–741, Sept. 1995.

[405] Y. Shama and B. Vojcic. "Soft output Viterbi algorithm for maximum likelihood detection of coded asynchronous communications." *Proc. 1997 Conf. Information Sciences and Systems*, pp. 730–735, Mar. 1997.

[406] S. Shamai and S. Verdú. "Worst-case power constrained noise for binary-input channels." *IEEE Trans. Information Theory*, 38:1494–1511, Sept. 1992.

[407] C. E. Shannon. "Communication in the presence of noise." *Proc. IRE*, 37:10–21, 1949.

[408] Z. Shi, W. Du, and P. Driessen. "A new multistage detector in synchronous code-division multiple-access systems." *IEEE Trans. Communications*, 44:538–541, May 1996.

[409] D. Shnidman. "A generalized Nyquist criterion and an optimum linear receiver for a pulse modulation system." *Bell System Technical Journal*, 46:2163–2177, Nov. 1967.

[410] M. Simon, J. Omura, R. Scholtz, and B. Levitt. *Spread Spectrum Communications Handbook*. McGraw-Hill, New York, 1994.

[411] Z. Siveski, Y. Bar-Ness, and D. Chen. "Error performance of synchronous

multiuser CDMA detector with multidimensional adaptive canceller." *European Trans. Telecommunications*, pp. 719–724, Nov.–Dec. 1994.

[412] B. Sklar. "Rayleigh fading channels in mobile digital communications systems – Part II: Mitigation." *IEEE Communications Magazine*, 35:102–112, July 1997.

[413] B. Sklar. "Rayleigh fading channels in mobile digital communications systems – Part I: Characterization." *IEEE Communications Magazine*, 35:136–146, Sept. 1997.

[414] D. Slock. "Blind joint equalization of multiple synchronous mobile users using oversampling and/or multiple antennas." *Proc. 28th Asilomar Conf. Signals, Systems and Computers*, pp. 1154–1158, Oct. 1994.

[415] D. Slock and T. Kailath. "Numerically stable fast recursive least-squares transversal filters." *Proc. Int. Conf. Acoustics, Speech, Signal Processing*, pp. 1365–1368, Apr. 1988.

[416] R. F. Smith and S. L. Miller. "Code timing estimation in a near-far environment for direct-sequence code-division multiple-access." *Proc. 1994 IEEE Military Communications Conf.*, pp. 47–51, 1994.

[417] V. Soon and L. Tong. "Blind equalization under cochannel interference." *Proc. 27th Conf. Information Sciences and Systems*, pp. 155–160, Mar. 1993.

[418] E. Sousa. "Performance of a direct sequence spread spectrum multiple access system utilizing unequal carrier frequencies." *IEICE Trans. Communications*, E76-B no. 8:906–912, 1993.

[419] M. Srinivasan, U. Madhow, and D. Sarwate. "Some results on parallel acquisition in CDMA systems." *Proc. 1997 IEEE Int. Symp. on Information Theory*, p. 353, July 1997.

[420] S. Stein. "Unified analysis of certain coherent and noncoherent binary communications systems." *IEEE Trans. Information Theory*, 10:43–51, Jan. 1964.

[421] S. Stein. "Fading channel issues in system engineering." *IEEE Journal on Selected Areas in Communications*, 5:68–89, Feb. 1987.

[422] Y. Steinberg and H. V. Poor. "Multiuser delay estimation." *Proc. 1993 Conf. Information Sciences and Systems*, pp. 635–638, 1993.

[423] Y. Steinberg and H. V. Poor. "Sequential amplitude estimation in multiuser communications." *IEEE Trans. Information Theory*, 38:11–20, Jan. 1994.

[424] B. Steiner and P. Jung. "Optimum and suboptimum channel estimation for the uplink of CDMA mobile radio systems with joint detection." *European Trans. Telecommunications*, 5:39–49, Jan. 1994.

[425] G. Stewart. "An updating algorithm for subspace tracking." *IEEE Trans. Signal Processing*, 40, no. 6:1535–1541, June 1992.

[426] I. G. Stiglitz. "Multiple-access considerations – a satellite example." *IEEE Trans. Communications*, 21:577–582, May 1973.

[427] M. Stojanovic and Z. Zvonar. "Multichannel processing of broad-band multiuser communication signals in shallow water acoustic channel." *IEEE Journal of Oceanic Engineering*, 21, no. 2:156–166, Apr. 1996.

[428] E. Strom and S. Miller. "Asynchronous DS-CDMA systems: Low-complexity near-far resistant receivers and parameter estimation." Technical report, Dept. of Elec. Eng., Univ. of Florida, Jan. 1994.

[429] E. Strom and S. Miller. "Optimum complexity reduction of minimum

mean square error DS-CDMA receivers." *Proc. 1994 IEEE Vehicular Technology Conf.*, pp. 568–572, June 1994.

[430] E. Strom, S. Parkvall, S. Miller, and B. Ottersten. "Propagation delay estimation in asynchronous direct-sequence code-division multiple access systems." *IEEE Trans. Communications*, 44:84–93, Jan. 1996.

[431] E. Strom, S. Parkvall, S. Miller, and B. Ottersten. DS-CDMA synchronization in time-varying fading channels. *IEEE Journal on Selected Areas in Communications*, 14, no. 8:1636–1642, Oct. 1996.

[432] G. L. Stuber. "Modulation methods." In J. D. Gibson, editor, *The Mobile Communications Handbook*, Chapter 33, pp. 526–539, CRC Press, 1996.

[433] G. L. Stuber. *Principles of Mobile Radio Communication*. Kluwer, Boston, 1996.

[434] V. Subramanian and U. Madhow. "Blind demodulation of direct-sequence CDMA signals using an antenna array." *Proc. 1996 Conf. on Information Sciences and Systems*, pp. 74–80, Mar. 1996.

[435] N. Suehiro and M. Hatori. "Modulatable orthogonal sequences and their application to SSMA systems." *IEEE Trans. Information Theory*, 34:93–100, Jan. 1988.

[436] C-M. Sun and A. Polydoros. "Performance evaluation of multicell direct-sequence microcellular systems." *European Trans. Telecommunications*, 6, no. 1:71–83, Jan.–Feb. 1995.

[437] P. Sung and K. Chen. "A linear minimum mean square error multiuser receiver in Rayleigh-fading channels." *IEEE Journal on Selected Areas in Communications*, 14, no. 8:1583–1594, Oct. 1996.

[438] S. Tachikawa. "Characteristics of m-ary spread spectrum multiple access communication systems using co-channel interference cancellation techniques." *IEEE Trans. Communications*, E76-B, no. 8:941–946, 1993.

[439] S. Tamura, S. Nakano, and K. Okazaki. "Optical code-multiplex transmission by Gold sequences." *J. Lightwave Technology*, LT-3:121–127, Feb. 1985.

[440] E. Telatar. "Capacity of multi-antenna Gaussian channels." Tech. Rep. BL011217-950615-07TM, AT&T Bell Laboratories, June 1995.

[441] Telecommunications Industry Association, TIA/EIA, Washigton, DC. *Cellular System Dual-mode Mobile Station-Base Station Compatibility Standard IS-54B*, 1992.

[442] Telecommunications Industry Association, TIA/EIA, Washington, DC. *Mobile Station-Base Station Compatibility Standard for Dual-Mode Wideband Spread Spectrum Cellular System IS-95A*, 1995.

[443] S. Talwar and A. Paulraj. "Blind separation of synchronous co-channel digital signals using an antenna array." *IEEE Trans. Signal Processing*, 44:1184–1197, May 1996 and 45:706–718, Mar. 1997.

[444] C. Tidestav, A. Ahlen, and M. Sternad. "Narrowband and broad-band multiuser detection using a multivariable DFE." *Proc. 6th IEEE Int. Symp. on Personal, Indoor and Mobile Radio Communications*, 2:732–736, Sept. 1995.

[445] U. Timor. "Improved decoding scheme for frequency-hopped multilevel FSK system." *Bell System Technical Journal*, 59:1839–1855, Dec. 1980.

[446] U. Timor. "Multistage decoding of frequency-hopped FSK system." *Bell System Technical Journal*, 60:471–483, Apr. 1981.

421

[447] L. Tong. "Blind asynchronous near-far resistant FIR receivers for CDMA communication." *Proc. 29th Asilomar Conf. on Signals, Systems and Computers*, Oct. 1995.

[448] L. Tong, G. Xu, and T. Kalaith. "Blind identification and equalization of multipath channels: A time domain approach." *IEEE Trans. Information Theory*, 40:340–349, Mar. 1994.

[449] M. Torlak and G. Xu. "Blind multiuser channel estimation in asynchronous CDMA systems." *IEEE Trans. Signal Processing*, 45, no. 1:137–147, Jan. 1997.

[450] M. Tsatsanis. "Inverse filtering criteria for CDMA systems." *IEEE Trans. Signal Processing*, 45, no. 1:102–112, Jan. 1997.

[451] M. Tsatsanis and G. Giannakis. "Coding induced cyclostationarity for blind channel equalization." *Proc. 29th Conf. on Information Sciences and Systems*, pp. 377–382, Mar. 1995.

[452] M. Tsatsanis and G. Giannakis. "Subspace methods for blind estimation of time-varying FIR channels." *IEEE Trans. Signal Processing*, 45, no. 12:3084–3093, Dec. 1997.

[453] M. Tsatsanis and G. Giannakis. "Blind estimation of direct sequence spread spectrum signals in multipath." *IEEE Trans. Signal Processing*, 45:1241–1252, May 1997.

[454] D. Tse and S. Hanly. "Multiuser demodulation: Effective interference, effective bandwidth and capacity." *Proc. 35th Allerton Conf. Communications, Control and Computing*, Sept.–Oct. 1997.

[455] J. Tugnait. "Blind spatio-temporal equalization and impulse response estimation for MIMO channels using a Godard cost function." *IEEE Trans. Signal Processing*, 45, no. 1:268–271, Jan. 1997.

[456] G. Turin. "Introduction to spread-spectrum antimultipath techniques and their application to urban digital radio." *Proc. IEEE*, pp. 328–353, Mar. 1980.

[457] S. Ulukus and R. Yates. "A blind adaptive decorrelating detector for CDMA systems." *Proc. 1997 IEEE Global Telecommunications Conf.*, pp. 664–668, Nov. 1997.

[458] S. Ulukus and R. Yates. "Adaptive power control and MMSE interference suppression." *Wireless Networks*, to appear, 1998.

[459] G. Ungerboeck. "Adaptive maximum likelihood receiver for carrier-modulated data transmission systems." *IEEE Trans. Communications*, 22(5):624–636, May 1974.

[460] F. Van Heeswyk, D. Falconer, and A. Sheikh. "Decorrelating detectors for quasi-synchronous CDMA." *Wireless Personal Communications*, pp. 129–147, nos. 1–2, 1996.

[461] P. Van Rooyen and F. Solms. "Maximum entropy investigation of the inter user interference distribution in a DS/SSMA system." *Proc. 1995 IEEE Personal, Indoor, Mobile Radio Communications Conf.*, pp. 1308–1312, Sept. 1995.

[462] H. Van Trees. *Detection, Estimation and Modulation Theory, I.* John Wiley, New York, 1968.

[463] L. Vandendorpe. "Multitone spread spectrum multiple access communications system in a multipath Rician fading channel." *IEEE Trans. Vehicular Technology*, 44:327–337, May 1995.

422

[464] L. Vandendorpe and O. van de Wiel. "Performance analysis of linear joint equalization and multiple access interference cancellation for multitone CDMA." *Wireless Personal Communications*, 3:17–36, 1996.

[465] V. Vanghi and B. Vojcic. "Soft interference cancellation in multiuser communications." *Wireless Personal Communications*, 3:118–128, 1996.

[466] M. Varanasi. "Noncoherent detection in asynchronous multiuser channels." *IEEE Trans. Information Theory*, 39, no. 1:157–176, Jan. 1993.

[467] M. Varanasi. "Group detection for synchronous Gaussian CDMA channels." *IEEE Trans. Information Theory*, 41:1083–1096, July 1995.

[468] M. Varanasi. "Parallel group detection for synchronous CDMA communication over frequency-selective Rayleigh fading channels." *IEEE Trans. Information Theory*, 42:116–128, Jan. 1996.

[469] M. Varanasi. "Power control for multiuser detection." *Proc. 1996 Conf. on Information Sciences and Systems*, pp. 866–874, Mar. 1996.

[470] M. Varanasi. "Optimizing symmetric energy and permuting users for decision feedback multiuser detection to user-wise outperform linear multiuser detection." *Proc. 1997 Conf. on Information Sciences and Systems*, pp. 492–497, Mar. 1997.

[471] M. Varanasi. "Decision feedback multiuser detection: A systematic approach." Submitted to *IEEE Trans. Information Theory*, Apr. 1997.

[472] M. Varanasi and B. Aazhang. "Near-optimum demodulation for coherent communications in asynchronous Gaussian CDMA channels." *Proc. 22nd Conf. on Information Sciences and Systems*, pp. 832–839, Mar. 1988.

[473] M. Varanasi and B. Aazhang. "Probability of error comparison of linear and iterative multiuser detectors." *Advances in Communications and Signal Processing*, Lecture Notes in Control and Information Sciences, 129:15–26, 1989.

[474] M. Varanasi and B. Aazhang. "Multistage detection in asynchronous code-division multiple-access communications." *IEEE Trans. Communications*, 38:509–519, Apr. 1990.

[475] M. Varanasi and B. Aazhang. "Optimally near-far resistant multiuser detection in differentially coherent synchronous channels." *IEEE Trans. Information Theory*, 37, no. 4:1006–1018, July 1991.

[476] M. Varanasi and B. Aazhang. "Near-optimum detection in synchronous code-division multiple-access systems." *IEEE Trans. Communications*, 39, no. 5:725–735, May 1991.

[477] M. Varanasi and A. Russ. "Noncoherent decorrelative detection for nonorthogonal multipulse modulation over the multiuser Gaussian channel." Submitted to *IEEE Trans. Communications*, July 1997.

[478] M. Varanasi and S. Vasudevan. "Multiuser detectors for synchronous CDMA communications over non-selective Rician fading channels." *IEEE Trans. Communications*, 42, no. 2/3/4:711–722, Feb./Mar./Apr. 1994.

[479] S. Vasudevan and M. Varanasi. "Optimum diversity combiner based multiuser detection for time-dispersive Rician fading CDMA channels."

IEEE Journal on Selected Areas in Communications, 12:580–592, May 1994.

[480] S. Vasudevan and M. Varanasi. "Achieving near-optimum asymptotic efficiency and fading resistance over the time-varying Rayleigh-faded CDMA channel." *IEEE Trans. Communications*, 44:1130–1143, Sept. 1996.

[481] A. Veen, S. Talwar, and A. Paulraj. "Blind identification of FIR channels carrying multiple finite alphabet signals." *Proc. 1995 IEEE Int. Conf. Acoustics, Speech and Signal Processing*, 2:1213–1216, May 1995.

[482] A. Veen, S. Talwar, and A. Paulraj. "A subspace approach to blind space-time signal processing for wireless communication systems." *IEEE Trans. Signal Processing*, 45, no. 1:173–190, Jan. 1997.

[483] J. Ventura-Traveset, G. Caire, E. Biglieri, and G. Taricco. "Impact of diversity reception on fading channels with coded modulation – Part III: Co-channel interference." *IEEE Trans. Communications*, 45, no. 7, pp. 809–818, July 1997.

[484] S. Verdú. "On fixed-interval minimum symbol error probability detection." Technical report UILU-ENG 83-2211, Coordinated Science Laboratory, University of Illinois, Urbana, IL, June 1983.

[485] S. Verdú. "Optimum sequence detection of asynchronous multiple-access communications." *Abstr. 1983 IEEE Int. Symp. on Information Theory*, p. 80, Sept. 1983.

[486] S. Verdú. "Minimum probability of error for asynchronous multiple access communication systems." *Proc. 1983 IEEE Military Communications Conf.*, 1:213–219, Nov. 1983.

[487] S. Verdú. *Optimum multiuser signal detection*. PhD thesis, University of Illinois at Urbana-Champaign, Aug. 1984.

[488] S. Verdú. "New bound on the error probability of maximum likelihood sequence detection of signals subject to intersymbol interference." *Proc. 1985 Conf. Information Sciences and Systems*, pp. 413–418, Mar. 1985.

[489] S. Verdú. "Near-far resistant receivers for DS/SSMA communications." U.S. Army Research Proposal, Contract DAAL03-87-K-0062, Princeton University, 1986.

[490] S. Verdú. "Minimum probability of error for asynchronous Gaussian multiple-access channels." *IEEE Trans. Information Theory*, 32:85–96, Jan. 1986.

[491] S. Verdú. "Optimum multiuser asymptotic efficiency." *IEEE Trans. Communications*, 34, no. 9:890–897, Sept. 1986.

[492] S. Verdú. "Multiple-access channels with point-process observations: Optimum demodulation." *IEEE Trans. Information Theory*, 32:642–651, Sept. 1986.

[493] S. Verdú. "Capacity region of Gaussian CDMA channels: The symbol-synchronous case." *Proc. 24th Allerton Conf. on Communications, Control and Computing*, pp. 1025–1034, Oct. 1986.

[494] S. Verdú. "Symbol-asynchronous Gaussian multiple-access channels." *Abstr. 1986 IEEE Int. Symp. on Information Theory*, p. 2, Oct. 1986.

[495] S. Verdú. "Maximum likelihood sequence detection for intersymbol

424

interference channels: A new upper bound on error probability." *IEEE Trans. Information Theory*, 33:62–68, Jan. 1987.

[496] S. Verdú. "Computational complexity of optimum multiuser detection." *Algorithmica*, 4:303–312, 1989.

[497] S. Verdú. "The capacity region of the symbol-asynchronous Gaussian multiple-access channel." *IEEE Trans. Information Theory*, 35:733–751, July 1989.

[498] S. Verdú. "Multiuser detection." In H. V. Poor and J. B. Thomas, editors, *Advances in Statistical Signal Processing: Signal Detection*, pp. 369–410, JAI Press, Greenwich, CT, 1993.

[499] S. Verdú. "Recent progress in multiuser detection." *Proc. 1988 Int. Conf. Advances in Communications and Control Systems*, 1:66–77, Oct. 1988, Reprinted in *Multiple Access Communications: Foundations for Emerging Technologies*, pp. 164–175, N. Abramson, Ed., IEEE Press, New York, 1993.

[500] S. Verdú. "Adaptive multiuser detection." In S. G. Glisic and P. A. Leppanen, editors, *Code Division Multiple Access Communications*, pp. 97–116, Kluwer Academic, Dordrecht, The Netherlands, 1995.

[501] S. Verdú. "Blind demodulation for multiuser channels." *Proc. 1995 IEEE Signal Processing/ATHOS Workshop on Higher-Order Statistics*, pp. 17–24, June 12–14, 1995.

[502] S. Verdú. "Demodulation in the presence of multiaccess interference: Progress and misconceptions." In D. Docampo, A. Figueiras, and F. Perez, editors, *Intelligent Methods in Signal Processing and Communications*, Chapter 2, pp. 15–46. Birkhauser, Boston, 1997.

[503] S. Verdú and H. V. Poor. "Abstract dynamic programming models under commutativity conditions." *SIAM J. Control and Optimization*, 24:990–1006, July 1987.

[504] S. Verdú and S. Shamai. "Multiuser detection with random spreading and error-correction codes: Fundamental limits." *Proc. 35th Allerton Conf. Communications, Control and Computing*, Sept.–Oct. 1997.

[505] S. Verdú and S. Shamai. "Spectral efficiency of CDMA with random spreading." Submitted to *IEEE Trans. Information Theory*, 1998.

[506] E. Visotsky and U. Madhow. "Multiuser detection for CDMA systems with nonlinear modulation." *Proc. 1997 IEEE Int. Symp. on Information Theory*, p. 355, July 1997. Also, M.S. thesis, University of Illinois, 1995.

[507] A. J. Viterbi. "Error bounds for convolutional codes and an asymptotically optimum decoding algorithm." *IEEE Trans. Information Theory*, 13:260–269, 1967.

[508] A. J. Viterbi. Review of "Statistical Theory of Signal Detection." *IEEE Trans. Information Theory*, 16:653, Sept. 1970.

[509] A. J. Viterbi. "Very low rate convolutional codes for maximum theoretical performance of spread-spectrum multiple-access." *IEEE Journal on Selected Areas in Communications*, 8:641–649, May 1990.

[510] A. J. Viterbi. *CDMA: Principles of Spread Spectrum Communications.* Addison-Wesley, Reading, MA, 1995.

[511] A. J. Viterbi, A. M. Viterbi, and E. Zehavi. "Other-cell interference in cellular power-controlled CDMA." *IEEE Trans. Communications*, 42, no. 2/3/4:1501–1504, Feb./Mar./Apr. 1994.

[512] B. Vojcic. "The role of data link layer in feedback interference cancellation." *Proc. 1996 Conf. Information Sciences and Systems*, pp. 205–210, Mar. 1996.

[513] B. Vojcic and W. Jang. "Transmitter precoding in synchronous multiuser communications." Submitted to *IEEE Trans. Communications*, 1997.

[514] B. Vojcic, L. Milstein, and R. Pickholtz. "Downlink DS CDMA performance over a mobile satellite channel." *IEEE Trans. Vehicular Technology*, 45, no. 3:551–560, 1996.

[515] B. Vojcic, R. Pickholtz, and L. Milstein. "Performance of DS-CDMA with imperfect power control operating over a low earth orbit satellite link." *IEEE Journal on Selected Areas in Communications*, 12:560–567. May 1994.

[516] B. Vojcic, R. Pickholtz, and L. Milstein. "DS-CDMA outage performance over a mobile satellite channel." *European Trans. Telecommunications*, 6, no. 1:63–70, 1995.

[517] B. Vojcic, Y. Shama, and R. Pickholtz. "Optimum soft output MAP detector for coded multiuser communications." *Proc. 1997 IEEE Int. Symp. on Information Theory*, p. 229, July 1997.

[518] S. W. Wales. "Technique for cochannel interference suppression in TDMA mobile radio systems." *IEE Proc.-Communications*, 142:106–114, Apr. 1995.

[519] J. L. Walsh. "A closed set of normal orthogonal functions." *American Journal of Mathematics*, 45:5–24, 1923.

[520] X. Wang and H. V. Poor. "Adaptive multiuser detection in non-Gaussian channels." In *Proc. 35th Allerton Conf. on Communications, Control and Computing*, Monticello, IL, Sept. 1997.

[521] X. Wang and H. V. Poor. "Adaptive joint multiuser detection and channel estimation for multipath fading CDMA channels." *Wireless Networks*, to appear, 1998.

[522] X. Wang and H. V. Poor. "Blind Multiuser detection: A subspace approach." *IEEE Trans. Information Theory*, 44:677–690, Mar. 1998.

[523] X. Wang and H. V. Poor. "Blind equalization and multiuser detection in dispersive CDMA channels." *IEEE Trans. Communications*, 46, no. 1:91–103, Jan. 1998.

[524] W. W. Ward. "The NOMAC and Rake systems." *The Lincoln Laboratory Journal*, 5:351–366, 1992.

[525] L. Wei and L. Rasmussen. "A near ideal noise whitening filter for an asynchronous time-varying CDMA system." *IEEE Trans. Communications*, 44, no. 10:1355–1361, Oct. 1996.

[526] L. Wei, L. Rasmussen, and R. Wyrwas. "Near optimum tree-search detection schemes for bit-synchronous multiuser CDMA systems over Gaussian and two-path Rayleigh-fading channels." *IEEE Trans. Communications*, 45, no. 6:691–700, June, 1997.

[527] L. Wei and C. Schlegel. "Synchronous DS-CDMA system with improved decorrelating decision-feedbak multiuser detection." *IEEE Trans. Vehicular Technology*, 43, no. 3:767–772, Aug. 1994.

[528] A. Weiss and B. Friedlander. "Fading effects on antenna arrays in cellular communications." *IEEE Trans. Signal Processing*, 45:1109–1117, May 1997.

[529] L. Welch. "Lower bounds on the maximum crosscorrelation of signals." *IEEE Trans. Information Theory*, 20:397–399, May 1974.

[530] B. Widrow. "Adaptive filters, I: Fundamentals." Technical Report 6764-6, Stanford Electronics Laboratory, Stanford University, Dec. 1966.

[531] E. P. Wigner. "On the distribution of the roots of certain symmetric matrices." *Annals of Mathematics*, 67:325–327, 1958.

[532] S. Wijayasuriya, G. Norton, and J. McGeehan. "Sliding window decorrelating algorithm for DS-CDMA receivers." *Electronics Letters*, 28:1596–1598, Aug. 13th 1992.

[533] S. Wijayasuriya and J. McGeehan. "Rake decorrelating receiver for DS-CDMA mobile radio networks." *Electronics Letters*, 29, no. 4:395–396, 1993.

[534] S. Wijayasuriya, J. McGeehan, and G. Norton. "Rake decorrelation as an alternative to rapid power control in DS-CDMA mobile radio." *Proc. 43rd IEEE Conf. on Vechicular Technology*, pp. 368–371, 1993.

[535] S. Wijayasuriya, G. Norton, and J. McGeehan. "A novel algorithm for dynamic updating of decorrelator coefficients in mobile DS-CDMA." *Proc. 4th Int. Symp. on Personal, Indoor and Mobile Radio Communications*, pp. 292–296, Sept. 1993.

[536] S. Wijayasuriya, G. Norton, and J. McGeehan. "A sliding window decorrelating receiver for multiuser DS-CDMA mobile radio networks." *IEEE Trans. Vehicular Technology*, 45:503–521, Aug. 1996.

[537] F. Wijk, G. Janssen, and R. Prasad. "Groupwise successive interference cancellation in a DS/CDMA system." *Proc. 6th IEEE Int. Symp. on Personal, Indoor and Mobile Radio Communications*, 2:742–746, Sept. 1995.

[538] J. Winters, J. Salz, and R. Gitlin. "The impact of antenna diversity on the capacity of wireless communication systems." *IEEE Trans. Communications*, 44:1740–1751, Feb. 1994.

[539] J. K. Wolf and B. Elspas. "Mutual interference due to correlated constant-envelope signals." In *Multiple Access to a Communication Satellite with a Hard-Limiting Repeater. Vol. II*, volume R-108, pp. 233–238. Institute for Defense Analyses, Jan. 1965.

[540] P. Wong and D. Cox. "Low-complexity diversity combining algorithms and circuit architectures for co-channel interference cancellation and frequency-selective fading mitigation." *IEEE Trans. Communications*, vol. 46, pp. 1107–1116, Sept. 1996.

[541] T. Wong, T. Lok, J. Lehnert, and M. Zoltowski. "A linear receiver for DS-SSMA with antenna arrays and blind adaptation." *IEEE Trans. Information Theory*, 44, pp. 659–676, Mar. 1998.

[542] G. Wornell. "Spread-signature CDMA: Efficient multiuser communication in presence of fading." *IEEE Trans. Information Theory*, 41, no. 6:1418–1438, Sept. 1995.

[543] J. Wozencraft and I. Jacobs. *Principles of Communication Engineering*. Wiley, New York, 1965.

[544] B. Wu and Q. Wang. "New suboptimal multiuser detectors for synchronous CDMA systems." *IEEE Trans. Communications*, 44:782–785, 1996.

[545] H. Wu and A. Duel-Hallen. "Performance of multiuser decision-feedback

detectors for flat fading synchronous CDMA channels." *Proc. 28th Conf. on Information Sciences and Systems*, pp. 133–138, Mar. 1994.

[546] H. Wu and A. Duel-Hallen. "Multiuser detection with differentially encoded data for mismatched flat Rayleigh fading CDMA channels." *Proc. 30th Conf. on Information Sciences and Systems*, pp. 332–337, Mar. 1996.

[547] H. Wu and A. Duel-Hallen. "On the performance of coherent and noncoherent multiuser detectors for mobile radio CDMA channels." *Proc. Int. Conf. on Universal Personal Communications*, pp. 78–80, Sept. 1996.

[548] Z. Xie, C. Rushforth, and R. Short. "Multi-user signal detection using sequential decoding." *IEEE Trans. Communications*, 38, no. 5:578–583, 1990.

[549] Z. Xie, C. Rushforth, R. Short, and T. Moon. "Joint signal detection and parameter estimation in multi-user communications." *IEEE Trans. Communications*, 41:1208–1216, Aug. 1993.

[550] Z. Xie, R. Short, and C. Rushforth. "A family of suboptimum detectors for coherent multiuser communications." *IEEE Journal on Selected Areas in Communications*, 8:683–690, May 1990.

[551] J. Yang and S. Roy. "Joint transmitter/receiver optimization for multi-input multi-output systems with decision feedback." *IEEE Trans. Information Theory*, 40:1334–1347, Sept. 1994.

[552] K. Yao. "Error probability of asynchronous spread spectrum multiple access communication systems." *IEEE Trans. Communications*, 25:803–809, Aug. 1977.

[553] W. Ye, X. Bernstein, and A. Haimovich. "Near-far resistance of space-time processing for wireless CDMA communications." *IEEE Communications Letters*, 1, no. 4, pp. 105–107, July 1997.

[554] C-C. Yeh and J. R. Barry. "Approximate minimum bit-error rate equalization for binary signaling." Submitted to *IEEE Trans. Communications*, 1998.

[555] Y. Yoon, R. Kohno, and H. Imai. "Cascaded co-channel interference cancelling and diversity combining for spread-spectrum multi-access over multipath fading channels." *IEICE Trans. Communications*, E76-B, no. 2, pp. 163–168, Feb. 1993.

[556] Y. Yoon, R. Kohno, and H. Imai. "A spread-spectrum multi-access system with co-channel interference cancellation over multipath fading channels." *IEEE Journal on Selected Areas in Communications*, 11, no. 7:1067–1075, Sept. 1993.

[557] Y. Yoon and H. Leib. "Matched filters with interference suppression capabilities for DS-CDMA." *IEEE Journal on Selected Areas in Communications*, 14, no. 8:1510–1421, Oct. 1996.

[558] B. Zaidel, S. Shamai, and H. Messer. "Performance of linear MMSE front-end combined with standard IS-95 uplink." *Wireless Networks*, to appear, 1998.

[559] X. Zhang and D. Brady. "Soft-decision multistage detection for asynchronous AWGN channels." *Proc. 31st Allerton Conf. on Communications, Control and Computing*, pp. 54–63, Sept. 1993.

[560] X. Zhang and D. Brady. "Asymptotic multiuser efficiency for decision-directed multiuser detection." *IEEE Trans. Information Theory*, 44: 502–515, Mar. 1998.

[561] D. Zheng, J. Li, S. Miller, and E. Strom. "An efficient code-timing estimator for DS-CDMA signals." *IEEE Trans. Signal Processing*, 45, no. 1:82–89, Jan. 1997.

[562] F. Zheng and S. Barton. "On the performance of near-far resistant CDMA detectors in the presence of synchronization errors." *IEEE Trans. Communications*, 43:3037–3045, Dec. 1995.

[563] F. Zheng and S. Barton. "Near-far resistant detection of CDMA signals via isolation bit insertion." *IEEE Trans. Communications*, 43:1313–1317, Feb. 1995.

[564] B. Zhu, N. Ansari, Z. Siveski, and Y. Bar-Ness. "Convergence and stability analysis of a synchronous adaptive CDMA receiver." *IEEE Trans. Communications*, 43:3073–3079, Dec. 1995.

[565] L. Zhu and U. Madhow. "Adaptive interference suppression for DS-CDMA over a Rayleigh fading channel." *Proc. 1997 Conf. Information Sciences and Systems*, pp. 98–103, Mar. 1997.

[566] Z. Zvonar. "Multiuser detection and diversity combining for wireless CDMA systems." In J. M. Holtzman and D. J. Goodman, editors, *Wireless and Mobile Communications*, pp. 51–65. Kluwer Academic Publishers, Dordrecht, The Netherlands, 1994.

[567] Z. Zvonar. "Multiuser detection in asynchronous CDMA frequency-selective fading channels." *Wireless Personal Communications*, 2, no. 4:373–392, 1995/1996.

[568] Z. Zvonar. "Combined multiuser detection and diversity reception for wireless CDMA systems." *IEEE Trans. Vehicular Technology*, 45, no. 1:205–211, Feb. 1996.

[569] Z. Zvonar and D. Brady. "A comparison on differentially coherent and coherent multiuser detection with imperfect phase estimates in a Rayleigh fading channel." *Proc. 1993 IEEE Int. Symp. Information Theory*, p. 48, Jan. 1993.

[570] Z. Zvonar and D. Brady. "Multiuser detection in single-path fading channels." *IEEE Trans. Communications*, 42:1729–1739, Feb. 1994.

[571] Z. Zvonar and D. Brady. "Suboptimal multiuser detector for frequency-selective Rayleigh fading synchronous CDMA channels." *IEEE Trans. Communications*, 43, no. 2:154–157, Feb. 1995.

[572] Z. Zvonar and D. Brady. "Differentially coherent multiuser detection in asynchronous CDMA flat Rayleigh fading channels." *IEEE Trans. Communications*, 43:1251–1255, Apr. 1995.

[573] Z. Zvonar and D. Brady. "Linear multipath-decorrelating receivers for CDMA frequency-selective fading channels." *IEEE Trans. Communications*, 44:650–653, June 1996.

[574] Z. Zvonar and M. Stojanovic. "Performance of antenna diversity multiuser receivers in CDMA channels with imperfect fading estimation." *Wireless Personal Communications*, 3, no. 1–2, pp. 91–110, 1996.

AUTHOR INDEX

Imai, H., 210, 269, 270, 384, 385, 387, 408, 417, 428
Ishii, N., 213, 408

Jacobs, I., 137, 402, 427
Jakes, W., 65, 405
Jalloul, L., 138, 405
Jang, W., 274, 426
Janssen, G., 385, 427
Jelinek, F., 211, 396
Johansson, A., 385, 405
Johnson, C., 327, 342, 405
Johnson, D., 66, 148, 327, 401, 406
Jones, T., 370, 385, 386, 395
Jung, P., 47, 65, 386, 406, 420
Juntti, M., 270, 271, 279, 406

Kaasila, V., 406
Kadaba, S. R., 406
Kahn, J., 201, 406
Kailath, T., 66, 78, 326, 398, 415, 420, 422
Kaiser, J., 64, 137, 406
Kajiwara, A., 65, 271, 406, 417
Kaleh, G. K., 270, 386, 406, 407, 414
Kalouptsidis, N., 326, 397
Kamel, R., 327, 407
Kammerlander, K., 65, 407
Kandala, S., 270, 385, 407
Kanterakis, E., 333, 412
Kasami, T., 65, 407
Kashihara, T. K., 326, 407
Kashyap, R. L., 65, 406
Kato, S., 384, 407, 408
Kaul, A., 270, 397
Kaveh, M., 270, 385, 401
Kavehrad, M., 138, 386, 407
Kawabe, N., 407
Kawahara, T., 270, 407
Kawahashi, A., 407
Kaye, A., 210, 325, 407
Kempf, P., 407
Kennedy, R., 386, 418
Kiaei, S., 399
Kim, H., 410
Kim, K., 271, 404, 407
Kim, S., 271, 407
Kim, Y., 271, 407
Klein, A., 10, 65, 270, 271, 397, 400, 406, 407
Kleinrock, L., 4, 408
Kobayashi, H., 210, 408
Koch, H., 231, 399
Kocic, M., 213, 402
Kohno, R., 65, 138, 210, 213, 269, 325, 328, 384, 385, 387, 408, 417, 428
Kolding, T., 384, 385, 415, 419
Komlós, J., 201, 406, 408
Kopp, L., 66, 397
Kubota, S., 384, 408

Kudoh, E., 137, 408
Kumar, P. V., 65, 326, 403, 408

Laforgia, D., 137, 409
Lancaster, P., 241, 409
Landau, H. J., 7, 409
Larsson, G., 385, 400
Laster, J., 66, 409
Latva-aho, M., 326, 328, 413
Learned, R., 65, 163, 211, 409
Lee, E., 326, 409
Lee, J., 271, 407
Lee, K., 409
Lee, W., 213, 327, 328, 409
Lehnert, J., 137, 327, 409, 411, 412, 427
Leib, H., 329, 428
Letaief, K. B., 137, 138, 409
Levitt, B., 29, 64, 65, 137, 419
Levy, A. J., 138, 409
Li, J., 213, 429
Liang, J-W., 270, 409
Lilleberg, J., 65, 413
Lim, T., 213, 271, 328, 409
Lindsey, A., 64, 410
Lindsey, W., 138, 410
Liu, H., 271, 328, 410
Liu, Q., 17, 211, 410
Ljung, L., 326, 400
Lo, N., 386, 400
Lok, T., 327, 427
Long, B., 410
Loo, C., 65, 410
Loubaton, P., 395
Luise, M., 64, 327, 385, 401
Lupas, R., 139, 211, 257, 269, 277, 325, 410
Lupas-Golaszewski, R., 139, 325, 410
Luvison, A., 137, 409
Lyu, D., 410

Müller, R. R., 212, 412
Madhow, U., 65, 211, 212, 213, 233, 325, 326, 327, 328, 385, 404, 410, 411, 420, 421, 425, 429
Magaña, M., 270, 413
Mailaender, L., 213, 270, 405
Mammela, A., 406
Mandayam, N., 270, 329, 333, 411, 419
Manolakis, D., 326, 397
Marcum, J., 131, 411
Massey, J. L., 65, 325, 411, 417
Mathews, V., 272, 418
Matsumoto, T., 137, 270, 407, 408
Mayyas, K., 327, 411
Mazo, J., 71, 221, 411
Mazzinni, G., 213, 385, 411
McDowell, A., 137, 411
McGeehan, J., 270, 271, 427
Meadors, H., 327, 402

SUBJECT INDEX

437